How to Use 21st Century Technology to Win Wars Without Combat

By Andrew Maximilian Molchan

How to Use 21st Century Technology to Win Wars Without Combat

Paperback Book ISBN 978-1-73285-173-3

Library of Congress Number 2019913156

Andrew M Molchan lives in Fort Lauderdale, Florida, USA. You can find Andrew M Molchan on Google. Kathleen G. Molchan was the Complier. This is an NAFLFD Corp Imprint.

Other Books by Andrew Maximilian Molchan

Military Strategy's 800 Rules and 400 Guidelines for the 21st Century Strategist - the 290,000 word Unabridged Edition

ISBN 978-0-692-13201-2 Copyright 2018

JOHN DAGON - God & the Devil Play Their Game Copyright 2005

JOHN DAGON - All the Time in the World Copyright 2003

The US Constitution is What Creates Being An American

Copyright 1976

Introduction

The 20 Objectives of This Book are:

To **(1)** stimulate thought, debate and the ongoing study of Grand Strategy for the purpose of **(2)** using the 21st Century's many new tools to win wars without using the old shooting war tools. **(3)** If shooting war tools must be used - this Book has Tactics and Strategy for how to win shooting wars quickly and decisively. **(4)** How to create and maintain long lasting periods of peace. **(5)** The expanded education of average American voters so the voters know what questions to ask Washington DC to avoid the Government's many past mistakes and **(6)** to NOT once again be con-gamed into more stupid wars in stupid places for stupid reasons. **(7)** To give Political Leaders, Teachers, the Clergy, the Press (everyone) a better understanding of military Grand Strategy to the point where they can **(8)** know the right questions to ask their Leaders, and **(9)** effectively cross-examine the powerful Washington DC Warmonger Lobby to the point of stopping them. **(10)** Teach voters to NOT automatically accept the usual glittering generalities about victory from either the Generals, the Press or the Warmonger Lobby. **(11)** To encourage the voters to demand specific battle outlines and plans from the Generals that are doable and with realistic time lines. **(12)** One of the objectives of this Book is to encourage the USA's voters to demand exact reasons for why Washington DC is going to win a war and **(13)** exactly why the other side is going to lose. **(14)** How to set "stop orders" so if your war objectives are not achieved at a certain point you end the shooting conflict. **(15)** The objective is to teach as many people as possible to recognize either victory or defeat (as early as possible) and bring shooting wars to a conclusion sooner rather than later. **(16)** The objective is to replace the old shooing war tools with today's many newly created non-shooting war tools. **(17)** The objective is to help create the new **United States Space Force** (USSF) and **(18)** to create the new **United States Communications Force** (USCF). **(19)** The objective is to create a new **Grand Strategy Team** of Experts. **(20)** The objective is to reform the now obsolete US Pentagon's organization so the USA can start winning everywhere rather than continuing to de facto lose everywhere.

The United States of America has been blessed with brave and courageous front line fighting soldiers from 1776 to now. However - the US Pentagon and Washington DC have NOT clearly and decisively won a significant shooting war since 1945. **The objective of this Book is to help give America's brave front line soldiers (and all citizens) the better quality of leadership that they deserve** and must have for the USA's long term survival.

How to Read this Book

The 737 axioms listed in this book cannot be listed in order of importance because there are 37 different types of warfare. **The axioms that are vitally important in one type of warfare might be completely irrelevant in another type of warfare.** The axioms and guidelines are listed by subject-issue for easy return study reference.

You can start reading anywhere in this Book. If you want to know how to win and end "Terrorist Wars" then start with Chapter 21. If you want to know about the USA-China War then start with Chapter 27. If you want to understand why wars happen then read Chapter 3. As situations happen in the real world read the appropriate Chapters.

There are 373 guidelines in this book. Many of the guideline strategy axioms have come from the past 4,000 years of military history. Each axiom can be taken as a stand-alone Truth and/or as a mini-essay about Strategy. Spend a few minutes every day to learn 2 axioms and in half year **you will know more about military grand strategy than 99% of the Generals who ever lived.** In today's world the understanding of strategy is NOT a luxury - it is your duty as a good citizen to understand what is happening so YOU as a citizen and voter can help reduce the war insanity.

Keep reading the Guidelines and you will eventually arrive at that "Ah ha" moment when all of the jigsaw puzzle axioms will come together in your mind and you will see the complete picture. **You will then be a better military grand strategist than 99% of the world's Generals who ever lived.** You will be able to look at every grand strategy issue and know the correct overall strategy action. If enough people acquire this understanding we can stop the US Warmonger Lobby and make the world a better place for everyone - everywhere.

Andrew M Molchan

Fort Lauderdale, Florida, USA

Contents

Chapter 1

Peace is the Ultimate Objective of Strategy

The Grand Strategy objective of every war should be exactly the same. The most advantageous longest lasting peace for your side and ideally the longest lasting peace for as many people in the world as possible.

(1) Prussian General Carl von Clausewitz (1780-1831) is Studied and Honored at Military Schools. However Clausewitz is Now Dangerously Obsolete. The Ultimate Objective of Strategy is NOT Winning Battles as Clausewitz Teaches. The Ultimate Objective of good Grand Strategy is Always the Same - an Advantageous Peace for Your Nation and ideally an Advantageous Peace for as Many People as Possible.

During the Vietnam War (1959-1975) the US Military won hundreds of battles. The US Military won every battle in Vietnam. However with a 1,000 to 1 advantage in gross firepower and a 1,000 to 1 advantage in money spent on the war the US Military lost the war. Why did the US Military lose?

The US Military's defeat in Vietnam was NOT for lack of trying. During 14 years of brutal combat the US Military dropped five-times the tonnage of bombs on Vietnam that it dropped on Nazi Germany in WW2. The US Military had the biggest Chemical Warfare attacks since WW1. The US Military killed 650,000 Vietnamese - many of them women, children and old people. The residue of the US Military's Chemical Warfare attacks during the Vietnam War continue to this day by killing animals and causing human birth defects.

The Vietnam War proved that winning combat battles is NOT the same as winning wars or winning the peace. **For the end of WW2 to now the US Military has won all of the significant shooting battles and lost all of the significant shooting wars.** Why? The exact reasons for "why" are in this Book. Plus the exact strategy to for how the USA can stop losing wars everywhere and start using modern 21st technology to win wars everywhere without using any of the shooting war tools.

Carl von Clausewitz in his writings wrote almost nothing about achieving a sustainable and lasting peace because he never knew anything about building a long lasting peace. However - **this book explains in great detail the Roman Empire's lost-art of building long lasting and highly successful peace.** The Romans maintained peace for hundreds of years in most of their Empire and with only very small military garrisons in 90% of the areas. This Book specifically explains how the USA can have the same degree of success the Romans had but in today's 21st Century.

Everyone who wants fewer shooting wars and much longer periods of peace needs to read and understand what is in this Book.

(2) The Reason for Studying Grand Strategy in All of its 37 Forms is NOT so you can creating more Shooting Wars. It is so you can Create Fewer and Shorter Shooting Wars and Transfer Warfare to the Non-Shooting Types of Warfare and Build Longer Periods of Peace.

Shooting wars must never be the objective. <u>A shooting war should only be reluctantly entered into after every other</u>

<u>non-schooting warfare tool for resolving disagreements has been tried and has failed.</u>

There are 37 main types of warfare and they have subtypes. As explained in this book **many of the 37 types of warfare are brand new to the last 20 and 30 years of world history.** This book makes a powerful fact based (and specific examples based) case for proving that the 21st Century's non-shooting warfare tools are now far superior to the old WW2 shooting war tools.

As explained (in detail in this book) the US Pentagon's mindset is dominated by obsolete WW2 concepts. WW2 will shortly be 2 Ages of Warfare in the past. **A WW2 mindset is ancient history and highly dangerous.** A WW2 mindset is the belief that <u>everything</u> connected to warfare is limited to only the 5 B's: Boots on the ground, Bombs, Bayonets, Battles and Blood.

The US Pentagon's (and Washington DC's) WW2 mindset is that if it is NOT the 5 B's then it is not warfare. The US Pentagon's obsolete mindset is the main reason for why the US Military has NOT clearly and decisively won any significant shooting war since 1945. **The US Pentagon will NEVER win any significant shooting war anywhere in the world until it brings its obsolete mindsets up into the 21st Century.**

This Book is all about how to create a 21st Century mindset and how the USA can go from losing everywhere to winning everywhere.

Chapter 2

Avoiding the Study of War is NOT Making Peace

Avoiding the study of war has never made wars go away. Avoiding the study of war has never made Nations or people safer. Avoiding the study of war will NOT make the 21st Century's wars go away. To study and master the art of War and Peace is to better protect your Nation and limit the destruction of wars. You are NOT studying war you are studying Peace by every means possible.

(3) For Success in Warfare, Business, Life and Making Peace you must Study Warfare in all of its 37 Forms because Warfare in All of Its Forms are Part of the Same Coin for Creating Peace.

It is one of the many paradoxes of life but you cannot fully understand how to make peace without also understanding how to make war. **Sir Liddell Hart** (1896-1970) one of the most widely read Grand Strategy analysts of the 20th Century said, "If you do not understand wars then you do not understand how to avoid them." The more we know about warfare in all of its forms the more we can have fewer shooting wars and the more we can have longer periods of productive peace.

(4) At the Core of Grand Strategy Planning is the Understanding that Nothing is Forever. Everything is always in a Constant Stage of Creation and Destruction.

The Grand Strategist understands that history never ends and thus learning never ends. There is no "settled." There is no "forever." There is no war that will "make everything right forever." No ally is forever and no enemy is forever. The details of every Grand Strategy plan are not forever. **A battle plan that was brilliant yesterday might be a disaster tomorrow.** Life is a constant learning and then re-learning process.

One of the many messages in this Book is that strategy has become evolutionary dynamic and must be under constant study and re-evaluation. For Centuries Strategy changed slowly or not at all. However since the start of the Industrial Age of Warfare (1885-1960) **Strategy has been changing at an ever increasing speed.** Today being a military General is like being a Computer/Software engineer. Most of what you learned 20 years ago is worth little.

(5) There will be Many Wars in the 21st Century. Some will be Big Wars. Wars are caused by: Social, Cultural, Economic, Political, Religious and Philosophical Tensions. Wars are caused by the Contradictions and the Incompatibilities of the Major Historical Forces in Play at the Time.

This Book examines all of the major causes of wars and goes into detail about how the mitigate them before they evolve into a shooting war.

(6) Grand Strategy's most Difficult Maneuver (but the Most Valuable) is Manipulating Somebody Else into Doing most of your Battlefield Fighting and Dying.

In shooting warfare the ideal is to have somebody else do the majority of the battlefield dying. The Grand Strategist should always be working to secure this great benefit for his Nation.

The biggest single reason for the USA being the biggest winner in WW2 was because the USA's Allies did over 97% of the combat dying during WW2. Russia alone lost 10,000,000 soldiers fighting **Adolf Hitler.** The UK and France lost over 1,000,000 fight **Hitler** - the USA lost 280,000 fighting **Hitler**.

(7) The Art of Grand Strategy is More than a Collection of Axioms but the Axioms are the Foundation for Learning Good Strategy.

Baron Antoine Henri Jomini (1779-1869) was a French General and Chief of Staff to one of **Napoleon's** best Generals - Field Marshal **Ney.** General **Jomini** later became the aide-de-camp to **Alexander the First of Russia.** General **Jomini** was the **Liddell Hart** of the 19th Century. **Jomini** wrote "It is true that theories cannot teach Generals with mathematical precision what they should do in every possible case. However it is also certain that **the axioms will always point out the errors which should be avoided** and this is a highly important consideration. The axioms thus become in the hands of skilled Generals who are commanding brave troops the means of almost certain success."

(8) Almost all National Leaders are Cheered by the Crowds when Wars Start. However if you Start Losing the War the Crowds Often Turn on the National Leaders who could Lose Everything.

At the start of WW1 (in 1914) all of the crowds in all of the European National capitals were cheering the war. They were all cheering their Leaders. However eventually the cheering crowds in: Russia, Germany, Austria-Hungary and Turkey turned into angry and ugly revolutionary crowds. The former ruling classes were lucky if they escaped with their lives (many did NOT).

(9) You Study Grand Strategy to Learn How to Use the Non-Shooting Tools to Win an Advantageous Peace.

There are 37 different types of warfare and the best Strategist will exhaust all of the non-shooting types of warfare before using the shooting war tools. The highest objective of Grand Strategy is to resolve disagreements without using any shooting war tools.

During the last 4,000 years about 70% of the combat wars ended up creating as many problems as they solved.

The USA (and its allies) resolving the Cold War (1946-1991) with the USSR was one of the greatest military achievements in the world's history of warfare. The USA and USSR never had a direct shooting war. The USA-USSR disagreements proved that **world class major disagreements can be resolved without shooting wars.**

(10) The Grand Strategist knows that Every War is only the Next War in History's Long List of Wars.

Most wars in modern times have been sold to the public as being "History's Decisive War" or "The War that Will End All Wars" or "The War to Defeat World Terrorism" or "The War to Defeat Islamofascism" etc and so forth. Most of the time it is the usual propaganda (in the usual way) to motivate the troops and the voters and "sell" that particular war.

(11) Being a Pacifist Nation is No Guarantee of Avoiding a War. Being a Pacifist is No Guarantee you will Not Die in a Shooting War.

A historically proven major mistake is to say "Our Nation will never attack anyone so we do NOT need to study or prepare for war." Or, "I don't want to know anything about wars and that means wars will not happen."

Refusing to learn anything about Cancer is no protection against Cancer killing you.

In WW2 Europe "Pacifist Nations" like Demark, the Netherlands, Norway, Czechoslovakia, Yugoslavia, Estonia. Latvia, Lithuania and so forth were aggressively attacked and occupied by an enemy.

In the 1930's China was attacked and occupied by the Japanese. In 1941 Pacifist Nations like Burma, Indonesia, Philippines and the USA were attacked. In November of 1941 75% of the US voters were strongly and adamantly anti-war.

History has proven that being a pacifist will not stop another Nation from attacking you. What stops others from attacking you is you completely understanding military grand strategy and being prepared.

Chapter 3

Understanding the Causes of War and Peace

You must have a complete and accurate understanding of the motivations and causes of the war you are fighting. This knowledge is absolutely essential for formulating the correct Strategy to build long lasting peace.

Your prime target for attack is always the same - your opponent's core strategy for achieving his core desires. There are many non-shooting types of warfare that could probably be used to neutralizing your opponent's core desires and objectives.

(12) The Most Common Cause of Wars are Changes. Most Wars are the Adjustments of the Tensions that Changes Create.

Many wars are caused by old historic power balances becoming obsolete and new rising power centers and world balances being created. Old declining powers (like the USA in the 21st Century) aggressively try to keep the status quo but the new raising powers demand recognition of their earned new status (and reset Bell Curves like China and the Muslim Middle East). The tensions between the old declining powers and the new raising power centers is a classic cause of major wars.

The classical mistake of the last 4,000 years has been Nations (or groups) who were losing power and believed that their lost power could be restored via aggressive wars. However once a Nation is on the downside of its Bell Curve then 95% of the time wars will increase the declining Nation's speed of decline even if that Nation wins their wars.

Another classic cause of wars has been Land. For thousands of years the possession of Land was the foundation of wealth and many wars were attempts to take control of land away from somebody else. **Many past wars were robbery on a grand scale.**

In the 21st Century the ownership of land is much less important than the ownership of advanced manufacturing technology, control of world markets and the successful use and control of world communications.

(13) In the last 3,000 Years 18 Factors have Caused Many Wars - Some of the Factors Are:

1. Changes from one historical age into another. This causes tensions and tensions cause wars.

2. Changes in religions and the clash of religions. The raise of Islam is an example.

3. Changes in the rise and fall of political ideologies. The idea of absolute power Kings was an idea that did NOT depart easily.

4. For thousands of years greatly increased population growths (especially without the ability to give the increased populations what they wanted and needed) produced wars of expansion.

5. If a population develops a big percentage of young males the probability of wars goes up. If a population is mainly old the probability of wars goes down.

6. Changes in National technology competence increases and decreases. This causes shifts in power centers and that can cause wars.

7. Homo sapiens have a natural tendency towards wanting to dominate others. There must be checks on that tendency or there will be wars.

8. Changes in trade and money flows and relative wealth.

9. Excessive wealth differences between the top wealthiest groups and the average worker.

10. Changes in relative economics and manufacturing. This once again changes the power equations and can cause wars.

11. Changes in National Bell Curves going up or down. Nations on the downside of their Bell Curves usually have more shooting wars then those on the upside.

12. Raising Nations with increased wealth and power demanding recognition that older declining Nations will not give them. From 1870 to 1914 Germany became a great industrial world power but England refused to acknowledge Germany's new power and status.

13. The rise of new political ideologies especially those that become semi-religions like Communism, Nazism and extreme Socialism.

14. Too much "Diversity" has been a cause of many wars.

15. Too many groups asserting "a right" to the same assets is a classic cause of wars. For example - who has a "right" to the West Bank?

16. Revenge and "getting even" from past wars or "wrongs" is a constant cause of wars.

17. The main cause of Civil Wars are when 2 big groups want the some factor but only one of them can have it - and neither side is willing to compromise. Civil Wars have many of the same characteristics of Religious Wars. Civil Wars and Religious Wars can be especially brutal and destructive.

18. A common cause of wars are Border Wars that are directly related to one or more of the above change factors.

It is absolutely vital that the Grand Strategy planners have a clear and accurate understanding as to what are the causes of wars. What are main causes of shooting wars and what are the correct solutions for establishing a new equilibrium without shooting war conflicts.

(14) **Expanding Human Populations have been a Historic Breeder of Wars.**

When most of the world's population were farmers an expanding population arrived at the point where more land was needed and other lands were occupied by somebody else. The results of expanding populations were many wars over who owned and controlled land.

For thousands of years wars were more frequent when there was (1) expanding populations that had (2) a young average age and (3) an increased percentage of males in the 17 to 29 age group. The above is a description of the Muslim Middle East's population in the early 21st Century.

Advancing population growth has much of the World on a collision course for wars in the 21st Century.

Most Americans are unaware of the exploding world population and how that puts tremendous pressure and tension on all of the systems. A few examples -

Mexico's population in 1930 - 16 million   Today - 130 million

Iran's population in 1930 - 14 million   Today - 82 million

Turkey's population in 1930 - 11 million   Today - 81 million

Too much testosterone and too few jobs for young men **is a proven formula** for creating conflicts and wars. There are too many young males with too few jobs particularly in the Muslim Middle East.

The USA is going to find nothing but trouble and muddled defeats in the Middle East. The good part is that **there is nothing of vital interest to the USA's survival in the Middle East.**

(15) Sometimes Having Too Much Money can Put your Nation into Trouble and Having Too Little Money can Keep your Nation Out of Trouble.

Economically weak Nations have difficulty having big wars because they cannot afford big wars and/or wars that go on forever. The ability of Washington DC to borrow Trillions of dollars from future US taxpayers makes it possible for Washington DC to finance today's never ending stupid wars in stupid places for stupid reasons.

What the American people need is a Law that says all shooting wars have to be 100% funded from increases in current taxes. That would bring a very fast end to the stupid wars.

Sometimes individuals are the same. Having too much money gets some people into serious trouble.

(16) The Grand Strategist Never Uses "Honor" and "Dignity" as valid Reasons for Starting a War.

Wars should never be started unless there are substantial and doable economic advantages for the Nation starting the war. Wars should be started only if there is a very high probability of a fast economical win that produces both significant short term and long term economic advantages for the citizens of your Nation and for the World.

(17) Vindictive Peace Agreements usually Produce more Long term Trouble then they Solve. One-Sided Treaties are often the Cause of the Next War.

In the last Century the Treaty of Versailles was like a large steel lid put on the top of a pot of boiling water that continued to have a fire under it. Many of the basic factors that caused WW1 continued to be in play after the war. Some of the factors that caused WW1 were made worse by the Treaty of Versailles. WW2 in Europe was Act 2 of WW1 and was only a question of "when." 21 years after WW1 ended all of WW1's European players were back into a World War. **WW2 is a reminder of the horrible costs of NOT knowing how to make a successful peace.**

One of the factors that caused WW1 was England refusing to acknowledge the growth in power of Germany and the Treaty of Versailles made those insults to Germany worse. The insults created a hunger for revenge that **Adolf Hitler** manipulated to gain power. **Adolf Hitler** in 1933 was elected into office in a fair democratic election.

(18) **Wars should Never Come as an Unexpected Surprise.**

There will be future wars for as long as there are Homo sapiens managing the world. The 3 main objectives of Grand Strategy are: (1) create more peace. (2) Keep an advantageous peace going for as many people as possible and for the longest possible time. (3) To postpone or sidestep wars for the longest possible time.

The Strategy should put maximum effort and work into postponing wars. If you can keep postpone wars then eventually you will have missed several wars.

(19) **Based on 3,000 years of History one of the Most Dangerous Times for Wars is when a Former Great Power is on the Downside of its Bell Curve (like the USA today) and Runs Into a soon to be Great Power that is on the Upside of its Bell Curve (like China today).**

The USA has been on the downside of its Bell Curve (relative to the world) since 1975. **The USA's downward angle in the last part of the 20th Century had been a very gentle downward angle but the speed of the decline and the downward degree of the angle is increasing in the 21st Century.**

China is on the upside of its reset Bell Curve (reset about 1979). Somewhere around 2045 China will eclipse the USA in both economic and military power. **The USA must accept the down step in status and learn from the British example after WW2.** The British did NOT make the classic mistake of destroying everything they had with stupid wars in attempts to stop history (the mistake Washington DC is currently making). England made the best arrangements possible and slowed its decline.

The bottom line is the average English family today lives much better than the average English family of 1910 when England had direct control of almost 25% of the world's population. The issue for the USA should NOT be how many US Troops are in somebody else's Nation killing the local people. The issue should be how happy is the average American? **One of the points of this Book is that average Americans can be happy without being in somebody else's Nation murdering the local people.**

(20) **Many Wars are Like Hurricanes and Earthquakes. They are Nature's Way to Even Out Pressures in the system. The skilled Grand Strategist looks for ways to Channel those Pressures into Areas and Adjustments that are NOT Wars.**

The correct action for once great Nations must be the acceptance the world's new realities. **One of the many things that history has clearly proven is that you cannot shoot the world back into yesterday.**

(21) The best Grand Strategist puts Great Effort and work into making sure His Side Constantly keeps Moving to the Correct Side of History's Changes.

Many wars are fought to **not** recognize the changes to what are world new power equations. **Trying to block history's major power changes with wars is a mistake**. Wars to return to the past like the USA-Vietnam War sometimes have short-term illusions of success but always fail in the long run. The Vietnam War from the first shoot fired by US Troops was on the wrong side of history and had zero chance of success.

A USA-Iran shooting war would also be on the very wrong side of history and have less than zero chance of success.

(22) One of the Most Valuable Art-of-War-and-Peace Talents is Knowing when NOT to Fight.

Chinese General **Sun Tzu** (544-496 BC) who is the **Aristotle** of Military Grand Strategy - wrote "The side that knows when to fight and when not to fight will over time be the winner. There are roadways not to be traveled and armies not to be attacked and walled cities not to be assaulted. Knowing when not to fight is one of the greatest talents in the Art of War."

Adolf Hitler should have read **Sun Tzu** before declaring war on Russia. The Japanese should have taken **Sun Tzu's** advice before bombing Pearl Harbor. The US Pentagon idiots should have taken **Sun Tzu's** advice before going to Vietnam. One of the places that should not be attacked with a shooting war is Iran.

Aristotle (384-322 BC) said about wars "We make war only that we may live in peace." **Aristotle** elaborated by saying that if your Nation is already living in peace then why would your Nation start a war to get what you already had?

Chapter 4

Understanding the Different Ages of Warfare

The great battlefield disasters of WW1 (that murdered millions of troops on all sides) were mainly caused by all of the Generals being one Age of Warfare behind in their mindsets and thinking.

You must be absolutely 100% aware of what Age of Warfare you are in and NOT have Strategy, tactics and mindsets from a bygone Age of Warfare.

The world is currently in the Information Age of Warfare (1960 to ?). The world is starting to transition into the Artificial Intelligence Age of Warfare (? to ?). However a large part of the US Pentagon's mindset and war planning remains stuck in the now bygone WW2 Industrial Age of Warfare (1885-1960).

(23) **When a new Age of Warfare arrives the Generals and Admirals must Advance their Thinking and Mindsets into the New Age of Warfare or risk National Military Disasters like the Vietnam War defeat.**

There have been 4 big ages of warfare in world history. The 4 major ages of warfare have different specific subgroups but the sub-grouping is a complete book by itself. The 4 larger groupings will suffice for this book's Grand Strategy teaching examples.

(24) **The Four Main Ages of Warfare were and Are:**

1. **The Age of Spear and Sword Warfare** (2700 BC to 1380 AD). The time period was from the first recorded in history war - the War between the kingdoms of Elam and Sumer in Mesopotamia 2700 BC. The center of gravity shifted to the next stage of warfare around 1380 AD. A few parts of Spear and Sword Warfare continued into WW2.

The year 1380 AD is approximate for when Black Powder cannons started to become decisively significant in Western World warfare. The Chinese invented Black Powder and first used Black Powder in warfare in 904 AD.

The use of **Black Powder cannons were definitely decisive in the 1453 AD capture of Constantinople.** A history changing war that forever destroyed the Eastern Roman Empire that had survived in a direct line for 1,777 years (324 BC to 1453 AD).

Constantinople's captured was accomplished with the brilliant construction of enormous Black Powder cannons by Sultan **Mehmed II** (1432-1481). The Cannons were 27 feet long and could fire a 600 pound projectile a mile. The fall of Constantinople made Artillery significant (and/or decisive) up to and including the Industrial Age of Warfare. **Black Power Cannons made the up to then #1 defensive weapons system of stone walls and stone castles obsolete.**

Many tactics from the Age of Spear and Sword Warfare persisted for hundreds of years after 1380. A few of the tactics of that age remain with us into the 21st Century. The US Army's current M4 Rifle comes with a bayonet (a short spear or sword arrangement) for charging the enemy.

2. **The Age of Black Powder Warfare** (1380 to 1885). During almost all of this time period Artillery and rifles were muzzle loading single shot and used Black Powder. One of the many drawbacks of both Black Powder Rifles and Cannons is that they do NOT work well in the rain. During most of this time period swords, spears and bayonets were in wide use.

Warships were basically the same from 1450 up to the US Civil War (1861-1865). For over 400 years warships were sail powered (and built out of wood) and had muzzle loading smooth bore Black Powder cannons. **A skilled Sea Captain of a warship in 1550 would have known what do to with a warship in 1850.**

The widely used Horse Cavalry of this Age of Warfare relied heavily on spear and sword tactics from the Age of Spear and Sword Warfare.

3. **The Industrial Age of Warfare**, (1885 to 1960). The year **1885 is when armies started adopting smokeless powder ammunition and firearms/cannons transitioned from single shot Black Powder muzzle loading weapons to multiple shots per minute breach loaded weapons.** Machine Guns firing 550 rounds a minute were invented and introduced into armies.

Black Powder is very dirty and will quickly foul a rifle or cannon barrel to the point where it becomes problematic and eventually impossible to reload unless cleaned. Smokeless power does not have that problem and made Machine Guns and rapid fire Artillery possible.

Around 1885 European Armies started adopting the modern Maxim Machine Guns (that used relatively clean smokeless powder ammunition) and European Armies acquired rapid firing cannons. The Cavalry in 1917 was reinvented as Tank Warfare. Warships that had basically been the same for hundreds of years (wooden ships with muzzle loading Black

Power cannons) saw an amazing revolutionary transition in size, construction, speed and firepower. Airplanes were invented and a brand new branch of warfare (Air Power) was created. Communications that for thousands of years had been extremely difficult on the battlefields became instantaneous with radios. In 1945 Atomic Bombs were invented.

With the start of the Industrial Age of Warfare the speed of warfare technology-change started moving much faster than the imagination and talents of the world's Generals. A major problem that persists to this day and is becoming worse.

4. **The Information Age of Warfare**, (1960 to ?). We are currently in the Information Age of Warfare. The General in a battle can often see the entire battle in real time. Winning or losing a battle has become very dependent on accurate and fast information.

For the first time in history one side in a conflict has the means to influence both the public and the troops of the opponent's force. For the first time in history communications and salesmanship have become decisive weapons systems.

In the Middle East's 21st Century many conflicts the factors of communications and salesmanship have become the decisive weapons systems.

It was information that defeated the US Military in Vietnam. The US Generals in the Vietnam War (1959-1975) never lost a battle. **The American Generals lost the Vietnam War NOT on the battlefields of Vietnam but on Main Street USA.**

In the Vietnam War the Viet Cong's most effective weapons were NBC, CBS and ABC's nightly TV news. The Pentagon's Generals never saw that one coming.

The information Age of Warfare has several brand new factors that never existed in any other age of warfare. One of those decisive factors (for the USA) is that a significant shooting war (in the current Information Age of Warfare) must be saleable to the US voters and continue to be saleable to the voters.

To be successful US Wars must (1) unequivocally be on the moral high ground and (2) have factors that directly affect the lives of average USA voters. Without the above 2 factors the Pentagon's stupid wars are going to continue to be mediocrity embarrassments that 75% of Americans do NOT support.

(25) There are No Clean or Sharp Lines between the Ages of Warfare.

Nobody rings a bell when one Age of Warfare predominately transitions into the next Age of Warfare. Start and stop dates for the Ages of Warfare are based on when the **center of gravity-mass moved into becoming predominately the next Age of Warfare.**

(26) Some Factors from all of the Past Ages of Warfare Continue to be Used into the Current Age of Warfare.

Most of the world's current military rifles come with a bayonet making them short spears from the age of Sword and Spear Warfare.

(27) For Thousands of Years Wars were Between Kings. What the Ordinary Person Thought meant Little or Nothing. All of that has Changed in the Current Information Age of Warfare.

In the current Information Age of Warfare the popular TV press will cheer the start of a significant shooting war (like a shooting war with Iran). However the popular TV press would very quickly turn against the war. **Part of the US Pentagon's war planning must be based on the fact that the USA's TV press has the power to create public opinion and thus turn voters against a war.**

In the USA the powerful Washington DC Warmonger Lobby might stamped (or pay $$$) Congress into starting a war. However - if it is NOT long term "salable" to the voters it will be another very expensive US Military defeat.

(28) The Next Age of Warfare will be The Artificial Intelligence Age of Warfare. The Advanced World is already Transitioning Into the Artificial Intelligence Age of Warfare.

The US Military has for some time been using drones to kill people. We are definitely on the road to the center of gravity transitioning mainly into the Artificial Intelligence Age of Warfare somewhere in the next 20 years?

The Artificial Intelligence Age of Warfare will produce profound changes in everything from Tactics to Grand Strategy. **Much of what US Platoons did in the Iraq War and the Vietnam War will be done by smaller highly intelligent flying and/or ground traveling machines with a mix of different weapons systems and abilities for many specialized jobs.**

(29) Mass Conscript Armies of the WW2 Type are Already Obsolete and are now a Dangerous Mindset from the Past.

Infantry Troops who can only handle a rifle are obsolete. Jet Fighters will be piloted by computers. Most of the immediate decisions that Generals and Admirals now make on the battlefields will be made by computers. Artificial Intelligence Warfare will continue to expand.

The new USA Strategy Team will attack worldwide with as many of the 37 different types of warfare as possible. **The definition of what is "a war" has already taken a quantum leap in complexity.**

The correct way to "see" the 21st Century is that almost everything is a war of one type or another type.

In the current age where advanced Nations have tactical nuclear weapons large concentrations of human troops is an obsolete and highly dangerous tactic that would produce mass deaths. Massive amounts of human troops are obsolete and that makes all of the WW2 and Vietnam Draft Laws obsolete.

Today's USA Generals must get their mindsets up into the 21st Century. One of the many new decisive issues is - **how to formulate new infantry tactics when casualties are no longer a major attack planning concern because most of the "Troops" will be smart machines?**

The USA must assume that Russian's and Chinese's strategy planners are working on the many new tactics changes.

(30) In the Soon to Arrive Artificial Intelligence Age of Warfare. The Total Number of "Troops" a Nation Has will NOT Depend on a Nation's Population Size but on a Nation's High Technology Manufacturing Capacity.

China and Japan are already the world leaders in computer and robotics manufacturing. Both China and Japan will be able to build large armies of intelligent robots in the future.

(31) Today's USA Silicon Valley Culture does Not Like the Idea of Working with the Pentagon's Culture. However many Silicon Valley Companies MUST Become Major USA Defense Contractors.

Silicon Valley's left-wing culture (and the Pentagon's WW2 mindset culture) MUST be reconciled so they can work together.

Part of that process to bring the Pentagon and Silicon Valley together must be the Pentagon's actions clearly being only on the moral high ground. **To secure Silicon Valley's full support the US Pentagon MUST support freedom and equal justice for everyone everywhere in the world.** The US Military cannot have shooting wars to keep everything the same as yesterday and continue to have selective amnesia about the Fascist governments it supports.

A much needed new National Defense Act should mandate very high USA Tariffs on: computers, smart phones, advanced chips, robotics and everything in high technology smart-machine manufacturing.

On the other side of the table the spoiled brats in Silicon valley have to understand that just because the USA is not 100% perfect that does not mean that it is 100% bad. Does Silicon valley really believe that a Communist Democrat Government would be an improvement? Does the USA really have to make the same Communist Government mistake that 70% of the world has already made over the last 100 years?

(32) The Time Between the Ages of Warfare is Becoming Less. The "Experience" of today's Military Officers has Less Value and Worth today Than in Past Ages.

Warfare for many thousands of years was in the Sword and Spear Age of Warfare (2,700 BC to 1380 AD). The slide into the Black Power Age of Warfare (1380-1885) was slow and there was significant use of Sword and Spear Age of Warfare tactics that continued for centuries after 1380.

Around 1885 the center of gravity transitioned into the Industrial Age of Warfare (1885-1960). Around 1960 the center of gravity transitioned into the Information Age of Warfare. When this book was written - warfare was starting to meet the leading edge of the Artificial Intelligence Age of Warfare.

For thousands of years the skills that a young soldier learned were useable when he was 60 years old and a General or Admiral. However - today's Tactics and Strategy and weapons systems are turning dangerously obsolete at an ever increasing speed.

Today **in the 21st Century "old age and experience" can be highly negative for US Military Officers and the USA.** Example. If you had received a PhD in Computer Engineering in 1985 80% what you had learned would have been obsolete by 2005.

(33) Today's Good Grand Strategists are NOT Followers of Prussian General Carl von Clausewitz who Believed that Destroying your Opponent's Army was the Ultimate Objective of Strategy.

General von Clausewitz (1780-1831) wrote that the opponent's Army should always be the prime objective. For **Clausewitz** the opponent's Capital City was the second most important objective. **Clausewitz's Strategy can be summarized as - if you win the main Battle then you win everything.**

For thousands of years the Clausewitz idea was correct. For thousands of years of warfare Nation's usually had the majority of their army at the location of a major battle. If you defeated your opponent's army on the battlefield then you usually defeated the majority of your opponent's military. The defeated side was forced to ask for peace. However **starting with the Industrial Age of Warfare (1885 to 1960) armies greatly increased in size** to include millions of men. It was impossible to defeat an army in only one battle.

Germany's Generals had a **Clausewitz** mindset in both WW1 and WW2 with disastrous consequences for Germany and the world. The USA's Generals had a **Clausewitz** mindset in the Vietnam War. **During the Vietnam War the USA's Generals won all of the battles and for years they deluded themselves into believing they were winning the war.**

(34) In the 21st Century Many Western Politicians Spin "Guerrilla Warfare" As Something New.

Some politicians are playing con games with "terrorist warfare." They say "terrorist warfare" and Guerrilla warfare are band new threats that demand restrictions on everyone's civil-liberty rights and justifies moral low ground actions. However - Guerrilla warfare and "terrorist warfare" (light and fast moving bands of men who hit a weak target by surprise and then move on quickly) is the original type of warfare from 10,000 years ago. Crooked politicians saying Americans have to give up their rights because of some new kind of terrorist warfare are con men and probably communists and/or fascists.

(35) The most Dangerous Time for Generals (and Nations) for making Disastrous and Decisive Mistakes is when the immediate Past Age of Warfare has been significantly Replaced by a New Age of Warfare.

The Industrial Age of Warfare came to an end around 1960 when the USA-Vietnam invasion went into high gear. All of the USA's Generals were one Age of Warfare behind in their mindsets and were completely blindsided.

(36) The Psychological Warfare Tool has Always Been a Major Part of Warfare.

In every war you try to intimidate your opponent to the greatest extent possible. **Exactly what will truly intimidate your opponent is part of the Art of War and Peace.** Are the USA's 10 Super Aircraft Carriers a intimidation to China? For the Chinese the Aircraft Carriers are a joke.

One of the earliest recorded highly successful uses of Psychological Warfare was by the Persian (Iranian) King Cyrus the Great and his Son Cambyses II. In 525 BC Persian King Cambyses conquered Egypt by using cats and National Character understand and Psychological Warfare. Cats were sacred to the Egyptians so the Persians tied cats onto the front of their shields when they attacked. The Egyptians would not shoot their arrows at the Persians for fear of killing a cat.

(37) One of the Biggest Changes and OPPORTUNITES that the Information Age of Warfare Created is the Ability to Communicate to your Opponent's Troops and Population.

For thousands of years (up to and including WW2) it was difficult or impossible to get information either into or out of an opponent's Army or population. However in the current Information Age of Warfare ideas can be sent out or received from almost anyone to almost anywhere. **Sending information and ideas and salesmanship to your opponent's civilian population and troops is now both doable and vital.**

Chapter 5

The Many Non-Shooting Ways to Win Wars

It is absolutely essential for you to have a correct and crystal clear understanding of (1) what types of wars you are fighting and (2) in what percentages.

Doing what the US Pentagon does - that is viewing all wars as the same and fighting them all in the same way is gross military incompetence. In 21st Century warfare one-size does NOT fit everything.

The US Pentagon has to stop being a one-trick pony and it has to become a 50-trick dragon.

In all bigger conflicts the two sides are fighting for a combination of several different reasons. One side's reasons for fighting are usually different from their opponent's reasons for fighting. However the US Pentagon since WW2 has suffered from "strategy confusion" because the US Pentagon always things in simplistic terms. For the US Pentagon they are the "good guys" and their opponents are the "bad guys" end of analysis.

In WW2 the US Pentagon's people really were the "good guys" and their opponents (Hitler and Tojo) really were the "bad guys." However since WW2 it has been much more complicated. A complication that the US Pentagon planners never seem able to understand. Sometimes you can build something of an argument for the Pentagon being the "bad guys."

For those that want to argue what are the facts? The FACT is the US Pentagon has NEVER publically and high profile endorsed equal rights, real freedoms, equal justice and equal opportunity for everyone in the world everywhere. If the Pentagon are the "good guys" then what are they waiting for?

(38) The good Strategist must Accurately Understand the True Motivations on All Sides so the Strategist can Tailor Fit His Strategy to most Effectively Deal with his Opponent's Strategy.

The best Grand Strategy is the reverse of **Carl von Clausewitz**. You do NOT attack your opponent's Army as the #1 objective. **Your #1 objective for attack is your opponent's strategy for securing what they most desire.**

(39) If the Desires of your Opponent are of Great Importance to your Opponent (but of Little Importance to You) then a Way To Win is to Give Your Opponent What They Want.

It is both stupid and immoral to be killing your own troops for something that is NOT important for your Nation's citizens. If giving your opponent something that is NOT vital to you will end a shooting war then do so.

(40) A Common Mistake is You Fighting one Category of War but Your Opponent is Fighting a Different Category of War.

An example of all of the grand strategy incompetents in Washington DC being disastrously blindsided was China's highly successful Economic War envelopment of a big part of the USA's manufacturing economy.

The US Pentagon's incompetents were wasting time, money and work building Super Aircraft Carriers for a completely delusional replay of WW2 in the Pacific while Chinese right in front of their eyes achieving a major victory.

China is doing it again with their Silk Road strategy. Building a major strategic asset covering the world's heartland. A weapons system that is not only high valuable it makes a money profit. Meanwhile back in the USA the Military Industrial Complex is dangling all kinds of bright color beads in front of the Generals and Admirals.

(41) Most Wars are a Combination of Several of the 37 Main Categories of Warfare that are:

Standard Army Warfare

Standard Air Force Warfare

Standard Naval Warfare

Space Satellite Warfare

ICBM Warfare

Tactical Nuclear Warfare

Chemical-Biological Warfare

Drone & Cruise Missile Warfare

Border Warfare

Economic Warfare

Psychological Warfare

Political Ideology Warfare

Religious Ideology Warfare

Guerrilla and Insurgent Warfare

Independence & Revolutionary Warfare

Intelligence/Information/Idea Warfare

Political and/or Semi-Religious Fanatical Warfare

Superiority of Organization Warfare

Messianic Leader Warfare

Annihilation Warfare

Ego Wars Fought for Worthless Status Symbols

Diplomatic and Trickery Warfare

Proxy Warfare

Spy and Espionage Warfare

Cultural Warfare

Civil War Warfare

Revenge and Getting Even Warfare

Crusading Warfare

Communications Warfare

Regime Change Warfare

Propaganda (in all of its forms) Warfare

Limited Objective Warfare

Total Population Warfare

Terrorist & Guerrilla Warfare

Salesmanship (in all of its forms) Warfare

Insane and Pointless Warfare

Artificial Intelligence Smart Machine Warfare

With big wars there will be 8 to 14 of the 37 types of warfare going on at the same time. Those percentage combinations are dynamic and will change as the war changes. A constant updated understanding of what wars you are fighting (and Not fighting) is vital for winning the overall war.

At the start of strategist's analysis there are four vital questions the strategist must asks himself. (1) Exactly what types of wars am I fighting? (2) Are the types of wars I am fighting the correct types of wars to effectively outflank the types of wars my opponent is fighting? (3) What types of wars do I need to be fighting to defeat my opponent's Strategy for achieving his core desires and objectives? (4) What are all of the non-shooting types of warfare that can be used before considering the shooting-combat types of warfare?

Correct Grand Strategy is you applying the types of warfare where you are the strongest against the areas and factors where your opponent is both the weakest and his vital interests are the most exposed to your doable alterations.

An example is China's highly successful Economic War against the USA.

(42) In the 21st Century you MUST Think Big about What Warfare Includes and Encompasses.

It is a major mistake (in this Century) to think of "war" as only armies with Tanks and Guns. Total war in this Century is **total** - it is everything your Nation possesses against everything your opponent's Nation possesses.

Doing the what US Pentagon does - only counting Tanks and Troops and Bombers etc - that is out of the dark ages.

(43) The Core of a Good Plan is Using the Types of Warfare Best Fitted to Counter the Types of Warfare that are Being Used to Attack You.

An attack from an opponent that is centered on Economic Warfare must be fought with a different Strategy than a war motivated by religious crusading. For example - Communism is a semi-religion and that is why the USA's Economic Warfare sanctions for 50 years never had a decisive affect against Cuba and North Korea. Economic Warfare was NOT the correct tool for the type of Crusading Warfare that Cuba, North Korea and now the Radical Muslim Groups are fighting.

The best way (and in some cases the only way) to fight an idea is with a superior idea. Decades ago the strategy should have been to open up Cuba and North Korea so the people could see that free markets are a better idea than Communism.

An example is West German enveloping East Germany without combat by exposing the East Germans to West Germany. West Germany defeated the idea of Communism with the better idea of Capitalism.

(44) One of the Grand Strategist's Basic Decisions (at the Start an Analysis) is Deciding between the 3 Groups of Warfare. (1) the Shooting types of Wars and (2) the Non-Shooting types of Wars and (3) Both Shooting and Non-Shooting types of Wars at the Same Time.

As many (as possible) of the Non-Shooting types of wars should always be used first unless there are overwhelming reasons to start with the shooting war tools.

During the 1946-1991 "Great Cold War" (between the USA and the USSR) both sides were evenly matched with nuclear weapons. The USA used Economic Warfare, Diplomatic Warfare, Cultural Warfare, Propaganda Warfare, Superiority of Organization Warfare, Salesmanship Warfare, Idea Warfare and several other non-shooting types of warfare to achieve a stunning victory WITHOUT using any of the shooting war tools against the USSR.

The USSR had the biggest Army (by numbers of men and equipment) in the world. If there was a shooting war between the USA and USSR half of the US population would have been killed. The Red Army probably would have been able to push its way into France. However the Communist USSR ceased to exist without one US soldier dying in combat with the USSR. This is real world proof that **the non-shooting types of warfare are vastly superior to the shooting types of warfare.**

The Salesmanship Warfare of President **Ronald Reagan** succeeded in finally changing the USSR. The complete historical process took 45 years of hard work and patience on the part of countless Americans. However **using the non-shooting tools to win the war saved many tens of millions of USA and Russian lives compared to a Shooting War.**

If you are an American (or Russian) reading this book you are probably here and alive because the USA and USSR did NOT have a shooting war and your parents did not die in that war.

(45) The Greatest Victory that a Grand Strategist can Achieve is Winning the Wars that Never Turned into Shooting Wars.

President **Ronald Reagan** (President 1981-1989) was the best Grand Strategy USA President of the 20th Century. President **Reagan** won the Cold War (1946-1991) without a USA-USSR shooting war.

President **Reagan** achieved what is the ultimate correct objective of all wars (shooting and non-shooting wars) - the most advantageous peace possible for as many people as possible.

Any moron can start a shooting war (and most shooting wars are started by morons) but it takes a person who has talent and skill to resolve a major conflict by using the non-shooting types of warfare.

(46) The Grand Strategist is very Wary of "Unifying Wars." A War to "Unify your Nation" is a Strategy that usually Works Only if the War is Short, Economical, Successful and Was Clearly on the Moral High Ground.

"Unifying Wars" are attempts by Leaders (who usually have serious domestic problems) to start a shooting war with the belief that the shooting war will unify the population and make the average person forget about the domestic problems. Sometimes Unification Wars work but usually in the long run they make everything worst - often very worse.

If you cannot clearly and swiftly win decisive real advantages in a shooting war then DO NOT be in that war. Do not start shooting wars (or keep shooting wars going) to take attention away from domestic problems.

(47) The Grand Strategist Guards against Wars Taking on A Life of Their Own.

For the US Pentagon (since WW2) wars taking on a life of their own has been the standard situation. **The more the US Pentagon is losing the more it insists on staying in a shooting war to prove that it is not losing.**

For post WW2 Washington DC - fighting wars becomes the justification for continuing to fight the wars. **It is like a gambler at a casino who is losing and the more he losses the more he is driven to continue gambling and losing.**

The Grand Strategist must constantly be asking "Exactly what are the reasons for fighting this war and are those reasons valid now? There must be an honest answer to the question of "Are we fighting this war only because we are fighting this war?"

(48) Non-Shooting Wars usually Cover Long Time Periods and You Must Have the Skills of (1) Perseverance (2) Patience and (3) Deep Faith in your Objectives Being on the Correct Side of History.

The non-shooting types of warfare usually take years (and often decades) and require the fortitude to keep working towards the objectives for a very long time. Unfortunately for the USA - the Chinese are masters at making and successfully completing long term plans. On the other hand Washington DC's "time lines" are usually short with a lack of long term planning.

(49) Wars Evolve (Especially Big Wars) and can Dramatically Change from Low Risk Relatively Safe Wars Into very High Risk National Survival Wars.

The Grand Strategist never assumes that the type of war he was fighting (when the war started) will be the same type of war he will be fighting when the war ends. Often it is NOT.

Japan started the Great Asian-Pacific War of 1931-1945 for very logical and doable economic reasons. Japan wanted to colonize a large part of China. For the Japanese (in 1931) a war to colonize China was a low-risk war with a possible enormous economic reward. However in the Summer of 1941 **Winston Churchill** and President **Roosevelt** conspired and put an oil embargo on Japan - and told Japan to (1) stop colonizing China and (2) to get out of China. **Churchill's** and **Roosevelt's** oil embargo was a Black Swan that Japan's planners of 1931 had never imagined or expected.

Churchill's and **Roosevelt's** oil embargo was a deliberate trap to trick the Japanese into attacking the USA and UK. **Japan's "safe war" in China mutated into a deadly war that destroyed a large part of Japan.** Three Million Japanese were killed and Japan lost all of its overseas territories of: Taiwan, Korea, Manchuria, the Kuril Islands, Sakhalin Island, Guam and other Pacific Islands.

The good Grand Strategist uses the shooting war tool only as the very last tool applied because absolutely nobody can be 100% sure where how big war will mutate and how it will end.

(50) The Grand Strategist is always On-Guard against the "Limited War" that keeps Growing and starts Drifting Upwards and Turns into a Major War and Possible a Major Disaster.

Many wars start as limited wars but start drifting upwards. Like a Cancer that starts very slowly and almost unnoticed but keeps growing until it is a deadly disaster. An example is the Vietnam War (1959-1975). The war started with one USA "advisor" being killed. The war at first grew by small steps but was always drifting upwards until it turned into the

USA's greatest military defeat up to this point. In total 2,709,918 US Military personnel fought in the Vietnam War. For comparison the total size of the British Army at the end of WW2 in 1945 was a little over 2,900,000.

From the end of WW2 to now the US Pentagon has been acting like a loser at a gambling casino - "One more troop SURGE and we'll win it all back." **Only incompetent and intellectually bankrupt Generals have never ending shooting wars.**

(51) History Fluctuates between Periods of Big Wars and Small Wars. During the Period of Small Wars many Analysts Make the Mistake of Thinking that Large Wars are Gone Forever. They have Always Been Wrong.

There will be shooting wars for as long as there are Homo sapiens. There are unfortunately many wars in the world's future and some of them will be big wars. NOW is the time to create the vitally important new Grand Strategy Team as outline in Chapter 31 because it will be needed.

(52) The US Military Must Have Major Efforts and Programs for Taking Its Mindset Out of WW2.

World War Two is already ancient history in warfare terms. Any Tactics and Strategy from WW2 will probably be more wrong than correct. For example, a D-Day type of landing against Russia or China would be a complete disaster for the USA. US Aircraft Carrier operations in a war with China would be a disaster for the USA. A USA Heavy Tank attack (in large numbers) against Russia or China would be a disaster for the USA.

Making battle plans in the 21st Century based on WW2's 1939-1945 examples is the same horrible mistake that WW1's Generals made during 1914-1917 when they based their WW1 battle plans on examples from the American Civil War (1860-1865) and the Franco-Prussian War of 1870.

(53) The USA's Political and Military Leaders who Think of Warfare as Only having Bombs, Bullets and Bayonets are Highly Dangers to America's Survival.

If your opponent is highly intelligent and skilled (like China) he will be attacking you (the USA) with several of the 37 different categories of non-shooting warfare at the same time.

The USA has been in an Economic War with China since the early 1980's. Most of Washington DC (and 98% of the US Pentagon) never realized that the USA was in a war with China until the Trump Presidency. A war that China had already won before Washington DC realized there was a war.

Chapter 6

How to Turn Generals into Peace Makers

Over 97% of the Generals who ever lived were NOT good at Grand Strategy. Generals need to be competent at winning battles and big Nations need many Generals (Germany in WW2 had over 3,500 Generals and Admirals). However a Nation needs only one excellent Grand Strategist.

Germany in both WW1 and WW2 did NOT have one excellent Grand Strategist. The USA has not had an excellent shooting war Grand Strategist since the end of WW2.

(54) If you are looking for the Rules of Military Grand Strategy one of the Places Where you will NOT Find Them is in the Writings of most of History's Famous Generals.

From **Alexander the Great** (356-323 BC) to General **Dwight Eisenhower** (1890-1969) most of the great commanders departed this world with very few written thoughts about how to win wars and win the peace.

President **Eisenhower** (President January 1953 to January 1961) did warn about the US Military becoming too large (and too bureaucratic) and too cozy with defense contractors selling over priced and questionable weapons systems for yesterday's wars. History has (unfortunately for the American people and the world) proven President **Eisenhower's** warnings to be correct.

(55) Where do you Find Good Grand Strategists? That has been a Vital Question for over 4,000 Years.

Most military Generals are NOT good at Grand Strategy. Why? Because all established military systems for the first 15 years of a military officer's career limited their decisions to small specific jobs all within a very narrow set of rules and regulations.

The usual iron-box mental conformity of military organizations produces a narrow tunnel vision small-minded view of warfare. However most Generals do not need to be good at Grand Strategy. If Generals are competent managers of the military during times of peace (and competent at winning shooting war battles) then that is adequate for over 95% of what almost all Generals are supposed to be doing.

The solution for solving the Pentagon's highly serious "Grand Strategy Skills Deficit" is to build the new Grand Strategy Team as outlined in detail in Chapter 31.

(56) Ninety-eight Percent of all senior Military Officers do NOT Need to have Inquiring Minds and Questioning Imaginations - or a deep Understanding of the Grand Strategy Big Picture.

National Leaders should always ask for the advice of their Generals while always remembering that most Generals throughout history have NOT been good Grand Strategists. Most Generals have too narrow of a world view - especially in this highly complicated 21st Century when warfare is multi-leveled and includes everything.

The National Leader should ask for the opinions of as many people as possible while understanding their analysis limitations. **The key for the Leader is asking questions. Cross examine the planers and Generals to see if their ideas can stand up to challenges.** What questions do you ask? This Book will teach you the correct questions to ask.

The conundrum is that the system that predictably produces dependable Generals is NOT the system that can produce good Grand Strategists. This is why in the 21st Century the USA must have a Grand Strategy Analytical Team. A group of 40 to 60 experts (plus their personal assistants) who are constantly doing Grand Strategy analysis in their areas of expertise.

(57) The Grand Strategist Never Attacks Only Because it is Possible to Attack. He Never Starts a War Only Because it is Possible to Start a War.

One of the biggest complications of Post WW2 Washington DC is the massive cost of the US Military. **The irrepressible mental deadly-contradiction is that because the USA has an enormously expensive military machine having that machine demands that Washington DC use their Trillion dollar expensive machine.**

It is like having a Rolls Royce in your garage that you are paying $100,000 a month to rent. The compulsion to drive it somewhere (for any reason) is overwhelming.

Looking back at Post WW2 US Military History **there is an argument for saying that the USA might have been better served with a smaller US Military.** Certainly if the Vietnam War could have been avoided that would have been enormously advantageous for both the USA and the world. The USA continues to carry the deep and permanent scares of the Vietnam War.

France did NOT stay in Vietnam an extra 10 years to keep fighting a losing war because France did NOT have the money. However the USA did have the borrowing power and did make the mistake. Sometimes having too much credit can get you into trouble as Washington DC has learned the hard way.

(58) Commanders with too Much Testosterone are Good on the Front Lines but Not in the Grand Strategy Planning Room.

Grand Strategy planning must be as devoid of emotion as possible. Military people love **Vince Lombardi** theatrics but **Lombardi** never won a significant shooting war - and neither has any US General since the end of WW2.

(59) The Grand Strategist Never Falls in Love with a Plan.

All military plans have Bell Curves and yesterday's excellent plan will someday be on its downside and obsolete. That is never a question of "if" and always a question of "when."

No matter how much you loved a plan (and no matter how much you want a plan to work) if it is a failure then you need the courage to admit defeat and do what is needed to save what can be saved.

(60) Abundant and Expensive Military Hardware (Tanks, Artillery, Combat Aircraft, Warships etc) Combined with a Fundamentally Flawed Grand Strategy and Mediocre Generals is a Lost War Waiting to Happen.

World history has many examples of Nations that created large military organizations that were big enough to stumble them into disastrous trouble but NOT intelligent (or big enough) to get them out of trouble (like the US Military in Vietnam).

The German Navy in 1914 was big enough to get Germany into a war with the British Navy but NOT big enough to win the war against the British Navy.

The Japanese Navy in 1941 was big enough to start a war with the USA and UK but NOT big enough to win that war. The US Navy today is big enough to start a war with Iran but NOT big enough to win a war with Iran.

(61) Generals and Admirals usually become Attached to Old Weapons Systems that ALWAYS eventually Become Obsolete.

Every weapons system (including the basic human infantry soldier) has a Bell Curve and eventually becomes obsolete - it is only a question of "when?"

The emotional and mental attachments that Generals have for obsolete weapons systems (and obsolete tactics) usually ends with battlefield disasters and sometimes horrible National defeats.

(62) Knowing and Understanding All of the Axioms in this Book will Make you a Better Strategist. However to Become an Outstanding Grand Strategist means Understanding almost Everything as it Relates to almost Everything.

Most of today's US Generals do NOT understand Grand Strategy because their minds are living in a **Clausewitz** world where they view most major factors as not related to military Strategy. For most of the USA's Generals if it does not at some point produce an explosion then it is not part of their world. However - **today's Grand Strategy includes everything as it relates to everything.**

Chapter 7

National Bell Curves and Peace Bell Curves

Every Great Power has three distinct Grand Strategy stages (1) the Upside time period. (2) The Topping Out time period and (3) the Downside time period. Significantly different National Grand Strategies are A MUST HAVE for each stage.

The location of where Nations are at on their Bell Curves is vital information for the Grand Strategy equation. The type of Strategy a Nation can have on the upside of its Bell Curve is very different from the type of Strategy it MUST followed on the downside of its Bell Curve.

The USA was on the Upside of its Bell Curve from 1775 to 1975 (200 years). The USA was running over the top of its Bell Curve during the late 1970's and into the 1990's. In the 21st Century the USA is on the Downside of its Bell Curve (relative to the world) and will be on the downside of its Bell Curve (relative to the world) for probably all of the 21st Century.

(63) **Every Great Power has Bell Curves. When Your Nation is on the Downside of its Bell Curve (as the USA now) You must do Everything Possible to Avoid Shooting Wars because ALL Shooting Wars will Increase the Downward Speed of the USA's Decline even if the USA Wins the Wars.**

(64) **There are Seven Vital USA Strategy Bell Curve Questions that must Constantly be Asked and Constantly Answered and Updated.**

The 7 Basic Bell Curve Questions are:

Where is the traditional US Air Force on its Bell Curve?
Where is the traditional USA Army on its Bell Curve?
Where is the traditional US Navy on its Bell Curve?
Where is the US Marine Corps on its Bell Curve?
Where is the US Space Force on its Bell Curve?

Where is the US Economy on its Bell Curve?
Where is the US Dollar on its Bell Curve?

Remember - **the correct strategy to be used for situations when a Nation is on the upside of its Bell Curve is usually very different from the correct strategy needed when a Nation is on the downside of its Bell Curve.**

In older Nations like the USA it is always a mixed bag. In the above list some factors (the US Space Force) are on the upside of their Bell Curves and some factors (the US Navy's big surface ships) are on their downsides.

The US Pentagon is failing to closing down the factors that are on their Bell Curve Down Sides. The Pentagon should be transferring the work, time and money that is going into obsolete factors and moving that money and effort into the factors that are on the Upside of their Bell Curves.

(65) Everything in God's Universe has a Bell Curve - From the Few Seconds Flash of Lighting in the Rainy Night Sky to the 5 Billion Year Life of our Sun.

The standard historical mistake (of the last 4,000 years) are great powers (that are on the downside of their Bell Curves) continuing the same actions and wars and mindsets as if they were continuing to be on the upside on their Bell Curves. At the time this book was published **the USA was making his classic mistake.**

Every corporation has a Bell Curve. The Strategy for Amazon Corporation is very different from the Strategy for Sears Corporation. Someday Amazon Corp will be today's Sears. Someday Google will be today's CBS.

(66) Once a Nation is on the Downside of its Bell Curve Military Aggression Increases the Downward Speed of the Bell Curve and Causes ever Increasing Levels of Internal Political Discord and Tensions.

Nations on the downside of their Bell Curves have increasing problems and tensions that often produce radical and dangerous political extremes of either right or left (or alternately both). The USA is currently having extreme internal political tensions and contradictions in large part because it is on the Down Side of its Bell Curve but Washington DC's culture refuses to believe it is on the Down Side.

(67) The Grand Strategist must always be Aware of (1) the World's Major Bell Curves that are In Play and (2) must also be Aware of the "Temperament of the Times" and "the Spirit of the Times" that is Motivating Immediate History.

The Grand Strategist is constantly asking himself "Am I on the right side of history - or has history marched on and left be behind?" Today in the 21st Century Washington DC and the US Pentagon are on the very wrong side of history in the Middle East.

The best immediate strategy for the USA in the Middle East is to disengage as much as possible and to build the new Grand Strategy Team as fast as possible.

(68) The Grand Strategist does NOT Confuse Short Term Trends with the Major Bell Curve Directions.

The lines of the Grand Bell Curves are never straight and even. They are like the stock market in a major up rally or in a major decline. **There are large vibrations up and down along the major trend direction. The Grand Strategist's long term planning must NOT be confused or distracted by the short term trends.**

Washington DC and the USA really need long term grand strategy objectives. There should be an overall plan for how the USA is going to get to the end of this Century in some type of form representing the best parts of today's USA. That plan should be under constant updating and re-evaluations.

(69) The US Dollar and US Banking System will Not be a Weapons System Forever. The USA absolutely Must Become more Realistic about its Relative World Declining Financial Importance.

Washington DC started sanctions against North Korea in 1950 and sanctions against Castro's Cuba in 1960. **The USA was on the upside of its Bell Curve during that time period and the sanctions DID NOT work.**

Currently the USA is on the downside of its Bell Curve (relative to the world) and Washington DC's economic sanctions will become less effective with each passing year. There is a desperate need in the USA for a new Grand Strategy Team and a new mindset.

Before the start of WW1 the British Empire could (and did) very successfully use the British Pound as a weapons system. However today's England can no longer use the Pound as a decisive weapons system. The same will eventually happen with the US Dollar - and probably in the next 20 to 30 years.

(70) In Washington DC Any Consideration of the Costs-Benefits Ratio of any Shooting War Anywhere is Viewed as both Irrelevant and Anti-American.

From the end of WW2 to the first decades of the 21st Century the attitude in Washington DC has been the belief that America had unlimited money, unlimited wealth and unlimited production capacity that would last for the next 1,000 years.

The disastrous mindset in Washington DC continues to be the delusion that the Federal Government can afford any shooting war anywhere for any reason. That is a major Grand Strategy mistake. When this Book was published the Federal Government had been running an overall deficit of over a Trillion Dollars (with a T) per year for over 10 years straight.

The USA stopped being an overall profitable Nation in the mid 1970's went it went into its Bell Curve downside. The USA needs at least a Trillion dollars of new Debt per year to balance the books. **Stupid costly wars in stupid places that have zero vital interests (or profits) for the average US citizens are the last things the USA needs.**

Washington DC has to change its current insane and suicidal attitude of believing that Washington DC has to be in the middle of everyone else's shooting wars anywhere and everywhere in the world.

The poor people of the USA have a 100% accurate and morally legitimate argument when they ask - "If the warmongers in Washington DC have endless money to be in every stupid worthless criminal war in every mud hole in the world then why can't they pay for basic health care for me?"

(71) The Bell Curves of Great Powers can be (and have been) Reset. However no Great Power in History has ever Reset its Bell Curve to the Upside without First either: (1) Crashing-and-Burning or (2) having a very Long Period of Stagnation and Rot that is de Facto a Slow Motion Form of Crashing.

Someday the USA will re-set its current declining National Bell Curve but historically NOT before the USA crashes and burns.

China reset its Bell Curve in the late 1970s. However the old China had to first crash-and-burn. The Opium wars of the 1840's sucked much of the silver coinage out of China (and into England) and starved the Chinese economy for hard money. Then China suffered the humiliations of a hundred years of Western and Japanese colonization. The horrors of the 1931-1945 Great Asian-Pacific war (17 Million Chinese dead) plus the Chinese Revolution and Civil War of 1921-1950. In quick step the next disaster was the Chinese fighting the USA in Korea 1951-1953. Then the horrors of the "Cultural Revolution" 1966-1976. **It was**

130 years of China's Bell Curve going down, down and down until it had completely crashed and burned. In 1976 (at the end of the Cultural Revolution) China's economy was in ruins - 75% of China's population was living in extreme poverty and 98% of the Chinese population was poor.

In 1810 China was 30% of the World's Economic Gross Product and in 1976 China was 5% of the World's Economic Gross Product - an over 80% decline in the Chinese economy relative to the world.

All of the world's great classic civilizations: Egyptian, Persian, Greek and Roman and so forth are gone and dead. Only the Chinese culture is basically similar to what it was 2,000 years ago.

All of the classic major civilizations that crashed and burned are civilizations that are gone forever. For example - today's Greece is nowhere close to being what Greek civilization was 2,500 years ago. There is today an area called "Greece" with people in that area but they are NOT the Greeks of 2,500 years ago.

The Muslim Middle East was in a state of stagnation for hundreds of years (like Japan before 1868). **The Middle East started resetting its Bell Curves in the 1950's.** Resetting the Muslim Middle East's Bell Curve has been uneven. However - **in the 21st Century the overall change to a new Middle East Bell Curve upside is in full force.** Some Muslim areas like Dubai and the Emirates are already very advanced.

(72) The Vital-Decisive Fact for USA Military Grand Strategy Planning is the Muslim World Being On the Upside of its Reset Bell Curve While the USA is On the Downside of its Bell Curve.

The 3,000 year historical fact is that Nations on the downside of their Bell Curves SELDOM win shooting wars against Nations (and areas) on the upside of their Bell Curves or reset Bell Curves. That means the probability of the USA clearly winning any significant shooting war in the Middle East is very low.

The probability of the Down Side USA decisively winning a shooting war against Up Side China - that probability is already zero.

(73) The Grand Strategist Never Opposes Change Only Because it is Change.

Nations on the downside of their Bell Curves have rancid bureaucracies that are also on the downside of their Bell Curves. **The "establishment" of downside Nations always unthinkingly falls into the wrong Strategy of trying to maintain the status quo** and constantly opposes the needed common-sense changes.

When your Nation is on the downside (relative to the world) of its Bell Curve (like the USA in the 21st Century) the Nation should actively be looking for constructive good management Leadership that can successfully adapt to the new realities.

(74) For Thousands of Years every Great Power on the Downside of its Bell Curve had groups of People who sincerely Believed that Wars would Restore the Nation's Greatness.

Washington DC currently suffers from this history proven disastrous sickness. There are many military morons in Washington DC who sincerely believe it was wars that made the USA great and more wars will make it greater.

In the last Century one of the main factors that made the USA great was the USA being the very LAST great power into both WW1 and WW2. Everyone else (England, UK, Russia, China, France etc) did 99% of the combat dying in WW1 and 97% of the combat dying in WW2. **In the last Century it was the relative lack of USA shooting war combat that made the USA the #1 greatest winner and greatest power.**

For thousands of years most wars were profitable for the winning side. However in today's world every US Pentagon shooting war is a guaranteed massive money loss. **Why would any sane Nation continue to do things that cost it massive amounts of money and gave it nothing?**

(75) No Nation or Society in Relative Decline to the World has Managed their Decline either Easily or Happily. The Classic Pattern is Increased Political Turmoil.

The textbook classic psychological Six Stages of finding out that you are dying (or something horrible has happened like the USA being on the downside of its National Bell Curve relative to the world) are:

(1) Shock, (2) Disbelief, (3) Anger, (4) Bargaining, (5) Self Pity and (6) Acceptance.

In the 2016 US Presidential election the US voters went for Mr. **Trump.** They voted for the #3 **"Bargaining"** Stage. The "Bargain" was that if America elected Mr. **Trump** then America "would be great again." However - the US voters who wanted **Hillary Clinton** to win went into - Stage #1 "**Shock**" or Stage #2 "**Disbelief**" followed by Stage #3 - "**Anger**."

(76) The USA in the 21st Century is in a Similar Position to the British Empire at the End of WW2.

The USA needs to follow the British Empire's post WW2 example. The British would fight independence movements in their Empire (many times with heavy handed repression and shooting wars) until it was clear that the other side had a greater will to win and the fighting would never stop.

The British (unlike the US Pentagon and Washington DC) did NOT keep wars going forever so they could keep on deluding themselves into believing they had never lost. Once it was clear that the "insurgents" would never surrender then England would do a 180 degree turnaround, declare victory and say that independence was what they always wanted. The British made the best deals possible with the new power structures and saved whatever they could save rather than lose everything.

The USA's history will follow England's history. After WW2 the left-wing Socialists came into power with the idea that if England could afford WW2 it could afford Socialism. Economic Socialism has been a disastrous failure everywhere in the world for over a 100 years and it failed in England. Conservative **Margaret Thatcher** came into power in 1975 when the USA was at the top of its Bell Curve.

A question seldom asked is **"Why is Communist China succeeding if Communism is so bad?"** Because China has a Communist Government System (Basically a one party dictatorship-monarchy) but NOT a Socialist economic system. China's economic system is almost pure 1890's Capitalism. If you want to build a factory go build it - no permits needed. No building regulations needed. No building restrictions. You have waste slug from the factory - dump it in the river. You have air pollution - who cares. The employees complain about working 12 hour days for low pay and no benefits - the Police will arrest the protestors. Unions - there are NO Unions. Liability - there is NONE. Worker's Compensation - NONE. Employee protection - NONE. Lawsuits - there are NONE. The cost of Lawyers per year - Nothing. **China has succeeding in taking over the world's manufactured-product markets because China is a Capitalist manufacturer's paradise.**

(77) Nations that are on the Upside of their Bell Curves Many Times Do Not Realize how Powerful they Have Become.

Upside Nations often do not realizing how strong they have become. On the other side - declining great powers (the USA today) often do not realize how powerful some Up Side Nations and areas have become.

In 1940 - the USA at first glance looked weak. In 1940 the USA had an almost non-existent Army. The US population looked discouraged after 10 years of economic depression. However - fundamentally the USA in 1940 was enormously strong. 1940 was the reverse of today when the USA looks strong but is living off of a Trillion Dollars of yearly new debt creation.

The Grand Strategist must always have a realistic and accurate analysis of who is becoming relatively stronger and who is becoming relatively weaker and he must factor those realities into his constantly updated Strategy planning.

(78) Classically when a Nation is on the Down Side of its Bell Curve most of that Nation's Bloated Bureaucracies have already become Bastions for Mediocrity with Reactionary Resistance to the Needed Changes.

The US Pentagon has 26,000 full time working people in the biggest single office building in the world. What do they do? Who knows - there has never been an independent audit. However we know what they do NOT do - and that is clearly and decisively win shooting wars. Or win ANY of the 37 types of wars.

The US Pentagon has lost all of its major shooting wars since the end of WW2. Maybe the Pentagon could learn how to lose wars with only 16,000 employees?

(79) Washington DC Must STOP Believing that Wars are the Creator of National Greatness.

World War Two (by a confluence of historical anomalies that will never happen again) projected the USA into being the world's biggest military powerhouse. WW2 was NOT the cause of America's greatness. America's already existing greatness was the cause of the USA easily becoming the world's military superpower. WW2 was in the middle of the Industrial Age of Warfare (1885-1960) and **the USA was already the greatest industrial power in the world before the start of WW2.**

The Vietnam War was a disaster in many ways. The Vietnam War greatly speeding up the USA's Bell Curve "Topping Out" process. The USA's Bell Curve upside was 1775 to 1975 - and then the "topping out" process during the 1980's and 1990's.

Nobody rings a Bell at the top of a Great Power's Bell Curve. However there are always symbolic happenings at the top. For the British Empire it was WW1 - after that it was downhill at ever increasing speeds.

The symbolic "Bells" for the USA's Top were (1) the US Government going off of the Gold Standard in 1971 because Washington DC could NOT pay its debts in Gold. (2) The US Military losing the Vietnam War in 1973. The US Military had a 1000 to 1 advantage in gross firepower and lost the Vietnam War. The US Military's defeat was a clear "message" of the decline starting.

The start of the USA versus Muslim World War (1991 to ?). Plus the Iraq War 2 (2003 to ?) and the Afghanistan War (2001 to ?). <u>They were all the Bells ringing</u> for the increased speed **of the USA's decline.** The above wars have all **increased the speed of the USA's Bell**

Curve decline. All future US shooting wars will increase the speed of the USA's decline relative to the world even if the US Pentagon by some chance happens to clearly win any of the wars.

(80) A Highly Dangerous Myth in the USA is the Belief that Wars are the Cure for Economic Recessions.

In the USA it is widely believed that WW2 was the CURE for the Great Depression of the 1930s. It is also widely believed (in the USA) that wars were the cause of America's National greatness. Both are 90% wrong and **highly dangerous American myths.**

The facts are that by the end of 1939 the "Great Depression" was 96% over in the USA.

American war-production massively increased from 1942 to 1945 because most of the factories were already there. During the worst of the 1930's Depression the factories were working at only 25% but the factories were there. Bringing all of the factories up to 100% was relatively easy. However today many of the factories are 100% gone.

USA Federal debt did greatly increase during WW2. USA Federal Debt went from $14 Billion in 1940 to $99 Billion in 1944. However **America's personal savings rate skyrocketed upwards. Average workers were saving 28% of their total income by 1944.** The debt of the US Government was 99% owned by USA citizens and for them the debt was a asset. Many US citizens ended WW2 much richer than when the war started. In addition - over 95% of all US citizens, companies, Local and State Governments had NO debt when the war ended. The Federal Government was 99 Billion dollars in debt but the US taxpayers were 99 Billion dollars richer.

Today the average American worker and company and taxpayers are all drowning in debt. State and Local Government are deep in debt. The US Federal Government has been increasing its debt by over a Trillion dollars per year for over 10 years (with no end in sight) when this Book was published.

There is proof positive that wars do NOT cure economic depressions. In 2008 when the US economy had its biggest crash since 1929 there were 2 US shooting wars going. The Iraq shooting war had been going since 1991 and the Afghanistan shooting war since 2001. However the US economy crashed.

(81) No Nation has Continued to be a Great Military Power Without Also Continuing to be a Great Economic Power.

In 1946 the USA's percentage share of the world's advanced hard goods product manufacturing was 37%. In addition to the USA being the world's biggest maker and exporter of advanced hard goods manufactured products the USA was also a major oil and food exporter and the world's biggest creditor.

In 2018 the USA had 14% of the world's advanced hard goods product manufacturing. What the current numbers mean is that **86% of the world's advanced product manufacturing is NOT in the USA.**

Relative to the world the USA is a minority player with (14%) of advanced manufactured products. That number continues to go DOWN every year. **Today the USA is also the world's biggest debtor Nation by gross amount of money owed.** The bottom line is the USA cannot afford stupid wars that are of zero value to the average American worker.

The Warmonger Lobby keeps screaming "America needs a far forward military strategy." No it does NOT. The Afghanistan Taliban is not going to build a Navy and invade California. ISIS is not going to build a Navy and invade Long Island. US Military Forces do NOT need to be "over there" to stop them from doing what they cannot do.

The Warmonger Lobby keeps screaming "America has to be engaged with the world." Yes it does - it needs to be much more engaged with the world and using all of the non-shooting types of warfare to the maximum degree possible.

The US Warmonger Lobby's definition of "being engaged with the world" means US Forces being in somebody else's Nation bombing the local population.

(82) Much of what Passes for USA-GNP economic "Growth" in 21st Century USA is really the Growth of Debt and NOT classic Real Economic Growth of Manufacturing and Production.

Some of what passes for "growth" in today's USA is the deliberate (and systematic) under-reporting of inflation that makes "real growth" look bigger than it is in reality. If real inflation is 6% but Washington DC reports inflation as only 3% then the "3% NOT reported becomes the "growth of the economy." However that "growth" is inflation under-reporting accounting fraud. Then the Government charges you higher income taxes on your "growth" in income but in reality the "purchasing power" of your "added income" has NOT grown. The final results are you end up paying higher and higher taxes on a purchasing power that is NOT increasing and in reality your overall real world purchasing power is thus declining because of the higher real taxes you are paying.

A highly important factor that is never measured (or reported) is the "Quality of the USA's Growth." Is USA Gross National Product going up because more factories are being built? Or because more loans are being made? Or because more foreign cars are being sold in the USA - and so forth? The fact is that BOTH the gross dollar percentage (relative to the world) and the "quality" of the USA's "growth" is deteriorating.

The provable FACT is that if you stopped all of the Debt Creation in the USA the USA would have a Depression worse than the Great Depression.

What is the solution? When a Nation (the USA in this case) falls into the trap of powering its economy with massive amounts of new debt **there is no easy way out**. Once you are on the back of the Tiger you have to stay there or you are dead.

The problems (or happiness) of average Americans are NOT helped by the USA's Warmonger Lobby always paying off Congress to have the USA in never ending stupid wars in stupid places for stupid reasons that are a moral embarrassment to all sane and rational Americans.

(83) When you Are on the Downside of your Bell Curve Make the Best Arrangements You can Make and Secure as many Benefits for your Nation as Possible.

Realistic and conservative polices can greatly slow the downward movement of the USA relative to the world. The USA needs to admit its decline is happening and do whatever is needed to slow the decline.

Chapter 8

The Mindsets Needed for Military and Peace Success

Only a few of history's well known Generals gave us any thoughts on Strategy. Only a few Generals in history have given the world any ideas about how to create peace. One of the exceptions was Field Marshal Helmuth von Moltke - the Elder (1800-1891) Chief of Staff of the German General Staff from 1871 to 1888. Moltke the Elder wrote "Strategy is more than a system of specific regulations, rules and particular practices and expedients; it is more than knowledge. It is the application of knowledge to practical life. The development of an original idea in accordance with the continually changing circumstances. It is the art of creative action under the pressure of the most difficult conditions."

The key words from the above are, **"The development of an original idea" to solve the new challenges of the new Ages of Warfare that advancing history will place in front of every current and future General and National Leader.** It is the exact opposite of what the US Pentagon has been doing for decades - and that is **fighting today's wars with yesterday's mindsets** and assuming that battlefield victory is the same as creating peace.

During WW1 (by December of 1914) the new challenge of warfare (on the Western Front in particular) was barbed wire protecting trenches that were full of machine guns that could fire 500 rounds a minute.

All of the Generals in 1914 (on all sides) had a Black Powder Age of Warfare mindset. It took years (and millions of dead soldiers) before the new challenges of trenches and machine guns were intellectually solved with new mindsets and new weapons.

In the current age (with a growing number of Nations that have nuclear weapons) the USA does NOT have the luxury of giving its living-in-yesterday Generals years of on-the-job battlefield training. Generals cannot do it wrong for years before they start to see the correct practices.

Field Marshal Moltke the Elder put a premium on the independent judgment of officers at all levels. **Moltke** taught that a dogmatic enforcement of any plan of operation was a deadly sin and great care should be taken to encourage initiative on the part of commanders at all levels. In the 21st Century that "initiative" in the US Military must be centered around the many non-shooting types of warfare.

(84) The Mindset of Your Opponent is always the #1 Prime Objective.

The correct Grand Strategy is understanding that your opponent's army is (almost always) a secondary objective. **The main objective for attack is always your opponent's core desires and his Strategy for achieving those desires.**

Your Grand Strategy is always the same in every war. Your Grand Strategy is (1) defeating your opponent's Grand Strategy - and then (2) building the most advantageous peace for as many people as possible.

The ideal situation is to alter your opponent's desires by using as many of the non-shooting tools of warfare as possible. You must also position yourself (and your assets) so your opponent's desires become clearly unachievable in a shooting war.

(85) The Essence of Grand Strategy is Accurately and completely Analyzing your Opponent's Grand Strategy. Then Formulating a Logical and Doable Path for How Your Strategy is Going to Defeat your Opponent's Strategy.

General Sun Tzu wrote:

1. "If you clearly and correctly understand your enemy's Grand Strategy, and you correctly understand your own Strategy, then you will win the war."

2. "If you do not understand your enemy's Grand Strategy, but only understand your own Strategy, then your chances of winning the war are 50-50."

3. "If you do not understand your enemy's Grand Strategy, and you do not understand your own Strategy, then you will lose the war."

The US Pentagon never bothers to analyze their opponent's desires and strategy to achieve those desires. For the US Pentagon every opponent is **Bonnie and Clyde** and every war is a re-run of WW2. This is one of the many reasons for WHY the US Pentagon has NOT clearly won any shooting big wars since WW2.

(86) The Highest Level of good Grand Strategy is the Dynamic Analysis of Everything in Real Time. The best Grand Strategist sees Strategy as a giant Mobile hanging from the ceiling. A Mobile with 3,000 different Hanging and Moving Parts of Different Sizes and Weights.

If the Mobile has 3,000 pieces and one piece becomes significantly lighter or heavier then the shape of the total Mobile changes. This is the challenge of Grand Strategy in the 21st Century.

(87) If Your Opponent is already Defeated in Their Mind then sometimes Victory can be Surprisingly Easy.

When WW2 started **Adolf Hitler** knew that the French were already defeated in their minds. **Hitler** understood that the French National mindset had never recovered from the horrors of WW1. In May of 1940 when the French Army was enveloped by German General **Heinz Guderian's** Tanks the French people's will to resist collapsed.

The USA's National mindset has never recovered from the Vietnam War disaster. The continuing US Military's de facto defeats in the Middle East are only **deepening the fragility of the USA's National mindset towards shooting wars.**

The USA's National mindset needs a long period of ZERO never ending wars. A long period where all of the shooting conflicts are ultra fast and clearly decisive wins for the USA. **The festering infection of never ending stupid wars that are poisoning the blood stream of the USA has to END.**

(88) Most of the Axioms of Grand Strategy are Simple. One of the most Important Rules of Grand Strategy is to NOT Ignore the Axioms just because They Seem to be Simple.

(89) There are Dozens of major Factors that go into the Grand Strategy Military Matrix Equation. One of the most Important Factors is the Talent of the Leadership.

Human talent is one of the biggest factors in the success or failure of most wars. It is also the single biggest factor in the success or failure of almost all businesses.

If you have a creative, courageous and talented Leader it makes no difference what business they are in the business will be a success.

In the current US Military versus the Muslim World War the USA needs to build a highly Talented Grand Strategy Team that can master and apply the 7 strategies.

1. Correctly and completely understand their opponent's Desires and Objectives.

2. Attack their opponent's Desires with as many of the 37 different types of warfare as can be applied.

3. The main effort (and biggest amounts of money) should be used in applying all of the non-shooting tools of warfare before any of the shooting tools are reluctantly applied.

4. The objective is NOT "victory" in the WW2 way. The objective is a workable peace for as many people as possible.

5. Have the courage to understand and use new weapons systems like the Internet and salesmanship. Have the Talent to think outside of all of the old WW2 boxes. **Washington DC must find the intellectual courage needed to use the new communications weapons systems to change the mindsets in the Muslim world's population.**

6. **WW2 is already ancient history and will soon be TWO ages of warfare in the past.** WW2 has few (if any) lessons for 21st Century success in the Middle East.

7. In future wars the US Pentagon must be on the clear moral high ground with ALL of its actions and polices. In the Middle East the US Pentagon must stop trying to play both sides of the fence. **The Pentagon's moral Schizophrenia has been a failure and will in the future continue to be a failure.** The USA and US Military must be supporting equal rights, equal fair justice and as much freedom as possible for everyone everywhere.

(90) The Grand Strategist Never Burns Bridges Only For the Fun of Burning Bridges.

Nobody can be 100% sure if they might need to go over some past bridge they thought they would never have to use again. Remember the rule - **Enemies are not forever and Allies are not forever. Never burn a bridge unless there are overwhelming reasons to do so.**

(91) The Grand Strategist Understands that most of the "Burning Issues" of Today are in the Long Term Irrelevant.

The Grand Strategist always keeps his eyes on the big picture and keeps the "noise" of current passions in historical perspective. The world has a million years of history in front of it - nothing today is important in the world's long-term big picture. **It is insane to be murdering people over trivia and in the long-term big picture it is all trivia.**

(92) The Strategist is always Thinking in TEN levels of Analysis. The same Challenge will have Different Solutions Depending on what Level you are Analyzing.

The 10 levels of analysis of operations are:

1. The Grand Strategy long-term world level
2. The immediate international level

3. The Grand Strategy long-term National level
4. The immediate National level
5. The long-term state or regional level
6. The immediate state or regional level
7. The local long-term level
8. The immediate local level
9. The long-term individual level
10. The immediate individual level

The classic mistake that has been made for thousands of years of warfare is to only operate on the immediate regional, local and individual levels. **The historical Hall Mark of truly great Leaders is they could see the big picture and think long term.**

After WW2 President **Truman** and Generals **MacArthur** and **Eisenhower** and **Marshall** did NOT make the mistake of trying to turn occupied Japan and German into West Banks. They did NOT make the WW1 Versailles Treaty mistake. They established freedom and equality and equal justice in both Germany and Japan and turned them into long term and highly valuable allies.

Unfortunately for the American people and the world the US Military no longer supports freedom, equal rights and equal justice for everyone everywhere.

(93) During the last 4,000 Years one of the Biggest Destroyers of Great Powers and Empires has been Too Many Wars in Too Many Places for Reasons that were NOT Vital to the Great Power's Core Existence.

Washington DC of the last 60 years has had too many wars in too many places for reasons that had zero value to the average American or to the world.

The Roman Empire fell to the "Barbarians" NOT because the Romans did not know how to defeat the Barbarians. Rome had been defeating the Barbarians for Centuries before the fall of Western Rome in 476 AD. Rome fell because in its last 150 years there were so many stupid wars in stupid places for stupid reasons the majority of Romans said "Who cares if the Barbarians take over. We are living in hell anyway."

(94) If you Have Been Defeated is a War You Need to Ask Three Questions: (1) Was the Leadership Lacking and Flawed? (2) Was the Strategy Lacking and Flawed? Or (3) were Both Lacking and Flawed?

The first step in problem solving is to admit that there is a problem. The US Military was defeated in Vietnam. The issue is WHY? You cannot solve a problem without first admitting there is a problem. The US Military has to admit that it has been de facto defeat everywhere in shooting wars since WW2.

In Alcoholics Anonymous the first step is to get the addicted person to admit they are addicted. Usually the addicted person will say - "I'm a social drinker." No - you are addicted and you are going to continue to go downhill until you admit to yourself that you are out of control and addicted and need help.

(95) The Grand Strategist has the Courage to Always Deal Honestly with the Facts as they Are.

The Grand Strategist never makes the mistake of seeing the facts as he would like them to be. He does NOT make the mistake of pretending the facts are what he wishes them to be with the common delusion that if he acts like his delusions are real then the facts will magically become what he wants them to be.

(96) The Grand Strategist After a Terrorist Attack Never Plays the Game of Being a TV News Hysterical Little Child.

In terrorist attacks (as when a theater is shot up or a building is blown up) the Grand Strategist is aware of the fact that the actual damage done to the Nation is next to nothing.

In terrorist situations the National leaders have to say "Your pin pricks have caused us emotional sorrow and grief but we are a large and mighty Nation. If your terrorist attack had been fifty times bigger it would have done nothing to defeat our will to eventual victory."

Every day in the USA 103 people are killed in automobile accidents: babies, children, teenagers, women and men. Every day in the USA 7,450 people die from all causes. The point is that the above is highly unfortunate but the Nation goes on.

(97) The Grand Strategist must be Constantly Reviewing all of his Basic Strategy Assumptions and always be Sure They Continue to be Accurate and True.

Never assume that a Strategy is correct only because it was the Strategy in the last war (or during the last 50 years) or last month. Always keep in mind the FACT that **Strategy and Tactics are evolving and changing at ever increasing speeds.**

(98) For Successful Military Grand Strategy Planning one of the First Things that the Grand Strategist must Conquer is his Own EGO.

For over a half Century the US Pentagon Strategy planners have automatically assumed that they would win every war because they were "the American Military and the American Military wins all of its wars."

The fact is that the US Pentagon's Strategy planners have NOT clearly and decisively won a significant shooting ware since 1945. **The Pentagon needs some healthy humility about its own lack of talent.** The proven Axiom since the end of WW2 has been "Where the US Military goes the US Military loses the shooting war."

(99) No Nation or Empire can be Strong Everywhere but the Nation that wants to Survive must Work Hard at NOT being Exceptionally Weak in any Vital Area.

Any Nation's Military with an understanding of basic Grand Strategy will NOT attack their opponent's strongest areas - but will attack first their opponent's weakest factors. **It is your weakest factors that are going to defeat you.**

Future war opponents of the USA are NOT going to play the Pentagon's games with the Pentagon's rules and attack where the Pentagon is the strongest. They are going to attack where the USA is the weakest.

(100) General Sun Tzu wrote that One of the Greatest Grand Strategy Military Weapons is Patience.

Sun Tzu advised that a Nation (given a choice) should only go to war when it is 100% ready to attack. In most cases it is much better if you have the patience to wait for your enemy to come to you.

(101) Never make the Tactical and Strategic Mistake of Believing that if Something Can Be Attacked then it Should be Attacked.

Never turn things upside down by saying that if an attack against a target is doable then because it is doable that means it is good Strategy and should be enacted .

Could the US Military bomb several things in Iran? Yes - that is easily doable. Any moron can start a war. Could the US Military win a shooting war with Iran? No. It would be another Vietnam but 10 times worse because today the USA is much more economically fragile then in the 1960's when the USA had a positive trade balance and was a creditor to the world.

(102) The Best Grand Strategist Never takes the Ability to Continue Fighting a War as Justification for Continuing to Fight a War.

Throughout history many wars became an end in themselves. The fighting became the justification for the fighting.

Washington DC's wars of the last 60 years (Vietnam, Iraq, Afghanistan) were (and are) wars that were continued mainly because they could be continued. **A war must NEVER become its own justification.**

(103) Wars are NOT Parlor Card Games where you Build Up Points with Battlefield Wins until you have Enough Points to Win the War.

One of the battlefield's proven realities (particularly in the last 100 years) is that winning battles keeps becoming less significant for the eventual outcome of wars. During the Vietnam War (1959-1975) the US Military won all of the battles (hundreds of won battles) - and was eventually defeated.

During the Vietnam War all of the USA's Generals were wrong. The decisive USA battle was NEVER in Vietnam. The decisive battle of the Vietnam War was 8,500 miles away from Vietnam on Main Street USA. The US Pentagon lost that battle.

If a USA-Iran shooting wars happens the decisive battle will once again be on Main Street USA and the US Pentagon will once again lose.

(104) Before a War Begins the Grand Strategist puts great amounts of Work and Planning into exactly How he is Going to Win the Last Battle of the War.

Your plan for winning the last battle of the war must be extensive, accurate, highly specific and absolutely doable before the first shot is fired.

In modern warfare it makes no difference who wins the first battle or who wins the first 100 battles. The only battle that is important is who is going to win the last battle.

During WW1 the German Army lost 360,000 troops killed winning the Battle for Russia (1914-1917). However Germany lost the war.

(105) The Grand Strategist Never Falls in Love with a Plan or Policy. If a Plan Stops Working and/or is Not Producing Significant Results then the Plan is Quickly Changed. The Motto of the Grand Strategist must be - "Fail Fast."

The overall rule is that **any long shooting-war is by definition a bad war.** If the shooting war tool is failing then you need an aggressive use of ALL of the non-shooting tools.

(106) There are Gold Nuggets in the History of Past Military Wars but Finding Them (and Appling them to Modern Challenges) Requires significant Strategy Talent and Work.

The overall Rule is that you have to work your way through tons of mud to find the Gold Nugget. Most "inventions" were the same - much highly frustrating hard work before a payoff.

(107) The Grand Strategist is Super Cautious when faced with Military Challenges that have Crossed the Line between the Different Ages of Warfare.

The Industrial Age of Warfare was from 1885 to 1960 and the Black Power Age of Warfare was from 1380 to 1885. However **from 1914-1917 all of the Generals in WW1 were using tactics that were mainly from the bygone Black Powder Age of Warfare.** The results were some of the biggest battlefield disasters in world history.

The world is transiting into the Artificial Intelligence Age of Warfare and when that happens it will be a SUPER DANGEROUS TIME for shooting-war Generals being mentally stuck in the past age of warfare like WW1's Generals. However - **it will be a super advantageous war winning time for the strategists who have their mindsets and actions firmly based in the next age of warfare.**

At this point in history the grand strategists who seem to be thinking about the next Age of Warfare appear to be mainly Chinese Generals and Not America's Generals.

(108) There is No End to History.

Nations (like most people) are always **looking for a final resting point where everything is settled and everything stays the same.** Especially if they are mainly on top like the USA today.

The "finish point that will last forever" is always a highly dangerous illusion. The Grand Strategist is clear in his mind that **the world's power equations are always changing and history never stops.** Today's great power is tomorrow's has-been and today's insignificant area is tomorrow's great power.

(109) The Grand Strategist understands that a Nation's Gross Military Firepower is only ONE of the Dozens of Vital Factors that go into the National Power Equation.

From the start of the Information Age of Warfare (around 1960) the factor of Gross Military Firepower has continued to decline in important in the Grand Strategy equation. However - **the ability to effectively communicate your ideas to the world continues to be more important with each passing year.**

Never forget that **during the Vietnam War the US Military had a 1,000 to 1 advantage in gross firepower and was defeated.** To win wars today the full matrix of ALL vital strategy factors must be predominately on your side. In the USA - two of the most important factors are (1) continued support for a shooting war from the TV News Press and (2) continued support from the voters.

(110) To Continue being a Great Power in the 21st Century (or to become a Great Power) you Must Have the Basic 7 Factors of Great World Power and Influence.

The 7 main factors are:

(1) A strong financial system that produces a National profit (i.e. positive trade balances and government surpluses).

(2) You need a Nation that produces more than it consumes and is not deeply in debt.

(3) You need a National people who are not afraid of work and pain. You cannot have a Nation addicted to pleasures.

(4) You need a National Vision of higher values for improving the world. Ideally you need a semi-religion like "Democracy."

(5) You need to be on the moral high ground and your citizens need a sense of mission for building a better and fairer world for everyone.

(6) You need a Nation that is on the upside of its Bell Curve.

(7) You need a Nation with an unshakable will-to-win.

The USA had all 7 of the above from 1941 to 1945 when it become the world's greatest power. However (and unfortunately) you could argue that today's USA has none of the above 7 in strong amounts.

(111) Never Build your Plans Around the Belief that if You Act as if a Desired Conclusion has Already Happened - Then It Must Happen.

Wishful thinking is NOT effective Grand Strategy. Never build your Grand Strategy ass-backwards around your wishful thinking. **Do not base your actions around the idea that your desires and dreams about the future are guaranteed to become a reality if you act like they are a reality.**

(112) The Grand Strategist is Always Doing MORE of what is Working and LESS of what is Not Working.

The above sounds like a simple truth that every child should know. However for over a half Century Washington DC has been doing MORE of what is failing. Standard Washington DC practice is to take proven failures and make them bigger.

(113) The Grand Strategist has the Courage to Know when a War has Failed and the Courage to End It and Move On.

The above Rule sounds simple but the US Pentagon (and Washington DC) are too weak to admit that something is a failure and a defeat. So the USA's average citizens (decade after decade) suffer endless wars in stupid places that have zero importance to the average American - while the USA's National Debt continues to become worse year after year.

Before WW2 Americans used to talk about "kitchen table logic." One of the rules used to be - "if you cannot afford it then do not buy it." **Stupid wars in stupid places is a bad habit that the USA can no longer afford.**

(114) One of the Best Talents the Grand Strategy Leader can Have is the Ability to Recognize People with Talent and then Put Them into Positions Where They Will Be The Most Effective.

The core problems are (1) you first need people with the talent to recognize other people with talent and then (2) the courage to promote them into meaningful positions. However - in today's Washington DC the mediocre hacks hire other mediocre hacks and the socialist idiots hire other socialist idiots.

(115) History has Proven that National Leaders (and Major Political Parties) can turn into Psychopaths, Schizophrenics and Megalomaniacs. Your Opponent (or Perhaps Your Ally) or a Major National Party inside of your Nation Can Become Dangerously Insane.

The key fact that you must understand is that Homo sapiens in the USA are NOT a different species from the Nazis, Stalinists, Tojo's Japanese, Cambodia's Pol Pot, the Chinese Red Guards etc - all Americans are Homo sapiens. The ONLY thing that has prevented the type of common world insanity (that has happened everywhere in the world at one time or another) from happening in the USA is the USA's Constitution. However - in **today's USA the left-wing radicals want to destroy the US Constitution.** Then the USA would be open to all of the insanities that have affected other Nations in history.

Without the US Constitution the USA is wide open to becoming the worst of the worst in world history. The USA would be open to self destroying itself like the USSR or Nazi Germany self destroyed themselves.

The US Constitution is NOT broken and the USA does NOT need America's left-wing Communists "fixing" the Constitution. If Communism is their paradise they should move to Cuba.

(116) Every War is Dynamic and every Big War has Black Swans - the Unexpected that can be Either Good or Bad for Your Overall Plans.

Wars never go the way that the original planners intended. There are always Black Swan surprises. The strategist most be very flexible (and fast) with his dynamic planning. You must never be a 100% slave to any big plan.

Remember the Axiom - **Your war plans are under your control ONLY until the first bullet is fired. After that anything can happen and probably will happen.**

On the good side sometimes there are unexpected Black Swans that are excellent opportunities for your side. You need the ability to change your set-plan and quickly exploit the unexpected opportunities for your advantage.

(117) The Grand Strategist does Not make the Mistake of Wanting to Control both the Grand Strategy and Every Detail of Front Line Tactics.

The Grand Strategist works towards the big picture objectives and gives his frontline commanders as much battle decision and management freedom as possible. If the Grand Strategist does NOT trust the ability of his frontline Generals to win battles (and be competent managers) then he needs new Generals.

(118) One of the Most Valuable Skills a Grand Strategist can have is the Ability to SEE what is NOT There and HEAR what is NOT Being Said.

What is NOT being said by your opponent is almost always some of the most important information. The best Strategist is excellent at hearing what is not being said and seeing what is not there. However **acquiring this decisive skill takes both extensive training and natural talent.**

The Grand Strategy Team when discussing their opponents should always be asking "What are they NOT saying?" And - "What is missing in the pictures we are seeing?"

(119) There is No Honor or Advantage in Continuing to be in a Losing War Once it is Clear that your Grand Strategy is Unworkable.

When it is clear that you are in a dead-end no win war then you must find the courage to stop murdering your own troops. If that includes ending the war on disadvantageous terms then that is what you do. If it means a new mix of Allies then that is what you do.

You do what is needed to stop your loses and move on. Remember the Axiom - **you cannot build the new without first discarding the old.**

(120) Grand Strategy Planning must be Devoid of Righteousness, Honor, Reputation Defense, Ego and Status Defense and your Subjective Views of What is Justice.

Righteousness, justice, honor and all of the other missionary, crusader and ego emotions are valuable to the Grand Strategist only to the degree they can be used to leverage the motivations of the Strategist's troops and voting citizens.

The best Grand Strategist is always 90% of the time thinking about "costs versus real benefits - and costs versus useable benefits."

The best Grand Strategist always tries to be on the Moral High Ground and thus does NOT have to Lie - but only tell the truth about his motives and what is happening.

(121) When the Major Historical Trends and events are going in Your Favor and Against your Opponent then one of the Best Strategies is to Postpone and/or "Slow Walk" major Military Actions and let the Historical Dynamics Weaken your Opponent.

The USA won one of the greatest victories in world history (the Cold War) by having the discipline to NOT have a shooting war. When you are on the right side of history then time is on your side.

(122) The Grand Strategist is always Asking Himself "Is this New Weapon going to be Useful to Those actually Doing the Fighting or is it an "Old Guard" Worthless Ego Toy from a Bygone Age of Warfare?"

The USA's military draft laws (that remain on the books today) are an aggravation producing relic from a bygone Age of Warfare. The Draft Laws should all be officially cancelled.

(123) The Grand Strategist Never Falls in Love with a Weapons System because Every Weapons System in the world will Become Obsolete.

It is best to be the first to obsolete an old weapons systems - and NOT the last to do so. Do NOT be the Polish Army in 1939 attacking German Armored columns with Horse Cavalry and Lances.

The US Army officially ended and disbanded Horse Cavalry in April of 1942.

(124) In the Last 120 years many Major Weapon Systems and Tactics have Became Obsolete Long Before most Generals and Admirals Were Able to See them as Obsolete.

Many Generals and Admirals believe that their fantasy world of fighting yesterday's wars will become a reality if the Generals and Admirals build weapons designed to fight yesterday's wars.

(125) One of the Grand Strategist's Objectives is to Force his Opponent into Making as Many Mistakes as Possible.

In shooting warfare the side that often won the war was the side the made the fewest big mistakes.

The **Napoleonic** way of winning battles is to seize and retain the initiative by using superior speed of action. That causes your opponent to be thrown off-balance and to make mistakes. Then with super speed and daring you exploit your opponent's mistakes.

During WW2 the British manipulated events and caused **Adolf Hitler** to make the decisive mistake of thinking the D-Day invasion would be at Calais, France rather than almost 200 miles away (by road) in Normandy. On D-Day the main German forces were in the wrong location (and thus off-balance) and never were able to gain the initiative.

(126) Carl von Clausewitz's Main Rule says that Winning Battles Means you Win the War. Today that Rule is Obsolete and Dangerously WRONG.

In modern warfare (and especially in wars that have many Guerrilla War characteristics) **winning battles means almost nothing towards winning the war.**

(127) The LAST Nation that Enters into a Major War is Usually One of the Biggest Winners.

The single biggest reason for the US being the biggest winner in both WW1 and WW2 was because <u>the USA was the very last major power in the World into those Wars.</u>

The first 2 Nations into WW2-Europe were Germany and Poland. Poland lost 19% of its total 35 million population and Germany lost 9% of its total 69 million population. The USA was the LAST major power entering into WW2 and lost 1/3 of 1% of its total population of 140 million.

During what was WW2 for the USA (1941-1945) the USA's total population increased in size with birth and immigration from 132 million to 140 million. The USA ended WW2 with 8 million more people then when it entered into WW2.

(128) The Grand Strategist is constantly Re-Running His Strategy Calculations to Make Sure that His Plans and Actions are Always on the Probability High Side.

Modern wars are dynamic and are becoming increasingly dynamic. The Strategy equation is always changing. Yesterday's excellent Strategy plan (or Ally) might be tomorrow's disastrously wrong Strategy plan (or Ally).

(129) The Grand Strategist Needs Subordinates to Help carry out his Plans and Appoints Subordinates Based on their Past Record of Being Correct, Effective and Intellectually Honest.

Subordinates must NEVER be chosen for their ability to tell the Grand Strategist (or Generals) what they want to hear. <u>The Grand Strategist guards against having "yes men" around him.</u>

One of the many problems with the old and bloated US Pentagon is little tolerance for "controversy." If something is impossible to do (and you are a younger Officer in the Pentagon) and say it is impossible. You will probably be labeled "a defeatist" and not promoted. <u>Old organizations ususally end up promote "yes boys" and they become increasing mediocre.</u>

The correct actions for a Leader is that if none of your subordinates ever suggested that you are wrong - and only tell you that you are always correct - then fire them.

(130) A Main Objective of both Tactics and Strategy is to Wherever and However Possible You must Try to Put Your Enemy on the HORNS OF A DILEMMA.

Putting your opponents on the Horns of a Dilemma is one of the most powerful battle tools for winning both battles and wars.

The historic way to put your opponent of the Horns of Dilemma is to simultaneously threatening several of your opponent's different positions at the same time. **You keep your opponent guessing about your specific attack objectives.**

To do the above you must first capture the initiative and keep the initiative. This is why capturing and keeping the initiative is usually decisive.

An example was US General **Sherman's** "March to the Sea" through Georgia and South Carolina (November-December 1864). General **William Sherman** (1820-1891) would continually threaten two or more objectives and the Confederate forces never knew which one he would attack. So the Confederates had to divide their forces. Then **General Sherman** would quickly concentrate his forces and attack one of the targets (destroy it) and then move deeper into the South and do it all over again.

The ultimate bold tactic in land warfare is to break away from you own lines (as **Sherman** did in the US Civil War and German General **Guderian** did in the 1940 WW2 Battle of France). Your Army operates in enemy territory like a ship at sea. However it is a big risk with historically a high probability of either stunning victories or being destroyed. **Napoleon** in 1812 broke his Army away and operated like a ship a Sea - and was destroyed.

Today's Terrorism is a variation of Guerrilla Warfare and **Guerrilla Warfare is a variation of putting your opponent on the Horns of Dilemma.** It is impossible for your opponent to be strong everywhere so the terrorist can choose his target and concentrate his surprise attack on a point of weakness.

The best solution for permanently winning Guerrilla Wars is to use General **Louis Lyautey's** methods of capturing the communications and salesmanship initiative (see Chapter 21). **You win Guerrilla Wars by drying up the Sea of popular local support that the Guerrillas swim in. That has been done and can be done.**

(131) If there is a Possible Attack against you (during wartime or peace) that Makes No Sense and would be Strategically Wrong and illogical for the opponent. The Grand Strategist NEVER Assumes that such an Attack will Not Happen Only Because it is illogical and Makes No Sense.

Many National Leaders have been insane and insane people do insane actions. Major powers like the USA need a large Grand Strategy Team that has one of its many specialties doing on-going psychoanalysis of every National Leader including all of the USA's Allies.

The Strategy Team must have an accurate understanding of who is sane and who is insane. A common complication is a Leader who is only insane in specific areas and on specific subjects while being brilliant in some other areas.

(132) The Grand Strategist must have a Mental-Set High Degree of Tolerance for Ambiguity.

The degree of ambiguity multiplies upwards as you go from: tactics to Grand Tactics to Strategy and Grand Strategy. A good Strategist and General must have the ability to operate decisively with only part of the information desired.

(133) If your Military Arrives at the Point Where They Believe that all of their Decisions are always Perfect then Defeat will Probably Not Far Away.

One of Washington DC's many problems is too much arrogance. Washington DC has NOT clearly won a significant shooting war since 1945. **Washington DC needs a little more humility about its war winning lack of talent** and lack of results.

Spending more money than anyone else does NOT guarantee winning wars as the USA's defeat in Vietnam proved.

(134) If Year After Year you Find Yourself Fighting basically the Same Battles in the Same Places and for the Same Reasons it means your Overall Grand Strategy is Defective and You Need to deeply Re-examine all of Your Basic Assumptions about Everything.

The current USA versus the Muslim World shooting war started in 1991 with operation Desert Storm against Iraq. In 2019 US Troops were in Iraq fighting "the enemy." The USA invasion of Afghanistan started in 2001 to supposedly destroy the Taliban. In 2019 the Taliban controlled about half of Afghanistan. The US Military has been basically fighting the same battles in the same areas of Afghanistan since 2001.

Long wars are a gross failure of military leadership and a gross failure of Strategy. Generals who fight never ending wars are incompetent by definition.

(135) When a Nation Changes its Grand Strategy it Changes its Relationship to the World and it Changes the Ultimate Destiny of that Nation.

If YOU change your mindset and actions significantly then YOU have changed your world. If a Nation changes its mindset and actions then it has changed its future.

(136) The Grand Strategist is always Asking Himself if the World's Realities are In-balance with his Nation's Grand Strategy.

A classic mistake with great powers (especially great powers on the downside of their Bell Curves like the USA) is that **the world changes but the Strategy of the declining great power does NOT change.**

(137) The Grand Strategist does NOT automatically Credit Officers with Outstanding Military Skills and Talents only because the Officers Won Easy Battles in Easy Wars.

The National Leader does NOT credit Generals who never won a shooting war (like the USA's Generals) with having great war winning talents.

(138) The Grand Strategist has an Achievement-Based Definition for "War Hero" as Being a Person who went Far Above the Call of Duty and was an Outstanding Example of Duty and Courage for Everyone in his Nation.

The true War Hero had to make an outstanding contribution to the morale of the war effort. His or her actions had to boost the pride of their Nation. **Their actions had to be an outstanding example of courage for their fellow soldiers and citizens.**

(139) The Least Desirable Battle Tactic is a Head-On-Frontal Attack Against an Opponent who is Expecting an Attack.

Direct head-on attacks should usually be avoided unless the attacking side is overwhelmingly stronger than the defenders and time is a crucial factor in the overall strategy big picture.

(140) The Ultimate in Grand Strategy Talents and Skills is to Convince (or Con Game) Somebody Else into Fighting Your Shooting Wars For You.

On the other hand the ultimate in military incompetence is to be tricked into fighting somebody else's wars as the USA's Generals have been doing for many decades.

Washington DC's Middle East "Allies" who say they "love America" and take the USA's free war materials and free money (while giving American troops zero combat help) are **exploitive phony Allies.**

Washington DC is a collection of naive-dupes for not being able to see the phony Allies who are (1) all take and zero give and (2) highly successful at tricking Washington DC into fighting their wars with the Middle East's Muslims.

(141) Knowing all of the Axioms of Grand Strategy and Grand Tactics will NOT by Themselves Make you a Competent General or National Leader.

Reading the axioms in this book is the easy part. However **having the talent and courage to enact the Axioms is the true test.** Knowing the Axioms of Strategy is like having a gun. If you do NOT have the courage to use the gun then it is less than useless because it could get you into trouble but not out of trouble.

On a personal bases never go out waving a gun at a burglar or whatever if you do NOT have the mindset to use that gun.

(142) The Grand Strategist Concentrates his Mind, Plans and Actions on the core of his Opponent's Strategy.

Your opponent's Strategy is always the main objective of your Strategy. You as a Grand Strategist must constantly be asking.
(1) "What are my opponent's overall objectives and desires?"

(2) " What is my opponent's Strategy for achieving his objectives and desires?"

(3) "What is the best way to attack his Strategy?"

(4) "How many of the 37 main types of warfare can I use to block and defeat my opponent's Strategy objectives?"

(143) One of the Most Valuable Talents (in both the Art of War and Business) is the Ability to always be Putting the Odds of Winning in Your Favor.

Business Strategy (and gambling strategy) and War Strategy are the same. You always try to position yourself so the odds of winning are on your side. **What you always want is a small loss if you lose but a very big reward if you win.** What you never want is a high risk of losing and only a small benefit if you win.

(144) Before Starting any War you MUST have the Basic Four Requirements.

(1) A specific list of exactly what would be highly valuable and materially advantageous to the people of your Nation by going into the war.

(2) A total analysis of the dozens of factors in the war-equation and a clear analytical conclusion that the odds of winning the objectives (and the odds of a short war) are on your side.

(3) There must be real and specific material advantages for YOUR NATION'S PEOPLE. Never try to "mold the world" into your subjective tastes when there are NO real advantages to your Nation's citizens for doing so.

Rule #3 is especially important for the USA (that is on the downside of its Bell Curve) and will never have the power to completely "mold the world" into its subjective (and often flawed) desires.

(4) You must be sure that when the dead bodies of your soldiers start arriving home you can retain the will-to-win support of your Nation's TV News reporters and the voters.

The USA's Generals must factor into their plans the fact that most of the USA's "Press" are left-wingers and anti-military. The Warmonger Lobby can and will use Lies and Con Games to start future wars but the Lies will last for only a limited time.

The will-to-win is always a major factor in the Grand Strategy analysis. There are no scientific ways to exactly measure the will-to-win and/or how the will-to-win will and does change during wars. **This is one of the areas where war and peace becomes Art and it depends on the God given talent of the Grand Strategist to make the correct analysis.**

(145) Nations that have Great Wealth (and an Abundance of War Weapons) Run the Risk of Believing that Winning Wars is All About having a Large Number of Assets.

Physical war machines and raw firepower are only one part of the Grand Strategy equation. Additional important parts are, talent, intelligence, the will-to-win and the many vital parts in the Grand Strategy equation that have little to do with wealth.

The USA in particular must remember that in the Vietnam War the USA had a 1,000 to 1 advantage in money spent on the war and the US Military was defeated.

During the American Revolution 1775-1783 the USA's population was only 2.5 million - 1/3 of that number were loyal to England and 1/3 did NOT care who won. So a USA de facto war population of 835,000 (mainly poor farmers) defeated the British Empire with a 1776 total world population of 40,000,000.

North Vietnam that had a population of 16,000,000 in 1960 eventually defeated the USA that in 1960 had a population of 181,000,000. The point is that **the grand strategy equations are highly complicated.**

(146) No Strategy plan is Forever. Every Strategy plan that Ever Existed (and exists today) will Become Obsolete - the Only Question is "When?"

No Strategy plan is forever. No ally is forever. No enemy is forever. No great power was forever. The USA will not be the biggest military power forever. The US Dollars will not be the World's #1 reserve currency forever. Everything changes.

Everything in the world has the word "temporary" in front of it and the Grand Strategist always plans accordingly.

The longer a Strategy Policy has been in play then the more you should assume it is on the downside of its Bell Curve (and is possible obsolete). There must be constant Strategy Policy reevaluations to make sure your main core-assumptions continue to be relevant.

Chapter 9

How to Have Short, Fast and Successful Wars

Every Strategy plan, every weapons system, every Ally and every program should undergo periodic vigorous cross examinations by skilled experts at regularly planned time periods.

The Grand Strategy Team should have pro and con juries to debate and cross examine everything. **NOTING should be automatically accepted as correct only because that was the way it was done for years.** What cannot pass a vigorous cross examination by skilled strategy experts needs to be updated (or discarded) and replaced with a better strategy with a long lasting peace as its ultimate objective.

Wars become long and endless and stupid when old bloated military organizations (like the US Pentagon) try to fight yesterdays wars with yesterday's tactics and mindsets. **The way to shorten wars is to fight today's wars with tomorrow's tactics and mindsets.**

(147) The Laws of Physics DO Apply to Military Actions. The Grand Strategist is Taking a Great Risk when He Ignores the Laws of Physics.

Clausewitz was correct when he wrote that (1) the determination to win and (2) the will to victory and (3) pride and confidence are vitally important in warfare. However **all of the football locker room bravado (and USA war room profanity) cannot by themselves change and overcome the basic laws of physics.**

If you are a bowman and your bow has a range of 350 yards (and your target is 375 yards away) all of the "determination" or "pride" or "profanity" in the world is NOT going to make your arrow hit its target.

(148) The Grand Strategist understands that Albert Einstein's Theory of Relativity is a Core Tool for Military and National Strategy Analysis and Planning.

The Theory of Relativity says that everything is in motion and the true picture of your position and speed can only be accurately understood in "relation" to everything around you.

When this book was published many Americans were looking only at the USA and saying, "We are better than we were 20 years ago." YES that is absolutely true compared to the USA of 20 years ago. However **the USA's military and industrial power in "relation" to the world's military and industrial power has been in decline since 1975.** The factual reality is that relative to the world the USA is in significant declining.

In 1950 much of the Middle East Muslim world consisted of Muslims living in tents (or in buildings much like they had 300 years ago). The USA on a relative basis was far more advanced in every vital area. However **around 1950 the Muslim World reset its Bell Curve onto an upwards trajectory.** Today a few parts of the Muslim World have already closed many of the gaps relative to the USA.

The USA's Strategists must assume that the Muslim Middle East's reset Bell Curve will probably continue upwards for all of this Century. The Strategist should also assume that the USA's Bell Curve (relative to the world) will probably continue downwards for all of this Century. **The USA's grand strategy MUST be based on and built around the world's realities.**

(149) As the Major Factors in War-and-Peace Evolve the Strategist Must Have the Mental Ability to Alter his Strategy Plans Significantly and Rapidly.

The Strategist makes the best plans possible for the situations at the time. He always keeps in mind the historical fact that wars and peace are dynamic and the life of a Strategy plan might be unexpectedly short.

Many times in history the General's strategy battle plans did NOT outlast the first day of battle. You have to be ready with a Plan B, a Plan C and a Plan D.

The brilliant Strategist Field Marshal **Helmuth von Moltke** the Elder (1800-1891) said - "A General can plan a battle with certainty only up to the first bullet being fired."

(150) Wars are like Street Cars. If you "Missed this War" Then Wait a Short Time and Another One will Soon Arrive.

It seems like the US Pentagon (of the last 60 years) has lived in great fear of missing a war. However the Pentagon since the end of WW2 has never decisively won any significant shooting war. The US Pentagon clearly needs a major reorganization.

During the last 2,000 years of history there have been wars somewhere in the world about 95% of the time. Nobody alive today ever lived in a world where there was not some types of shooting war, revolution and Guerrilla action somewhere in the world. We do NOT have to worry about a "war shortage."

For as long as there are Homo sapiens there will be wars. You study warfare to understand how to make wars shorter and less destructive. **You are studying this Book (and the 37 kinds of warfare) to win wars without ever using the shooting combat tools of war.**

(151) Every Nation and Culture has its Own Collection of Delusions and Falsehoods. The Grand Strategist Puts great Effort and Work into NOT Becoming a Victim of his Own Group's (or Culture's or Nation's) Myths and Self-Deceptions.

Many times in history groups and political parties and Leaders have become victims of their own propaganda, lies and self-created mental-delusions. Often to the great detriment of the groups or Nations that are self deceiving themselves.

The Grand Strategist is always asking himself - "Am I fooling myself? Have I become a victim of my own propaganda? Am I connected to the real world? Or - am I living in my own fantasy world?"

(152) World History is full of Examples of Where the Leaders Started their Planning with the Conclusions they Wanted to Happen and then Worked Backwards. That is a Common Mistake that Often has Disastrous Consequences.

Many groups do their Strategy planning ass-backwards. **They start with the conclusion they desire and then build their actions around the ending they desired being a reality.**

In 2003 the Washington DC so-called "war planners" started the USA-Iraq War 2 with the conclusion-delusions that the war would be fast, easy and cost little in USA lives and money.

Washington DC proceeded with the war as if their fast and easy delusion-fantasies were guaranteed. Washington DC's and the Pentagon's "planners" had decades earlier made the same mistakes with the Vietnam War (1959-1975). They were 100% sure that the Vietnam War would be fast, easy and economical. WHY? Because that is what their delusions wanted.

Currently the Warmonger Lobby believes that a USA shooting attack on Iran would be a short and easy war and a full surrender of Iran to the USA. They are very wrong.

(153) The Grand Strategist Never Assumes that his Wars and Plans are Absolutely Correct Because they are His Wars and His Plans.

The fact is that NO USA Military Officer (active duty or retired or alive) has ever been part of a USA planning team that decisively and clearly won a significant shooting war. **Based on the proven outcomes of all of the shooting conflicts since the end of WW2 any "Pentagon Shooting War Strategy" is probably going to be 70% wrong.**

Leo Tolstoy (1828-1910) Russia's greatest writer was a Russian Officer in the Crimean War. He wrote the greatest War Novel in the world - ***War and Peace*****.** Tolstoy wrote **"The most repulsive and dangerous of all politicians and military leaders are those who imagine that they possess the absolute truth based on a military science of their own invention."**

(154) The Grand Strategist is like a Good Boxer.

To be a good boxer (and good Strategist) you have to be light on your feet. You cannot have a lot of mental fat. You might be old but your mind cannot be old. (Some 30 year olds have old minds and some 70 year olds have young minds).

You have to always be moving and always maneuvering. You need a good defense. **You need to know where your opponent's thinking and actions are going to be in the future and NOT where his thinking and actions were in the past.**

You need to immediately see the opportunities and then hit fast and hard to exploit the opportunities. <u>You put time on your side by increasing the speed of everything you are doing.</u> You must fight in a way that will psychologically dominate your opponent.

Most importantly - you have to capture the initiative and force your opponent to always be reacting to what you are doing.

(155) For Thousands of Years the Possession of Land was a major Objective for most Wars. Land was Directly Equal to Power and Wealth until the Start of the Industrial Age of Warfare around 1885.

Historically the most common cause of wars have been Border Wars over land. In the 21st Century there continue to be Border Wars over land. However **land is no longer the basis of military power.** Intellectual ability, intellectual courage and the will to work very hard are the new engines of National and military power. <u>The Japanese proved the above in the last Century</u>.

In the 21st Century occupying land continues to be less and less important while occupying the space in people's minds continues to become all important.

(156) The 2 Most Used ways to Change the Power Equation between Nations are - (1) Weaken your Opponent with Shooting Warfare. (2) Make Yourself Stronger (compared to your opponent) with Economic and Diplomatic Warfare.

Making yourself stronger with the Economic and Diplomatic non-shooting tools of warfare always has a higher degree of long term success. It has been the Strategy of China since 1980. Since 1980 China has had a 250% growth in its economy with zero shooting wars anywhere.

Attacking your opponent with the shooting war tool (with the objective of making your opponent weaker) is the WW1 Verdun Mistake. **You will bleed and weaken your opponent but your opponent will also bleed and weaken you.** The mistake that all shooting warmongers make is the belief that their side will NOT be weakened in the process of attacking their opponent.

The Nations that are the biggest winners in Attrition Shooting Wars are always the Nations who were NOT in the shooting war (like the USA in WW1 and WW2 being the very LAST major power into the wars).

If "American wants to be great again" then it needs to RETURN to NOT being in shooting wars anywhere.

(157) The Grand Strategist understands that in the current Information Age of Warfare there are Brand New Roads to Success that did Not Exist in Past Ages of Warfare.

The new and DECISIVE military weapons system in the 21st Century is Salesmanship. Salesmanship and Internet-Warfare are now decisive weapons for either victory or defeat.

The US Pentagon needs to find its **Leni Riefenstahl,** its military **David Ogilvy** and its creative **Steven Spielberg.** The great irony is American Culture and Civilization has the best sales and persuasion talent in the world but the US Pentagon and Washington DC have a fanatical death grip on an obsolete tunnel vision stupid WW2 mindset.

If the USA wants to slow its decline then the talent of (1) Hollywood and (2) Silicon Valley and (3) the Pentagon ALL have be on the same sheet of paper. That is going to take considerable adjustments in attitudes and actions on everyone's part but IT IS ABSOLUTELY NECESSARY FOR THE USA's long term survival.

(158) Sometimes the Causes of Conflicts are illogical and you need to Understand the Logic of illogic Before you can Correctly Understand Your Opponent's illogical Strategy.

If your opponent has an illogical Strategy then sometimes the correct Strategy is an illogical counter-Strategy. **The best Grand Strategy Team must have the talent to understand the logic of illogic and the logic of insanity.**

(159) History is Starting to Transition from the current Information Age of Warfare into the Artificial Intelligence Age of Warfare. The World is Going Into an Exceptionally Dangerous Time for Military Planning. However Where there is Danger there is Also Opportunity.

The biggest danger period for Strategy and Tactical disastrous mistakes is when history is transitioning from one Age of Warfare into the next Age of Warfare. **The transition will present decisive opportunities for those Nations and groups who can put themselves out in front of the new Age of Warfare weapons changes and most importantly the new mindset changes.**

China has made no secret of its intentions to dominate the next Age of Warfare - the Artificial Intelligence Age of Warfare.

Historically old organizations like the US Pentagon fight the changes they need the most.

(160) The World will Continue to have Many Wars. If a Regional Small War (or Revolution or Civil War) <u>can be Contained</u> then Often the Solution is to Let the Local People Fight It Out and Get the Disagreements Out of Their Blood and Out of Their Systems.

Washington DC has to STOP destroying the USA by trying to end any and all small wars everywhere in the world. **Small wars relieve some of the tensions and adjust unbalances and contradictions.** Sometimes wars are the ONLY cure for the National insanities that periodically happen everywhere in the world.

History has proven that many times it is better to have some smaller wars that resolve contradictions and disagreements rather than bottling up all of the pressures until there is a large explosion that brings big powers into the war and creates a major conflict.

Sometimes small wars are like many small earthquakes that relieve tensions in the system and avoid a major earthquake.

Unfortunately history has hundreds of times proven that many disagreements in the system are resolved only with violence. However history has also proven that sooner or later people become tired of killing each other and then the disagreements find a resolution and a new balance is established in the system.

The ultimate solution is a world legal system to resolve disagreements without shooting wars but that is not going to happen anytime in the near future. Washington DC **does NOT** support the International Court of Justice and for decades has fought against the Court.

(161) The Grand Strategist does NOT Tolerate having People Around Him who Twist Language and Destroy the Honest and Correct Meaning of Words.

When you subvert the meaning of honest words your main thinking-tool (language) is corrupted. The corrupted and poisoned thinking has in the past created wars that turned into National disasters for one side or all sides.

The Information Age of Warfare has turned the systematic corruption of words into a highly dangerous poison.

George Orwell (1903-1950) in his famous book ***1984*** (published in 1949) was correct when he wrote that the first thing any Dictator will do is corrupt the meaning of words.

The Chinese genius-philosopher **Confucius** (551-479 BC) had it exactly correct 25 centuries ago when he wrote **"The beginning of wisdom is to call things by their proper name."**

We could update **Confucius** for the 21st Century and say - "The beginning of political honesty is to call things by their proper names."

One of the vital steps for good problem solving is for you to keep the meaning of your words honest because then you can keep your thinking honest. Then you have a better chance of seeing the correct conclusions. On the other hand - **once you corrupt the words you use to think with then your thinking is corrupted and your conclusions are almost always highly flawed and possible disastrous.**

(162) Most Wars Start with All of the Sides Believing that They are Going to Get Everything They Wanted.

About 70% of wars in the last 4,000 years have ended with neither side getting everything they wanted. **Many wars ended with no side getting anything they wanted.**

From 1946 to now the USA has fought many wars. How many of them ended with the majority of the American people happy? The Korean War? No. The Vietnam War? No. The Iraq War? No. The Afghanistan War? No. The US fighting in Syria? No. The answer is "No" to every USA big shooting war since 1945.

The USA's people are getting shooting-war zero satisfaction from the US Pentagon that every year spends 36% of ALL of the military budget money in the world. That is a BAD deal for the American people.

Total world military spending for every Nation in the world is 1.9 Trillion dollars per year. The US Pentagon (every year) spends $700+ Billion of the $1.9 Trillion.

What is the Pentagon giving the American people for $700+ Billion per year? **The average US worker is paying for never ending shooting wars that start nowhere and go nowhere.** Every 10 years the US Pentagon spends $7 Trillion dollars and what the average American gets is stupid wars in stupid places for stupid reasons.

General Sun Tzu had it correct 25 Centuries ago when he wrote, "There are some areas where you should not be fighting." Knowing the areas where the USA should NOT be fighting (the Middle East) is one of the many strategy planning talents that is missing at the US Pentagon.

(163) The Strategy of Always Postponing your Next War has an Excellent Track Record. If you Keep Postponing Wars you will Eventually Completely Miss a War.

When the choice is between a (1) an imperfect cold war peace that is aggravating and frustrating or (2) a shooting war. **Based on the US Military's proven record of losing shooting wars from the end of WW2 to now the US Military should religiously avoid shooting wars in favor of Cold Wars.**

(164) War Prisoners Should be Treated as Humanly as Possible.

If you acquire a reputation for harsh treatment of war prisoners then your opponent's troops will be LESS likely to surrender and more likely to keep fighting to the death.

The American Military's once high moral standards of WW1 and WW2 have been deeply corrupted. There were no American Abu Ghraids or Guantanamos in WW2. During WW2 there were 425,000 German military war prisoners in the USA. Most went home with good feelings towards the USA and some returned and became US citizens.

Abu Ghraid and Guantanamo were giant steps morally downwards and backwards for the US Military and the world public relations image of the USA. The US Pentagon has adopted some of the highly corrupt fascist practices of one of its Allies and keeps dragging all of America deeper into the moral low ground. The disaster for average Americans is made worse because the Ally that the US Military is in love with is less than worthless to the average US voter. Millions of USA voters are ashamed to be Americans because of what the US Military does and that is a big contributor to many enormous political and social problems currently in the USA.

If the US Pentagon had anyone that understood grand strategy they would be able to see the problem. However - for grand strategy a big part of the Pentagon is living in the dark ages.

Washington DC and the US Pentagon MUST learn that in the Information Age of Warfare ALL of the world knows what is actually happening. The US Military is supporting and protecting anti-freedom fascists and anti-equal justice psychopaths and everyone in the world is aware of the facts.

Rather than the US Military becoming angry at those telling the truth why not stop supporting fascism and start supporting equal rights, equal justice and equal freedom for everyone everywhere in the world? The USA-USSR Cold War is long OVER. The USA no longer needs every fascist in the world.

(165) You must Never Accept Any Plan Only Because the Plan Came From a "Proven Expert." Many Times "Proven Experts" Have Been Very Wrong.

Napoleon Bonaparte brilliantly won many of the greatest battles of the Black Powder Age of Warfare. However at the end of the day the Russian Army was in Paris (1814) and **Napoleon** had become a disastrous mistake for France.

(History Footnote) **Napoleon** put his tallest soldiers into his elite Imperial Guard. There have been estimates that the average height of today's French males is one inch lower because of all of the French tall soldiers that died in the Napoleonic Wars and never had children.

(166) The Grand Strategist Never Forgets that in Every War the People that you have to Make Peace With are the People Who Are Shooting At You.

In the USA's never ending wars in the Middle East it is NOT about what Washington DC's politicians want - or what Washington DC's often worthless allies want. **The decisive question is what do the people who are shooting at the US Troops want? You do NOT make peace with yourself or your Allies. <u>You MUST make peace with the people who are shooting at you.</u>**

Dead people do not sign peace agreements. Unless you can kill 100% all of your opponents you are eventually going to have to build a peace agreement with those who are shooting at you.

The US Military cannot successfully kill 1.8 Billion Muslims so eventually Washington DC is going to have to make peace based on what the Muslim world wants.

(167) The Grand Strategist does NOT Start Wars or Continue Wars to achieve a Subjective Definition of "Perfection" in a World that Never was and Never will be Perfect.

The standard rule is that if something big (like a marriage or job or National policy) is about 85% good that is as good as it gets. Demanding complete perfection in big issues is insanity. **Starting shooting wars to achieve some subjective dream of idealized perfection is insanity.**

(168) The Grand Strategist will Psychoanalyze his Opponent's National Character for: Mental Blind Spots, National Delusions, Social Myths and their collected National Character Weakness Factors. The Grand Strategist will Take the above Analysis and Put it Into his Matrix of Grand Strategy Analysis for use Against his Opponent in Any and Every Way Possible.

Modern Military Grand Strategy is everything in your Nation against everything in everyone else's Nation.

(169) The Grand Strategist is Constantly Searching for New Roads to his Specific and Decisive Grand Strategy Objectives.

The more the Grand Strategist understands the entire big picture the more possible roads he can use for reaching his objectives. The more roads used the greater the chances for success. **If there are five roads to your decisive objective and your opponent successfully blocks you at four of them but not the fifth - then you win.**

(170) In every Conflict one of the Most Important Grand Strategy Questions to Correctly Analyze is if Time is on Your Side or if Time is on Your Opponent's Side.

If time is on your side then you can afford to wait and have a Strategy of blocking your opponent (like the USA-USSR Cold-War). **When time is NOT on your side you need significantly productive actions and as fast as possible.**

In the current USA versus the Muslim World War time is NOT on the side of the USA. The USA is DECLINING in power relative to the world. The USA needs to be doing whatever is necessary to build as many Muslim Allies as possible while the USA retains factors of value for leverage.

(171) The USA is a Young Nation and Americans View 3 or 4 Years as a "Long Time." However - for Many of the USA's Opponents 20 or 30 Years is NOT a "Long Time."

The Warmonger Lobby keep saying "A US attack against Iran would be easy." So the US Military lands some troops by the City of Chah Bahar on the coast of Iran - then what? Do the US Troops travel the 900 miles by road over mountains and desert to Tehran? Or does the US Military stay on the cost of Iran?

If US Forces stayed on the coast the Iranians would know that the US Troops had trapped themselves and were going nowhere. So the US Troops were there for 30 years - for the Iranians (Persians) with 2,700 years of history it would be - "so what - we know the ending of this story?"

For the Chinese 50 years is NOT "a long time to wait." The Chinese are waiting. The Chinese are waiting for the USA's massive debts to destroy the international value of the US Dollar and then China wins.

(172) There are NO Ultimate Settled Weapons Systems and there is NO Ultimate Settled Nation that will be a World Power Forever.

The idea that some war or action will "settle history forever" is a major recurring delusion. History will never ended. Homo sapiens will be 100% gone - the USA will be gone - and history will continue. Washington DC must find the courage to STOP trying to stop history and to enjoy life more and attack other Nations less.

(173) The Grand Strategist Works Towards Positioning his Nation's Assets and Advantages in such a Way so if there Was a Shooting War He would Win.

The Chinese are using their massive flow of US Dollars to buy strategic assets (particularly) on the Eurasian Continent - but also all over the world. The Chinese are positioning themselves (and their worldwide assets) so if the Washington DC idiots do start a shooting war with China the Chinese will win.

(174) How Easily (or NOT Easily) a Nation or Group Surrenders Depends in Large Part on what Type of War you are Fighting.

In many religious wars (and with semi-religions like **Stalinist** Communism or **Hitler's** Nazism) many times one side or both sides are willing to die fighting for what they believe. **One of the very last types of wars you want to be part of is a Religious War.**

On the other hand - wars motivated by mainly economic gains can sometimes be quickly ended if the desired economic gains of your opponent are blocked and/or you give them what they want.

The Japanese in WW2 were always fighting for economic gains - there were always fighting an Economics War. When Japan's hopes for China Colonial economic gains were blocked by Russia attacking Japan then the Japanese surrendered. On the other hand Nazism was a semi religion and the Nazis never surrendered until **Hitler** (their God) was dead.

(175) Individuals, Groups and Nations Should Never Assume they are Going to Win only Because they are Historically and Morally Correct.

The reverse of the above Rule is you should never assume the other side will lose only because they are morally wrong.

Sometimes you can do the right thing for your Nation (and be on the morally correct side) but nevertheless end up dead.

When German **Kaiser Wilhelm** abdicated (November 28, 1918) and Germany surrendering in WW1. Many German politicians during that time were Jewish: **Rosa Luxemburg, Karl Liebknecht, Kurt Eisner, Paul Levi** and so forth.

Germany by late 1918 was decisively defeated with zero chance of a military victory. The war had to end. Ending the war was absolutely the right thing to do for Germany and the German people. However starting in the 1920's the Nazis propagated the false idea that Germany had been "stabbed in the back by the Jews." One of the eventual results of that lie was the Jewish Holocaust of WW2 and the murder of almost six million Jews.

Most Jews in Germany during WW1 were loyal Germans. Ending the war was something that was desperately needed for all of Germany. However **many times in history people and groups were murdered for doing what was morally and ethically correct for their Nation.** History's biggest example is **Jesus Christ** who was trying to save Israel from destruction by radicals. A modern example is **Gandhi**.

(176) Never Start a War that you know is Beyond Your Capacities and Resources and Based only on the Hope that your Opponent will Grow Tired.

Wars are easy to start - however (depending on the type of war) can be very hard to end. **If your opponent achieves a stalemate and your opponent does NOT want to negotiate - then your options are grim.** This is what happened to the US Military in Vietnam and what would definitely happen in a USA-Iran Shooting War.

(177) The Grand Strategist Never Starts a War for the Reason of Revenge.

Some wars can and do become never ending circles of revenge wars that go on for Centuries.

(178) The Dream of Generals for Thousands of Years has been the Dream of having the Entire War Come to a Conclusion in a Single Titanic Battle.

For thousands of years of wars most of a Nation's Army was on the battlefield. The side that decisively won the main battle won the war because most of the opponent's Army had been destroyed. However - starting with the Industrial Age of warfare (1885-1960) Nations could build Armies with Millions of troops and winning wars with one battle became impossible. The **Clausewitz** Rule that says if you win the battle then you win the war - that Rule is now dangerous obsolete.

(179) The good Grand Strategist has the Ego-Stability and Courage to Admit to Having Made Mistakes.

The US Pentagon (and Washington DC) do NOT have the courage to admit to mistakes. Conveniently **if there was NEVER a mistake then there is No Need for a correction.**

Washington DC never admitting to any mistakes is the reason for WHY the US Pentagon (and Washington DC) keep doing the same old failures in the same old ways over and over decade after decade. <u>They refuse to admit that there are mistakes and with "no mistakes" there is no need to make changes</u>.

(180) Never Put Yourself Into the Middle of Somebody Else's Fight if You Do NOT Have to be There.

It is amazing how many National leaders in history want into the middle of a war that was NOT in any way vital to their Nation. If your Nation does NOT have a vital interest in the outcome of a fight then what is your sane reason for being there?

From 1775 into the 1960's the USA grew into the world's greatest super power with essentially ZERO involvement in the Middle East. So WHY is it now so essential to have US Troops constantly (since 1991) fighting in the Middle East? The author of this Book can see NO logical strategy reason.

A major reason for the US Military's never ending wars is because when you have a military machine that costs $7 Trillion dollars every 10 years you MUST have shooting wars somewhere to justify its existence. In addition - with that much money the Pentagon can afford to have live-action playgrounds.

(181) If You had to Bring Grand Strategy Down to only One Word that Word would be "Adaptability."

Millions of things in the history of the world arrived at the point where they could NOT (or would not) adapt to the challenges that **God** put before them. There are millions of once living things that are gone forever.

Many Nations and cultures that are alive now will somewhere in the future fail to adapt and be gone forever. Their governments, cultures, language, religions, National character - everything will be gone. It has happened many times in the past - ancient Greece, ancient Egypt, the Minoans, ancient Rome, the Mayans, Aztecs, Incas, Angkor Wat Cambodians etc etc. Someday the USA as we know it will also be gone.

Successfully adapt or DIE - this is God's Grand Strategy Rule #1 for everything in the Universe.

(182) The Grand Strategist does NOT Let His Nation be Suckered into the Middle of Situations that have the Potential of Turning into a Nuclear War.

One of the most likely places in the world for the future use of nuclear weapons is the Middle East. The USA should be working hard to make itself as insolated and removed as possible from the Middle East.

(183) In the History of the World no New Weapons System Has Been an Exclusive Forever.

The invention of the atomic bomb was the most challenging weapon system in history (up to that point). However the USA had an exclusive on Atomic Bombs for only four years. On August 28th 1949 Russia exploded an atomic bomb.

(184) The Ability to Start A War must Never Become the Main Reason for Starting a War.

In the history of warfare many wars were started mainly for the reason that they could be started. Incompetent Generals will say "We have the ability to bomb those targets and to block those ports." The ability to bomb a Nation is never by itself the correct reason for bombing a Nation and starting a war.

Everyone has the ability to jump off of a high building. Would you jump off of a high building only because it was possible?

(185) The Ability to Start a War Guarantees NOTHING as to Who Will Eventually Win the War. In the Last 100 Years Most of the Wars Were Lost by the Side that Started the Wars.

Germany started a war in 1914 but did NOT win.

Germany started a war in 1939 but did NOT win.

Japan started a war with China in 1931 but did NOT win.

Japan started a war with the USA and UK in 1941 but did NOT win.

North Korea started a war with America in 1950 but was stalemated.

Saddam Hussein started a war with Iran (1980-1988 that killed 400,000 Iranians) and Saddam lost.

Saddam Hussein started a war with Kuwait (1990) and was defeated.

The Muslims started a war with Israel (1967 Six Day War) and they were defeated.

The Muslims started a war with Israel (1973 Yom Kippur War) and they were once again defeated.

We have the American wars: Vietnam, Iraq and Afghanistan all started (or greatly expanded) by Washington DC and all lost or muddled by the US Military.

Any military moron can start a war (and most wars are started by military morons) but starting wars is not winning wars.

The US Warmonger Lobby keeps saying that if the USA attacks Iran that is a guarantee the USA will win. Historically that is NOT True.

(186) Wars have Dozens of Motivations. It is absolutely Vital for the Grand Strategy Team to Correctly Understand Exactly what is Motivating your Opponent to Fight. If you Refuse to Understand the True Motivations of your Opponent then How do You Defeat or Neutralize those Motivations?

(189) In the History of Warfare Many Wars were Won NOT by What One Side Did Brilliantly Correct but by What the Other Side did Wrong.

Many times in history the winning side in a war was the side that made the fewest mistakes.

(190) Good Grand Strategy Planning Must always be Built on the Absolute Truth even if the Truth is What You do NOT Want To Hear.

You must never let your Strategy planning become a victim of your own cultural lies and your own propaganda and subjective delusions.

(191) The Grand Strategist is NOT a believer in "Democracy" by Itself having Mystical/Magical War Winning Powers.

The idea that Democracies have special war winning powers is a delusion. A war between a democracy and some other form of government by itself guarantees nothing.

Dictatorships have some decisive advantages over Democracies because Dictatorships do NOT have to worry about voters.

In WW2 most of the Nations that **Adolf Hitler** easily defeated: France, Poland, Czechoslovakia, Norway, Netherlands, Belgium, Denmark, Yugoslavia and Greece were Democracies. **The Nation during WW2 that did 65%-70% of the combat that defeated Nazi Germany was a ruthless dictatorship - Joseph Stalin's Russia.**

From 1959 to 1975 the USA (a Democracy) had a war with the North Vietnamese Communist Dictatorship - the Democracy lost and the Communist Dictatorship won.

(192) The Leaders of Nations Must Understand that They are NOT Managing a Flower Shop.

If you need to be a "nice guy" all of the time then open a flower shop. However - sometimes the Leader of a Nation has to open the car door and leave the sick puppy on the side of the road and then drive away.

Sometimes the Leader of a Great Power needs the courage to never stop for the sick puppy on the side of the road. **A car full of sick puppies does not increase a Nation's competitive advantages in a world that has never in the long run rewarded weakness and a lack of self survival.**

(193) The Grand Strategist is Constantly Scenario Building and Never Limits his Scenario Building to Only Scenarios that are Logical and Sane.

The military Grand Strategist is constantly scenario building and is constantly updating his past scenarios as conditions and events change for his Nation.

(194) Often the Core Problem for the Grand Strategist is How To Cure an Insane Nation Without Going to War.

Insane Nations are like insane people - they can be cured. Germany went from being 70% insane under **Adolf Hitler** to being one of Europe's most stable and safe Nations.

No Nation in history has been 100% sane and rational all of the time. Washington DC in the 21st Century is far from being 100% sane and rational.

(195) The Grand Strategist Believes in the Absolute Certainty of Almost Nothing.

Three of the things that are 100% Certain in this world is that (1) **God** does NOT forever reward those who cannot meet His constant challenges. (2) Everything has a Bell Curve. (3) Everything will eventually end including the United States of America. It is only a question of "how soon." With endless stupid wars for stupid reasons the end will be sooner rather than later.

(196) The Grand Strategist Understands that in Warfare there is almost Never an Ideal Time for Doing Something. There is Never an Ideal Time to Maneuver or to Attack or an Ideal Time to ask for Peace. If the Grand Strategist Believes he has more than 65% of the Odds-On-His-Side then He Proceeds.

If you wait for the "ideal time" to do almost anything you will be waiting forever. Especially in shooting wars - **you must capture the initiative and keep the initiative.** You capture the initiative by moving fast and boldly and forcing your opponent to always be reacting to you. If something is not working you do it differently. If that is not working you try another plan.

(197) The Grand Strategist does NOT either Over-Exaggerate or Under-Estimate the Importance of Events. He Works Hard to keep Everything in Proportion so his Thinking and Actions can Stay in the Correct Proportions.

When the World Trade Center in New York City was attacked (September 11th 2001) the USA's reaction was mass hysteria.

America's hysterical over reaction was part of **Osama bin Laden's** plan to stampede Washington DC into a war with Iraq.

President **George W. Bush** was successfully stampeded into a stupid and needless war with Iraq. **Osama bin Laden** leveraged his 9/11 damage against the USA by 200% on the American death toll count and well over 9,000% with the money lost by the USA.

Nineteen **Osama bin Laden** non-professional soldiers (with no guns and only box cutters as weapons) eventually cost the USA over a Trillion dollars - with the costs continuing to go up every day. Each one of Osama's 19 "volunteers" created $55 Billion dollars in costs to the USA. Washington DC's hysterical over reaction (and being con gamed into the Iraq War) turned **Osama bin Laden's** 9/11 attack into **the most cost/benefits successful attack in world military history.**

(198) General von Clausewitz (1789-1831) put Too Much Emphasis on Determination and the Will-to-Win's Ability to Overcome All Problems.

The will to Victory is definitely one of the most highly significant factors in the overall matrix of Grand Strategy. However it cannot overcome significant deficiencies in other vital factors.

All of the determination available to a human is not a guarantee of overcoming the laws of physics and mass differences in population, resources and geo-political positioning factors.

(199) The USA's current Culture's Emphasis on "Pride" is a Highly Dangerous Mindset for War Planning.

The current USA cultural emphasis on "pride" has become highly popular especially among Americans of "diversity." In the real world unjustified "pride" can be highly dangerous. **"Pride" by itself does not equal skills.** It does NOT make any difference how much "pride" you have - if you are NOT a highly trained 747 pilot and you use only your "pride" to land a 747 - you are going to crash and burn.

(200) The Competent Grand Strategist Must Learn from His Mistakes and Must Learn From the Mistakes of Others.

There is an old saying that applies to most Generals in history. **"They learned nothing new and they forgot nothing old."**

Tactics and strategy are changing and evolving at an ever increasing speed. The essential talents for today's professional soldiers are (1) the ability to let go of the past. (2) The ability to keep learning. (3) The ability to keep updating your mindsets into where wars will be in the future and not where they were at in the past.

(201) When Leaders Propose Wars the Voters must Loudly Demand that the Specific Details of Exactly what of Real Value the Average Citizen will Gain From the Proposed War.

America's voters have a National Duty to the USA (and to the world) to protect the USA from future Vietnam Wars and future Iraq/Afghanistan Wars. When the Washington DC Warmonger Lobby propose new wars (like attacking Iran) the voters must demand a list of specific and detailed real benefits of value that will be forthcoming to the average American worker.

If the USA is the aggressor then the question of going to war should be put to a National vote. The voting machines can easily be set up and all you need to vote is a drivers license.

Somebody in the US Congress should introduce such a Bill. For example - a vote on if there should be a shooting war with Iran?

(202) Recovery from a War is Never an Absolute Given. Many times the Losing Side Loses Major Assets Forever. Losing Wars can Sometimes Mean Complete Cultural Death.

For 1,777 years (324 BC to 1453 AD) the Eastern Roman Empire (Constantinople) had successfully resisted the attacks of sword and spear warfare. However the Ottoman Turks embraced the new warfare technology of Black Powder and built what (at that time) were the biggest cannons in the world. Some of the cannons were 27 feet long and could fire a 600 pound projectile one mile. In 1453 the Turks used their cannons to attack the Roman walls of Constantinople. The walls were 26 kilometers long and enclosed the entire city. At vital areas the walls were 30 feet thick.

The Byzantine Generals lacked the courage and imagination to understand the new weapons system of large cannons. Their Generals were living in the past. That was (and is) very typical for old, established and over-regulated militaries.

For 3,000 years stone walls had been a major weapons system. The Byzantines believed that their walls would continue to stop any attacker like they had stopped every attacker for over a 1,000 years. The Byzantine Generals had NOT developed large cannons as counter Artillery against the Turk's cannons. The Turk's great cannons shot holes in Constantinople's walls.

A 100,000 strong Turkish army swarmed into Constantinople. It was a bloodbath. Emperor Constantine was killed. **The Eastern Roman Empire's Christian religion, language, writing, laws, customs, art, culture and government were forever destroyed as living entities.**

The Byzantine Empire lost everything - including their Roman names. The last living vestige of the direct line to the Roman Empire (753 BC to 1453 AD) was destroyed forever. 2,200 years of direct-line Roman history came to a dead end.

Wars are never risk free or cost free. Wars have sometimes cost the loser EVERYTHING. This is why only an idiot goes happily into a shooting war.

(203) In Shooting Wars the Grand Strategist must have the Courage and Skills to manage Three Basic Outcomes.

(1) He needs the skills to know when **he has won** the shooting war and needs to know how to reduce the fighting and limit the loss of life on all sides.

(2) He needs the courage to know when **he has lost** the shooting war and needs to know how to make the best peace possible.

(3) He needs the courage to know when **the war has become a stalemate** and he needs to know how to extricate himself from the fighting.

A peace you are not totally happy with is better than a never ending war that keeps costing you dead soldiers, lost money and makes you a joke in the eyes of the world.

(204) The Grand Strategist is NEVER a Crusader. "Crusader Talk" is a Propaganda Tool for the masses and must NEVER be a Thinking Tool for Either a Competent Leader or Strategists.

The Grand Strategist is never an ideologue. Never a believer that the current war is the war to settle all issues and end all wars. <u>The Grand Strategist is constantly recalculating the costs vs. benefits equations</u> and constantly asking "Exactly and specifically what are the real and usable assets in this war for the people of my Nation?"

When a National leader starts talking about "Crusades" that is the time to become deeply frightened.

(205) Economic Sanctions against Dictators are Seldom Effective because Dictatorships are NOT Run by Popular Vote.

The "sanctions" that the USA has applied to North Korea have been a failure for over 60 years because of 3 core facts. (1) The average North Korean worker has a very limited idea of what material things (compared to the outside world) he does NOT have. (2) The average North Korean worker has zero ways to do anything to change this government. (3) Any "protestor" in North Korea is immediately arrested and sent to jail for years.

The situation in Cuba is only a little better. The Ruling-Dictators who have all of the power (and their core inner-groups) always have every world luxury they want and sanctions do NOT affect their life styles.

There is an ancient saying - "You cannot have rich people unless you also have poor people." It is poor people who make being rich worth something. The USA's sanctions make the Dictators who are in control richer in their own Nations.

(206) The Grand Strategist Treasures his Engineering Brigades. Sometimes they are More Valuable than Combat Brigades.

Good engineering brigades have been vital since the start of organized warfare. In the 21st Century the objective is to make life better in areas where you are trying to build peace. That means the engineering brigades are of decisive important in a Guerrilla type of war.

Those American who think a war with Iran would be easy and fast should know that Persian (Iranian) Emperor **Darius the Great** (522-485 BC) had his engineers build a 2 Kilometer (1.2 miles) pontoon bridge over the Bosporus so his army could invade Greece.

In 2,700 years of warfare the Iranians have been bold and aggressive fighters. The Romans never defeated the Persians - but the Persians defeated several invading Roman armies. Until the Roman's learned the hard way and stopped invading Iran. A USA-Iran shooting war has a 95% probability of being another US Pentagon defeat.

One of the defeats was Roman Triumvirate General **Marcus Licinius Crassus** who in 53 BC invaded Persia. His Army was destroy and **Crassus** and his son **Publius** were captured. The Persians told **Crassus** - "You came to Persia to find gold and we will give you gold." They forced his mouth open and filled it with molten gold. Then did the same to his son.

(207) The Key to Winning Guerrilla Wars is the fact that Guerrilla Groups Cannot Hide from their Own Minds. If you Change Their Minds and Thinking you have Won the War.

Guerrilla Wars in particular must be fought with a Grand Strategy centered squarely on (1) the main reasons for why your opponent is fighting and (2) centered on changing the reasons for your opponent fighting.

Remember the Axiom - **Victory in every war is always in exactly the same place - it is always in the minds of your opponents.**

It cannot be said too many times. You do NOT directly attack your opponent's army. You must attack the reasons for why your opponent is fighting. **If you remove those reasons (or alter those reasons) then you have defeated your opponent in the only place where Guerrillas can be completely defeated - in their minds.**

Once the insurgents have had their ideas changed you have won because **Guerrilla groups cannot hide from their own minds. This is WHY "salesmanship" has become the #1 weapon for long-term decisively winning Guerrilla Wars.**

(208) The Gold Standard of Strategy and Tactics is the Envelopment of Your Opponent.

Through all of the Ages of Warfare - from the Age of Sword and Spear Warfare to the current Information Age of Warfare. From Grand Strategy down to platoon tactics the <u>King-of-Maneuvers</u> has always been the envelopment of your opponent.

With your Grand Strategy war planning - if there is NOT a doable way to envelop your opponent then you must be super cautious about entering into that war.

With a USA-Iran War the US Military has zero doable ways to envelop Iran. The US Military would thus lose a USA-Iran shooting war for exactly the same grand strategy reasons that the US Military lost the Vietnam War.

(209) The Grand Strategist Never Automatically or Unthinkingly Assumes that Time is Going to Be on His Side in Every Conflict.

The Strategist for any Nation planning a war should have a well thought out and accurately researched plan for victory in under a year's time. Ideally a victory in under 6 months. **If the war plan does not have a high probability for victory (peace) in well under a year then the war should not be started** unless there are overwhelming circumstances for doing so.

(210) The Grand Strategist Avoids Mindless and Irrelevant Activities.

A well-lived-life does NOT consist of trying to put the maximum amount of pleasure and products into your body. A well-lived-life is NOT trying to do as little work as possible. A well-lived-life is NOT being "high" on some kind of drugs all of the time. **Too much comfort makes you weak and lazy. Too much comfort makes a Nation weak and lazy.**

The best Strategists will devote their lives to intellectual objectives of value. They will always work towards elevating their understanding of everything important. Do not waste your life on trivia. **Think big and have life objectives that are meaningful for you and for your family and for the world.**

President **Trump** is correct when he says - "It takes the same amount of time and energy to think small as it does to think big - so you should always think big."

(211) **The Grand Strategist is Highly Skilled at Tolerating Working Conditions that are Organized Confusion.**

Under the best of battle conditions you usually only have about 70% of the information you would like. If you cannot tolerate confusion then stay out of the war room.

(212) **The best Grand Strategists Should View Going into Shooting Wars as a Failure.**

Warfare has many non-shooting tools. The good Strategist will use as many of the non-shooting tools as possible before reluctantly using the shooting war tool.

Nations that are already Great Powers (the USA) should put great effort into avoiding battles and wars. Why? One of the many reasons is because **when you are on the top (like the USA) there is nowhere to go except down.**

The USA is already #1 in the world so why is the USA in the war-casino rolling the dice on shooting-wars that the USA does NOT have to be fighting? **If you already have 50 million dollars why are you walking into the casino?**

The best USA Strategy is to engage all of the non-battle system tools of Strategy (economic, diplomatic, cultural, salesmanship etc) and envelop an opponent's Strategy to a degree where a shooting conflict with the opponent would be a clear win for the USA.

(213) **The Strategist MUST Understand the National Character of His Opponent.**

The Grand Strategist understands the National Character of his own people - their strengths and weaknesses. He understands the National Character of his opponent's population - their strengths and weaknesses. The Strategist tries to leverage the strength of his Nation's "National Character" against the weakness of his opponent's "National Character" while doing what can be done so his opponent does not do the same to him.

Understanding the mindset of Muslims requires work for those with a Western mindset. However the job is NOT that difficult. A large part of the essence of understanding Muslims is in the Holy Koran. **The USA's Strategy planners should have all read the Holy Koran several times over.**

One of the key parts of the Koran is the commandment to attack the people, especially non-Muslim people, who are attacking Muslims. What that means is the more the USA attacks Muslims the more the world's 1.8 Billion Muslims will resist and attack the USA.

The Holy Koran in 22:39 states - **"Permission to take up arms is hereby given to those who are attacked."**

(214) Anyone at the US Pentagon who Says that Warfare in the Middle East is Simple (and has Easy Solutions) is Highly Dangerous and Should be Removed from Any Planning or Higher Command Position.

(215) The Grand Strategist Never "Gambles the Farm" Unless He is 100% Absolutely Certain of Winning.

One of President **Donald Trump's** rules is "Never go into a deal - no matter how good that deal looks - if the deal going bad will take you down."

(216) One of the USA's Biggest Flaws is that it has NO Long-Term National Grand Strategy Plan. The USA Must have a National Strategy Plan for Where it Wants To Be at the End of Every Decade - and Where it Wants to Be at the End of the 21st Century.

If you have NO roadmap for your Nation how do you know in which direction to go? How do you know when you have won? How do you know when you have lost? The USA at this time is a ship without a rudder steaming in the dark. This is why the USA desperately needs the new Grand Strategy Team that is outline in this book.

(217) The Grand Strategist does NOT Waste his Energy on Anger. He Works at Keeping his Mind as Focused as Possible on the Core Decisive Factors.

The Grand Strategist knows that he (like all humans) has a limited amount of mental and emotional energy. Time spent on anger is time NOT spent on realistic rational planning.

(218) Every Organization sooner-or-later has in-fighting. The Strategist takes pains to Not Destroy One or the Other Rival Sides. He will use Persuasion and Salesmanship plus whatever is Needed to Bring Everyone Back into an Effective Working Unit with the Same Overall Objectives.

The Strategist will do only what damage is necessary to settle his inter-group conflicts. **The last thing the good Strategist wants is a staff that is afraid to say anything that the Leader might not like** because then the Leader will end up with a staff that is only a collection of yes-boys. However - if a particular person is incompetent or unworkable or more trouble than their contributions are worth the Leader should quickly terminate the problem person.

(219) The Grand Strategist Must Never nonchalantly "Brush Off" Big Number Differences in either Population size or Geographical size between his Side and his Opponent's Side.

In the 1960's the US Pentagon failed at enveloping North Vietnam and was defeated. So how is the US Pentagon in the 21st Century going to use force-of-arms to successfully envelop the Muslim World that is 25 times larger than Vietnam?

For the USA there are NO military doable large solutions in the Middle East. The US Military can use its Allies to stamp down local problems like ISIS in Syria - and push ISIS around on the map. However long lasting large solutions are either going to be via non-shooting tools or there will be no long lasting USA solutions.

General **Sun Tzu's** axiom is - **You have to know what you cannot do.** The US Alligator is NOT going to swallow the Middle East Elephant.

(220) Do NOT Start a War You Have No Chance of Winning.

The above axiom qualifies as an oxymoron. However **history is full of examples of Nations starting wars that they had no chance whatsoever of winning.**

(221) The Grand Strategist Never Orders his Troops to do Things that are Impossible to Do.

The above axiom is important for ALL managers of people doing all jobs everywhere.

The fastest way to "burn out" troops (or people in a private businesses) is to order them to do impossible things.

There is a fine (but all important) line between a General **George Patton** pushing his troops and demanding they give their maximum 100% effort and a bad officer demanding the impossible.

The competent Leader has the talent to know where the line of do-ability is located and pushes his men to that line but does NOT go over that line and burn out his subordinates.

In "folk wisdom" it is called "Do not bite off more than you can chew." Do not consistently order your troops to do the impossible. The same applies to individual lives - push yourself to the maximum but do NOT try to do what is impossible.

(222) In the Middle East (and the World) Colonial Rule is a DEAD Idea. Attempts to Establish or Protect Colonialism are Doomed to Failure.

Washington DC keeps making the same mistakes. One of the biggest standard-mistakes is Washington DC's support for oppressive Colonial Rule. **In the 21st Century nothing in the world is more on the wrong side of history then supporting oppressive Colonialism.**

Washington DC's support for 21st Century Apartheid is the moral low ground, stupid, insane and is a world public relations disaster for the USA and the American people.

(223) The Grand Strategist Tries to Leverage the Forces of History for His Advantage.

The best Strategist works towards harnessing the momentum of history and the momentum of current world trends.

In both war and business if you are on the correct side of history and "have the wind at your back" winning wars is almost assured. However - if you are on the wrong side of history winning wars becomes almost impossible.

(224) Simultaneously Use as Many of the 37 Types of Warfare as Possible to Capture the Initiative and Keep the Initiative.

The Grand Strategist should be using as many of the 37 different types of warfare as possible to capture the initiative everywhere and keep the initiative everywhere. **You always start with the non-shooting types of warfare first.**

The best Strategist is never a one-trick-pony who only uses the same narrow old military tactics. He uses as many of the 37 different types of Warfare as possible to capture the initiative and force his opponent to always be in a catch-up mode.

Once a war has "stalled" and you have lost the initiative. As the US Military did in Vietnam (and has done in Iraq and Afghanistan and everywhere) **then you are basically back into Trench Warfare of 1915 WW1 France.** Everything becomes a brutal slugfest that goes nowhere.

(225) To Win Wars the Leader must Avoid the Incompetent Hacks, Showboats and Game Players.

The good Strategist does NOT want - (1) ass kissing hacks who tell the Leader everything he wants to hear and nothing about what he should hear. (2) Showboat bombastic profanity-fountains who believe their profanity compensates for them being stupid and incompetent.

A half century ago the author of this book did an experiment. He collect over 400 profanity words and phrases. After working with them for a while his conclusion was that **profanity is completely useless for constructive thinking and creative problem solving.** Profanity is meaningless adjectives that disguises the fact that the talker is usually saying nothing other than he or she is angry and does not like something.

The characteristic of all manipulative politicians is they use tons of adjectives but **if you can think while you lessen you can understand that they are talking many words but saying nothing.**

(226) The Strategist does NOT Hid Failure and Incompetence by Promoting Those Who are Failures and Incompetents.

The Washington DC practice (for far over a half Century) has been to reward the ass kissers and incompetents even if they are gross failures. **The Washington DC mindset is that rewarding and promoting failures proves a failure never happened** and thus proves that Washington DC never makes mistakes.

(227) The Grand Strategist puts Much Effort and Work into Keeping Secret his True Objectives and his True Strategy for Achieving his Objectives.

Secrecy is an absolute must for military success. Your plans and intentions must be kept as secret as possible. **A constant stream of disinformation to keep your opponents guessing is essential.** You must work hard at NOT being predictable.

For decades one of the many serious weaknesses of Washington DC is that it has been totally predictable. **Osama bin Laden** was absolutely sure as to what President **George W. Bush** would do after the 9/11 attack - and that is why the attack happened. The 9/11 attack was the bait on the trap - and Washington DC swallowed the bait - hook, line and sinker.

(228) Never Make Fun of your Opponent's Fighting Ability Unless You are Absolutely Certain of the Facts.

It is best to have a 100% accurate assessment of your opponent's fighting ability. **Without total accuracy it is usually a little safer to overestimate your opponent's ability than to underestimate your opponent's ability.**

At the start of the Vietnam War US Troops were told they were fighting "Gooks in black pajamas." At the start of the Iraq War American troops were fighting "Rag Heads."

At the start of the Vietnam War (1959-1975) the US Pentagon war planners should have been asking themselves, "The Viet Minh just defeated the French Army - so are they really going to be that easy to defeat?"

At the start of the USA-Afghanistan war the US Pentagon's so-called strategy planners should have been asking themselves - "The Afghan irregulars defeated the Russian Red Army and before that the British Army - so are they really going to be that easy to defeat?"

The British-Afghan War of 1842 is the only war in British History where the British lost a complete Army wiped out in battle (4,700 soldiers) - with only one (1) survivor - one of the British Army doctors who was kept alive to treat the wounded Afghans.

The British Army's defeat in Afghanistan should have been a sobering fact for the US Pentagon planners who in 2001 confidently assured everyone that the US-Afghan War was going to be "a piece of cake" and "over in less than a year."

The US-Afghanistan War started in 2001 and when this book was published in 2019 about 50% of Afghanistan was controlled by the Taliban. Afghanistan is one more US Military defeat. Yes there is definitely room for improvement at the US Pentagon.

Today in Washington DC the very powerful Warmonger Lobby keeps pushing for a USA-Iran shooting war. The US Pentagon's war planners should be asking themselves - "the Persians have been tough, determined and highly successful warriors for 2,700 years. Are the Iranians really going to surrender to the USA after the US Military drops a few bombs on them?"

The Iranians have been thinking about (and planning for) a USA-Iran War since 1979. There are no "surprise factors" on the US Pentagon's side. The US Military has already lost the factor of surprise. **We can be 100% sure that the Iranians have their own strategy plans that are NOT based on Iran playing the US Pentagon's game by the US Pentagon's rule book.**

(229) The Grand Strategist is often Faced with a Decision Between (1) a "Close in Strategy" of Blocking an Opponent from Invading his Homeland or (2) a "Far Forward" Strategy of "Taking the War to your Opponent" and Fighting Them Far Away.

Today's Radical Muslim Groups in the Middle East cannot invade the USA. **A far forward Strategy to block an "invasion" that cannot happen is by definition stupid.**

The more the US Military kills local Muslims inside of their Nation the more "terrorists" the US Military will create for future bigger shooting conflicts.

US Troops shooting up foreign lands has become a vicious cycle. US Forces killing Muslims in somebody else's Muslim Country has become the Pentagon's justification for the continued US Forces fighting and killing Muslims in somebody else's Nation. The whole process has become self defeating.

(230) The Grand Strategist always Listens Carefully to Everything his Opponent Says because Sometimes it is the Truth (and a Valuable Key) to What Your Strategy Must Be.

One of the hardest parts of the Art of War and Peace is knowing when your opponent is bluffing and giving out disinformation and when they are telling the truth.

In Korea (on June 25th 1950) North Korean forces crossed the 38th Parallel and attacked the US forces in Korea. The North Korean Army pushed the South Koreans (and US Forces) into a small pocket around the far southern Korean port of Pusan. US forces counter attacked with the landing at Inchon (September 10-19, 1950) in the middle of the Korean peninsula. General **Douglas MacArthur's** strategy was to trap and envelop the North Korean forces that were in the far South of the peninsula. Many of the North Korean forces escaped the trap but **the disorganized North Koreans had lost the initiative** and were pushed back to almost the Chinese-North Korean border.

Mao Zedong (1893-1976) publically **many times warned** and said that if the American Army approached the Chinese border then he (**Mao**) would send in Chinese Red Army troops.

General **MacArthur** ignored **Mao's** many warnings. American troops approached the Chinese border and **Mao Zedong** as promised sent in Red Army "volunteers" that pushed the American Forces back to Seoul South Korea - with a loss of half a million South Korean civilians murdered by the North Korean/Chinese forces during the battle for Seoul.

The cease fire armistice of July 1953 ended the active war at positions near the 38th Parallel on almost the exact lines where the war had started.

If General **MacArthur** had believed **Mao Zedong** and stopped his advance 75 miles from the Chinese border that would have never been enough land for a viable North Korean State to exist. However **USA Generals have the weakness of always looking for a complete 100% victory.**

The smart Strategist understands that a 75% win is better than a 100% defeat. Or a horrible blood war like Korea (with 44,499 dead US Troops and 566,434 wounded) in a war that ended almost exactly where it started and created many serious problems that exist to this day.

(231) The Grand Strategist knows that He must First Understand all of His Mental and Organizational Weaknesses to Best Exploit his Opponent's Mental and Organizational Weaknesses.

Adolf Hitler (1889-1945) had a genius for perfectly understanding his opponent's weaknesses while being almost totally blind to his own fatal weaknesses and mental flaws.

Probably any person who has lived a fair amount of time (and been aware of things around them) has known somebody who could see everyone else's weaknesses perfectly but see none of their own.

(232) When the Grand Strategist has a Choice of (1) Directly Attacking his Opponent's Strongest Points or (2) Braving Environmental-Physical Challenges in a Roundabout Indirect Attack on his Opponent's Forces. World military-history has clearly Shown it is usually Better to Risk the Environmental-Physical Challenges in a Roundabout Indirect Attack.

When the Grand Strategist is faced with a decision of directly attacking his opponent's troops or indirectly attacking via a route that is environmentally/physically very difficult it has historically been more successful to brave the environmental challenges (like crossing a desert, or bringing troops over mountains, or building a railroad system to connect all of Eurasia as the Chinese are now doing, or taking over much of the USA's retail products as China has already done).

Human opponents who are defending a prepared position (and are mentally set for your attack) can be dynamic, highly adaptive, amazingly resilient and never a guaranteed predictable. However - the environmental challenges (once completely and correctly analyzed) are then <u>highly predictable and almost never change.</u>

The most symbolically decisive battle of the last half of the 20th Century was the Battle of Dien Bien Phu in Vietnam (March 13 - May 7th 1954). The French made a deep thrust into Viet Minh territory and built a fortress at Dien Bien Phu. The fortress was surrounded by jungle covered mountains and the French Strategists were "absolutely certain" that nobody could put Artillery up onto the mountains without the French shooting them off the mountains before they were emplaced.

The Dien Bien Phu fort was deliberate "bait." The French plan was to have the Viet Minh attack their fort and then the French would use the power of the defensive and destroy the Viet Minh.

Viet Minh General **Vo Ngauyen Giap** (1911-2013) was NOT going to be suckered into playing the French Army's game with the French Army's rulebook.

(Side Note) When the US Military invaded Vietnam (in 1960) General **Giap** did NOT play the US Pentagon's game with their rulebook. **The Chinese are also NOT going to play (and are NOT playing) any of the US Pentagon's games with the Pentagon's rulebooks.**

At Dien Bien Phu General **Giap** knew that positioning Artillery up onto the mountains around the French fort would be a super-human physical challenge - but not impossible.

<u>General **Giap** chose the environmental challenge over a direct infantry attack on the French fortress.</u> That was the correct Strategy. The French wanted a direct attack by the Viet Minh infantry against their dug-in French machine guns and Artillery. Attacking the fort would have been the Viet Minh slaughter the French wanted. The French Strategy was to have a mini Verdun in reverse.

Rather than a direct infantry attack against prepared positions General **Giap** had his Army start with an environmental attack. General **Giap's** Viet Minh Army (with super human efforts) dragged and hauled their heavy 120mm and 105mm Artillery pieces up the sides of the jungle covered mountains. Usually during rain storms (and at night) so the French artillery could not see them. Then the Viet Minh dug their Artillery into caves overlooking the French fort and French artillery. When the placement of the Viet Minh Artillery was finally completed the French fortress of Dien Bien Phu became a shooting gallery for the Viet Minh's Artillery.

Every earthen fortification at Dien Bien Phu was destroyed by General **Giap's** Artillery. With the fortress defenses shattered the Viet Minh infantry enveloped the fort and attacked. **The day that the French garrison at Dien Bien Phu surrendered (May 7th 1954) was the day that White man colonialism in Asia became the walking dead.** The over 400 years of White men successfully colonizing Asia came to an end.

Other White man colonies in Asia would hold out for a while longer but their day was over. **The Age of White man colonialism was DEAD in the world.**

Tragically for the American people and the world the US Military (and Washington DC) never learned anything from the French defeat - or from the US Pentagon's defeat in Vietnam. Washington DC and the USA Military are struggling to protect the status quo of the West Bank - a pre WW2 vestige of brutal colonialism in Asia that is doomed to failure.

In Vietnam the US Pentagon followed their usual post WW2 pattern of taking a proven failure (France's defeat) and making the proven failure much bigger. The US Pentagon wanted to prove that it could also lose in Vietnam like the French had lost in Vietnam - but have a much bigger defeat.

The US Pentagon's Generals also proved they could lose in Afghanistan like the Russians (1979 -1989) had lost their war in Afghanistan.

The US Pentagon's Generals also proved they could lose in Afghanistan like the British had lost their two wars in Afghanistan (1839-1842 and 1878-1880).

The US Pentagon's Generals also proved that they could have a muddled stalemate in Iraq like the British had during the 1930's war in Iraq.

Maybe someday the US Pentagon's Generals will prove that they too can lose a war against Iran like Iraq's **Saddam Hussein** lost the Iraq-Iran war of 1980-1988 (300,000 Iraqi soldiers killed for a 50 mile advance into Iran).

How clear do the repeated defeats have to be for the Pentagon's brain-dead Generals to get the message? **White man colonial rule is DEAD in Asia. Washington DC trying to protect White man colonial rule in the Middle East is the moral low ground (it is military insanity) and on the very-wrong side of history.**

The Washington DC Warmonger Lobby says that the West Bank slave colony is an "independent nation." However nobody else in the world is buying that lie. Does anyone in Washington DC understand that in today's Internet world everyone knows what is happening? Yes - the Pentagon and Washington DC believe their own lies but nobody else in the world does.

(233) The Proven and Reproven Rule that Applies to the US Military (from the End of WW2 to Now) is that Where Other Armies have Failed and been Defeated the US Military will Also Fail and be Defeated.

The very powerful Warmonger Lobby keeps pushing for a USA-Iran War. One of the FACTS that the US Pentagon's proven flawed planners should think about is the fact that in the last 2,700 years only 2 groups have successfully invaded Iran (Persia) and succeeded in staying for a long time - and both of those groups had extenuating conditions.

One group was the Muslim Arabs in 651 AD during the Arab's Crusade for expanding Islam. You could argue that "conquest" of Persia was NOT a true military conquest of one side decisively defeating the other side but was a "conversion" of Iran from its ancient religion of Zoroastrianism to Islam.

Persian Zoroastrianism believed in a Deity of divine wisdom and believed in a heaven and hell and the free will of individuals. In essence it was not radically different from Islam (or Christianity).

The only other group in 2,700 years to successfully invade Persia (and stay for a long time) was **Alexander the Great** in 333 BC (over 2,300 years ago). The Persians greatly outnumbered **Alexander's** Army but the Persians de facto "surrendered" to **Alexander the Great.** Why? Because the Persians wanted the creative superior Greeks to rule them.

At that time in history the Greeks had the correct reputation of being the smartest people in the Western world. **The Persians did not surrender their culture to the Greeks - the Persians used the Greeks to improved their culture.**

Three hundred years later the Romans tried to defeat the Persians several times and their Roman Armies were destroyed several times.

There was a USA-Iran 2015 Peace Agreement. The author of this book believes that Peace Agreement was a opportunity to get inside of Iran and apply the many non-shooting types of warfare. However - that opportunity was changed back to sanctions that in 2015 had already been a 35 year failure.

Washington DC not living up to its 2015 agreement Iran was a horrible mistake. The Peace Agreement was an open door to get inside of Iran and make real changes. The USA could have been the Greeks and taken over Iran from the inside out. The ONLY 2,700 year proven way to succeed in Iran.

(234) One of the many Characteristics of a Great Power in Decline (like the USA) is that the People Become Soft and Weak and No Longer have the Will to Solve the Challenges that History and God Puts in Front of Them.

The USA has been on the downside (relative to the world) of its Bell Curve since 1975. Interestingly **the USA was at the top if its 200 year long Bell Curve at the same time when China was at the bottom of its 150 year long downward Bell Curve.** Does that mean a 100 years from now the USA will be at the bottom of its Bell Curve when China is at its top? Possible.

Some of the classic characteristics of Great Powers on the downside of their Bell Curves are addictions to: pleasure, comfort, food, dinning, drink, drugs, sports, sex, gambling and entertainment. Money is made not by making products but by manipulating money and credit. Does that sound like today's America?

There is a good argument for saying that right now. At this time - the USA is in the Gold Age of its history.

In today's USA a large part of the population believes that everything should be a Disneyland. If it is not perfect then something is wrong and somebody is a criminal.

The paradox is that President **Trump** is working hard (and enduring much pain and criticism) trying to save the American "gravy-train" that makes life so easy (relative to the world) for so many tens of millions of Americans.

(235) The Best Grand Strategist has to be Born with a Talent for Grand Strategy. A Decisive Factor for every Nation is Being Able to Recognize the Gift in the Few People Who Have It.

No school can teach a person how to be an **Alexander the Great** or a **Saladin** or a **Genghis Khan**. Often true **God** given talent is hard to recognize (like **Vincent Van Gogh**) because the talented person's actions are often related to the future when the "official judges" are usually dead-heads living in the past.

The US Military has a death grip on its WW2 mindset. How is the bloated, rancid and war losing US Pentagon going to recognize a gifted 21st Century Grand Strategist? There are probably some good 21st Century Strategists in Silicon Valley but Silicon Valley's culture and the US Pentagon's culture are oil and water. That must change because Silicon Valley must become major defense contractors.

In the past the lines between "civilian mindset" and "military mindset" were sharp and clear. In the future those lines must blend and merge. The shooting war tools become more obsolete with each year and the non-shooting war tools become more decisive with each passing year.

The USA needs (as soon as possible) the new US Space Force. The Pentagon is already "dragging its feet." President Trump should find the courage to kick-start everything by turning NASA into the US Space Force. The US Pentagon will scream -"It would NOT look like a military organization." Wrong. It would not look like yesterday's obsolete military organization but it would look like tomorrows military organization.

The USA also must have the new United States Communications Force (USCF). It is hard to imagine the high degree of resistance that is going to run into at the Pentagon.

(236) With Negotiations and Peace Talks Sun Tzu Worried that You Should Beware of People who (1) Offer you too Much and too Readily or (2) they Give-In to Your Demands too Easily. If They Offer too Much (or Give-In too Easily) it could Indicate They Never Intended to Keep Their Promises.

The above is also a Mafia trick. The Mafia would offer a non-Mafia criminal a large amount of money to do a job. When the person who did the job went to collect his money he would be killed.

Before it became illegal in the USA - a company would hire a key employee of a rival company at a much higher salary. The company that hired the victim would then learn all of the secrets and contacts of the rival company. After learning everything of value they would fire the person they hired.

It is very easy to offer somebody a giant monthly salary increase over what they had been making if your plan was to pay the high salary for only a short time.

Adolf Hitler easily agreed to the non-aggression terms of the 1938 Munich Agreement because he never intended to keep his promises. All Hitler wanted was more time to build up his military.

(237) The Voters in any Nation Should be Highly Skeptical of Leaders who Tell Them that They must Give Up Their Freedoms to "Fight for Freedom."

Voters should be skeptical of National Leaders (and Allies) who impose fascist controls and oppress freedoms to supposedly "protect freedom and democracy."

Shooting wars do require more government controls because of the greater risks during active wars. However wars are often used to establish dictatorships that remain in place for many years (or many decades) after the war has ended.

(238) **Sometimes Winning Wars is the Same as Winning a Game of Tennis. If you can Stay in the Game Until your Opponent Makes a Decisive Mistake - Then You Win.**

For the first 2 years of WW2 in Europe the English held their own against **Hitler**. The English were not winning but they were not defeated. Then **Adolf Hitler** made a decisive mistake by declaring war on Russia in June of 1941. Then - **Hitler** made another decisive mistake by declaring war against the USA on December 11th 1941. At that point **Hitler** had defeated himself.

(239) **The Strategist Guards Against the "Winner's Curse." The inability to Move On From the Last Big War that Was a Big Win.**

Historically the Nations that lose wars learn from their mistakes and the winners do NOT learn. However - that has NOT been the situation with the US Pentagon. The US Pentagon's WW2 mindset believes that the Pentagon has never lost a war and therefore there is nothing to improve.

What about Vietnam? For the Pentagon the Vietnam was NOT a defeat - it was a war the Pentagon declined to continue.

A half Century after the Vietnam War defeat Washington DC continues to protect and prop-up White Man colonialism in Asia. Washington DC and the US Pentagon learned nothing from their Vietnam defeat.

(240) **Sun Tzu wrote that there are Six Mistakes the Leader Must Not Tolerate. If they Exist they Must be Corrected as Soon as Possible.**

1. **DISORDER.** It comes from unprincipled officers and/or leadership that is weak and undisciplined and/or a Strategy that is indulgent towards delusions and basically an illogical Strategy.

2. **DESERTIONS.** They come when officers are (1) unfocused on their duties and (2) have muddled objectives and (3) give impossible to-do orders. The officers (4) spend too much time and emotional energy confronting low ranking troops about trivial and unimportant issues. (5) The battle tactics are seen by the troops as obsolete, bankrupt and failing.

3. **COLLAPSE.** It will happen even if the officers remain strong but the troops become defeated in their minds.

4. **DISINTEGRATION.** It will happen if the officers let their anger rule their orders and meet their opponent hatefully. It will happen if the commanding officers make decisions based on emotions rather than cool headed calculations and planning.

5. **INSUBORDINATION.** It happens when troops are strong but senior officers are weak. It comes when the troops no long trust their Leaders. **The National Leader must always appear confident even if he has to pretend.**

6. **FLIGHT**. A disorganized retreat comes from attacking an opponent who is far superior to you and/or has a superior Strategy that confuses and disorganizes your troops. It comes from being attacked by a far larger force with a greater will to victory.

The above 6 Rules also apply to Political Parties and to Business Organizations.

(241) The Grand Strategist Never Takes the Courage of His Troops and Army for Granted and as Something that Will Always Be There.

Field Marshal of France **Maurice de Saxe** (1696-1750) wrote, "Of all of the elements of war the one that is most necessary to study is courage. The courage of the troops must be reborn daily. The true skill of a General consists in knowing how to guarantee courage by his dispositions and his positions and those traits of genius that characterize great captains."

Doing what the US Pentagon has been doing for the last 40 years. Having US Troops in de facto trench warfare around the Middle East. That is historically proven to be bad for overall moral and is one of the reasons for WHY so many US Troops returning from the Middle East have so many problems.

Chapter 10

The Axioms of Strategy and Tactics for War and Peace

There have always been military "police actions" and very limited wars that are sometimes needed. However one of the Most Important Rules is that you should never enter into a substantial sized shooting war if you have no way to envelop your opponent and/or force them to make peace.

(242) The Ultimate Objective of all of the 37 Different Types of Warfare is Always the Same - an Advantageous Peace for Your Nation.

There are many roads to an advantageous peace. However - shooting wars should always be the last road taken. At the proposed start of any war the Grand Strategist is always asking "Exactly and specifically what advantageous long lasting peace for my Nation is going to come from this war?" The answers must be honest, truthful and worth the costs of the war. **If there are not good and achievable real advantages for your Nation's citizens then you should NOT voluntarily enter into a shooting war.**

(243) The Ultimate Attack Objective of Grand Strategy in Every Shooting War is Always the Same - an Attack on the Strategy of Your Opponent for Achieving His Desires and Dreams.

You must first correctly discern your opponent's core Grand Strategy desires. Then you must list all of the ways that the 37 types of warfare might be used to attack your opponent's Grand Strategy desires and objectives.

The larger the number of different ways you use to attack your opponent's Strategy the more roads you have to the same ultimate objective. **If you attack your opponents Strategy from six different main roads and your opponent is successful in defending and/or blocking five of those roads (but NOT the sixth road) - then you win.**

Your Grand Strategy has to be focused and centered on the mind of your opponent. What are your opponent's dreams, hopes, fears and Strategy for achieving his objectives? Most of your attacks should be a combination of both indirect and direct attacks.

NEVER limit your thinking or attack Strategy. There are many different types of non-shooting warfare and the good Grand Strategist will <u>simultaneously</u> use and apply as many of the non-shooting different types of warfare as possible.

(244) Every Road of Attack Used must have Grand Strategy Significance Towards the Ultimate Objective of Weakening your Opponent's Core Assets that are Being Used Towards Achieving his Core Desires.

Many attacks in the history of warfare had little or no Grand Strategy significance (and were done only because they could be done). Many attacks (and wars) were useless dead-ends before they started.

If you DO NOT have a crystal clear definition of your objectives and why they are decisive then you are wandering in the dark.

During WW1 Germany's General **Erick Ludendorff** (1865-1937) launched his great March 1918 offensive. <u>Tactically it was brilliant and changed basic infantry tactics (in all armies) up to this day.</u> However - in Grand Strategy terms General **Ludendorff's great offensive was a dead-end from the start.** There never were any doable Grand Strategy objectives of decisive significance within **Ludendorff's** reach. It was the classic **Clausewitz** mistake of attacking your opponent's army to only win battles that mean nothing in the big picture. It was a tragic waste of lives.

Today in the Middle East the US Forces are pushing the Radical Muslim Groups around on the map of the Middle East. US Forces are NOT attacking the strategy or the motivations of the Radical Muslim Groups. **The US Military (as always) wins the battles but are they winning any of the war? The answer is NO.**

For 60+ years the USA's Generals have all been **Ludendorffs** having battles because they can have battles. However most of the battles had no direction, no propose other than having battles and no grand strategy meaningful objectives.

During Iraq War 2 (2003 to ?) the General **David Petraeus** great Troop SURGE of 2007 was (in the opinion of the author of this book) one more example of General **Ludendorff's** stupid Surge of 1918. Neither **Petraeus** or **Ludendorff** were attacking the opponent's Strategy. All they were doing was pushing the opponent around on the map at a great cost in blood and treasure.

The United States has entered a stage in its history where it can no longer afford to have grand strategy incompetents in high command positions.

(245) Small Forces Can Defeat much Larger Armies.

In 1519 **Hernando Cortes** landed in Mexico with 500 soldiers, 14 small cannons and 17 horses. The Aztec Empire had 5 million people covering 80,000 square miles. **Cortes** made an alliance with the Tlaxcaltec Indians who had been conquered by the Aztecs and had been treated badly. The Tlaxcaltec Indians (with Spanish military guidance and tactics) defeated the Aztecs and made it possible for **Cortes** to conquer all of Mexico and Central America. An area that contained massive amounts of Gold and Silver that made Spain the richest Nation in Europe.

The reverse of **Cortes** was the US Military in Vietnam. During the Vietnam War the US Pentagon had a 1000-to-1 advantage in both gross firepower and money being spent on the war and the US Pentagon was defeated.

(246) If You do Not Understand your Opponent's Strategy then You are "Shooting in the Dark."

For many decades Washington DC and the US Pentagon have NOT had crystal clear and highly defined objectives (that are doable and valuable) for their shooting war blunders. The Pentagon has been a rudderless ship in the dark where "the battle" is an end it itself and most of the time the only objective.

(247) The Basic Essence of Shooting War Grand Tactics on the Battlefield can be Summarized by Three Actions:

(1) **You constantly maneuver your forces in a way that threatens as many of your opponent's positions as possible.** You force your opponent to garrison with troops as many different points as possible. Then you quickly concentrate your attacking forces and achieve superiority of force (and a win) at a decisive point.

(2) **You concentrate your superior forces (hopefully in a surprise attack) onto one of your opponent's assets** that are weakly defended and is of importance to your overall attack against your opponent's Strategy.

(3) **You attack your opponent's weakest points first** and continue to eliminate your opponent's positions from the weakest points first while always advancing towards your opponent's core positions.

Incompetent Generals and Admirals cling to the **Clausewitz dream** of one titanic battle that will decide everything. The one great "Troop Surge" for final everlasting victory. However - **the most gifted Grand Strategist can win a major war without ever having fought a major shooting battle.**

General **Douglas MacArthur** defeated the Japanese in the Pacific and brought his troops to the doorstep of Japan with NO titanic battles by WW1 and WW2 standards. General **MacArthur's** biggest battle of WW2 (in the Pacific) was on Okinawa (April-June 1945). 12,526 US Military were killed in battle. For comparison the Russians during the Battle of Stalingrad lost 478,741 killed.

(248) To Win Significant Struggles in the 21st Century you Must Successfully Capture the Minds of your Opponents.

The non-shooting weapons (in the 21st Century) must be used first and with great vigor because the **traditional military shooting war combat weapons are becoming increasingly obsolete at an increasing speed.**

Today the **Clausewitz** axiom of capturing your opponent's Capital City means nothing because many of the USA's opponents (like the Radical Muslim Groups) have no Capital City. Their "Capital" is the Internet that now covers the total Earth. Their "Capital" is in the minds of their supporters who can now communicate as if there were all living in one Capital City (even though they are physically living in many Nations of the world).

(249) There is No Honor or Sanity in "Fighting to the Death" and Destroying your Nation.

By August of 1943 both Germany and Japan had zero-chance of winning a continuing war. Their defeat was only a question of time.

Both Germany and Japan made the Grand Strategy mistake of letting their egos do their thinking and they continued to cling to their delusions. They continued to fight and lose until both Germany and Japan were in large part burning ruins. **In Grand Strategy it is vital for a people to know when they are defeated** - and to make the best peace possible.

BUSINESS. Everyone starts a business with the dream of it being a big success. However 75% of new businesses fail within three years. When the numbers are NOT adding up you need the courage to admit that you are defeated. **Save what you can and move on.**

PERSONAL. Many people enter into marriages (or personal relations) that stop working. If there are fundamental problems and you cannot resolve the problems then it is best to admit defeat (and save what you can of your life) and move on. **As long as there is life there is hope for a future.**

(250) When an Attack (or War) Fails Never have Another Attack (or War) along the Same Lines and in the Same Way. 90% of the Time Your Result will be a Bigger Failure.

Dozens of times (during WW1) the French and British Generals would attack the German lines and the attack would fail with great loss of life. The French and British Generals would then have the SAME type of attack in the same location with a 50% "Surge" of more troops. However - the Germans had always increased their defenses with 50% more machine guns and the attacks failed with a bigger loss of British and French life.

Around 1950 Washington DC became addicted to taking proven failures and making the proven failures bigger. For the end of WW2 to now Washington DC and the Pentagon have continued to be surprised and shocked when they made a proven failure bigger and then ended up with a bigger failure.

(251) Do NOT Put great amounts of Money into Building a Weapons System only Because you Have the Money to Build a Weapons System.

The US Military must stop build yesterday's weapons (like super Aircraft Carriers) only because they can build them.

Every USA weapons system needs a detailed and specific set of reasons for exactly and specifically why that system will be needed in future wars. NOT why it was needed in yesterday's wars.

Does the USA really need 10 super Aircraft Carriers to protect everyone else's ocean trade routes? The USA only has a few cargo Merchant Marine ships over 20,000 tons. The USA has NO overseas Colonial Empire. So **who is the US Navy protecting?** Mainly Chinese and Japanese ships that bring products to the USA to increase the USA's negative trade balance.

If 90% of the foreign ships could reach US Ports long term that would be one of the best things that could happen to the USA. The USA would eventually return to 1940 when 90% of everything in US stores was made in the USA. The average American citizen would be happier.

(252) The Grand Strategist Understands that Most of the "Burning Issues" of Today are Long-Term NOT Worth the Costs of a Shooting War.

During the Vietnam War (1959-1975) the American people were told over and over again that all of Southeast Asia would become Communist if North Vietnam took over South Vietnam.

Since 1975 North Vietnam has controlled South Vietnam and all of Southeast Asia DID NOT turn communist. What happened is North Vietnam became more Capitalist. The justification for the USA invading Vietnam and murdering 650,000 Vietnamese civilians - those reasons were lies, falsehoods and a con game.

(253) The Grand Strategist Never Assumes that the Destruction of a Regime is going to Automatically be Followed and Replaced with Something that is Better than the Original.

Without a clear, detailed and carefully reasoned and doable plan of action **the type of government that follows a destroyed regime might be far worse than the government that was destroyed.**

During WW1 German **Kaiser Wilhelm** (1859-1941) was forced to abdicate and his government was eventually replaced with **Adolf Hitler** and the Nazis.

In Russia - **Tsar Nicholas II** (1868-1918) was replaced with **Joseph Stalin**.

The US Military destroyed the regime of **Saddam Hussein** (1937-2006) and **Saddam** was replaced in part by ISIS and al Qaeda and other Radical Muslim Groups.

The Libyan regime of **Muammar Gaddafi** (1942-2011) was replaced with a collection of warlords that created a safe haven for Radical Muslim Groups.

(254) Forts and Walls are of Temporary Value. <u>Walls and Forts can Buy You Time</u> but have Never been a Final Solution for a Permanent Defense.

1. The Great Wall of China failed to keep out the Mongols in the early 13th Century.

2. The walls around ancient Rome failed.

3. In 1453 the walls around Constantinople failed.

4. Hadrian's wall in England failed.

5. The Hindenburg line in 1918 WW1 failed.

6. The Maginot Line in 1940 WW2 failed.

7. **Adolf Hitler's** Atlantic Wall in 1944 failed.

8. Japan's wall of Pacific fortified Islands in WW2 failed.

9. The French 1160 km long "Morice Line" electrical wall during the Algeria Revolution failed in 1962.

10. The Berlin Wall in 1989 failed.

11. The Zionist's Wall around the West Bank & Gaza will fail.

12. The Trump Wall along the US-Mexican border will fail at keeping out many illegal immigrants and drugs.

12A. The solution for the USA is NOT a wall but a National ID Card for everyone inside of the USA. Cards with time limits could be issued for Students and those on work projects etc.

The ancient Greek Spartans deliberately did NOT build any walls around their cities because 2,500 years ago they realized that a mobile defense is superior to Walls. The USA's "mobile defense" is a National ID Card.

For thousands of years Walls and defensive lines only bought time. **The long-term solution is a USA National ID Card.**

(255) Advancing ICBM Technology Made most Underground USA Nuclear Missile Silos Obsolete in the 1980's. The Same High Precision Target Hitting Technology has Made the USA's Super Aircraft Carriers floating Coffins in a Shooting War with Russia or China.

Russia and China know the location of all of the USA's ten Aircraft Carriers all of the time. Today's ICBM's have pinpoint accuracy. Russia's and China's new power-dive ICBMs could come straight down onto the top of a US Aircraft Carrier at 24,000 miles per hour (a bullet from a high powered rifle travels at 2,300 mph). **Far out at Sea the USA's 10 Aircraft Carriers have ZERO effective defense against ICBMs.**

In a shooting war with China or Russia the US Navy could lose its 10 Aircraft Carriers in the first week. That would be a devastating blow to the USA's morale and a great morale booster to the USA's opponents.

The defense of the US Aircraft Carriers must come from the new **US Space Force.** At the start of a shooting war with Russia and/or China the **US Space Force** would have to destroy all of the opponent's navigational and targeting satellites before the US Carriers could be targeted. There would be a "window" of **only hours** to destroy Russia's and China's satellites.

The Chinese Navy has a back-up system. There are over 100,000 Chinese fishing boats all over the world. They can be quickly disguised as being from other Nations and the Chinese fishing boats can (and would) radio the locations of the US Navy's Ships.

The Bell Curve for surface Navies is on its downside. **The US Navy's Super Aircraft Carries in a 21st Century war with China would be WW1 Horse Cavalry charging machine guns.**

(256) A Good Strategist is Always in a Constant State of Reviewing and Updating his Strategy and Tactics.

World history never stops and the revisions of your Strategy and tactics must never stop. **Yesterday's excellent Strategy plan is often tomorrow's disaster waiting to happen.**

(257) Constantly Checking Your Plans (including a re-verification of all Past Assumptions) is the Mother of Victory.

The Strategist builds his many war plan scenarios from the best possible outcome at one end to the worst possible outcome at the other end. **Every scenario must have a "next step plan"** - how best to speedily follow-up a major success (or in case of a failure) a doable "retreat-exit plan."

The objective is to never have paralyzing confusion at the Command Center because of some unexpected turn of events. All likely possibilities must have at least a rough outline of actions that were pre-thought out.

Laborious analysis and planning is both the mother of victory and the protector against disastrous defeat. **A war should be planned to the point where the war is essentially won before the first bullet is fired**.

(258) Long Wars must be Avoided because they Create Learning Opportunities for Your Opponent to Acquire Greater Combat and Leadership Skills.

All wars create battle experienced officers and leaders (on all sides) who have increased battlefield skills.

The difference between a US soldier learning combat skills on a Middle East battlefield (and a Muslim fighter learning combat skills on the same battlefield) is that after 12 or 15 months the US soldier will be gone and (85% of the time) will never return to the Middle East. However **the Muslim soldier's home is where he is fighting and he will stay to meet the new US soldiers who arrive with no battle experience.**

About 15% of US Troops serving in the Middle East have multiple tours and they become first-rate combat soldiers. However this group has some serious issues when eventually adjusting to civilian life.

The more the US Military mucks around in the Middle East (with the US Military's never ending shooting wars) the more the Muslims will improve their fighting abilities and improve their tactics and strategy leadership skills.

(259) The Grand Strategist Fears an Opponent who is an Excellent Organizer Far More than he Fears a "Showboat" Loudmouth Opponent.

At one end of the Leadership spectrum are "showboat" National Leaders who are excellent at giving enthusiastic speeches and exciting the masses. At the other end of the spectrum are National Leaders with little showmanship but they are excellent organizers, team builders and project completers. The management slang from the old days was "rat catchers versus showboats."

History has proven that the leaders who are excellent organizers are the ones to fear far more than the "showboats."

Business is the same. A manager with excellent organization talents and job completion skills (the ability to do whatever is needed to get the job done) will in the long run win out against a flashy showboat.

Sometimes a Leader can be both - like General **George Patton** - and **Donald Trump** - and until 1943 **Adolf Hitler** In Soviet Russia (during the 1920's) dull and plodding **Joseph Stalin** was a rat catcher (an excellent organizer) - and flashy exciting **Leon Trotsky** was a showboat - **Trotsky** ended up dead.

(260) A Nation that is Growing in High Culture and Sophistication is ALSO Growing in Advanced Military Skills and Leadership Power.

When a culture becomes more sophisticated then the more expanded thinking and culture spreads into that Nation's militarily planning, organization and capabilities. Especially now in the 21st Century when **good Military Strategy includes everything.**

The Middle East is on the upside of its reset Bell Curve and becoming more culturally sophisticated with each year. The military capacity and skills in the Middle East are growing. Places like Dubai City and the Emirates are already World Class. In 2018 Dubia Airport had 89,149,387 passengers and Chicago's O'Hare Airport (once the #1 in the world) had 83,399,186 passengers. (Note - today the #1 airport in the world is Atlanta USA with 107 million passengers a year).

(261) Significant Sized Wars almost Always take Twists and Turns Neither Side Expected and End in Ways Neither Sided Foresaw.

When WW2 in Europe started (September 1st 1939) there was no person in the world smart enough to see what Europe would actually look like in September of 1945.

In 1962 if you had told all of the USA's Generals the eventual outcome of the Vietnam War (a US Military defeat) none of them would have believed you.

(262) One of the USA's Biggest National Complications and Problem Producers is the US Pentagon's $700+ Billion Dollar Per Year Cost.

Washington DC is constantly on the "Horns of a Dilemma." The US Pentagon costs a staggering $700+ Billion dollars per year ($7 Trillion dollars every 10 years) but **the US Pentagon's ability to forcibly change history to fit Washington DC's subjective delusions has been proven and reproven to be impotent.**

To justify a $700 billion yearly budget there is always tremendous pressure to use the USA Military to solve perceived problems with Military Force <u>when Military Force is the wrong solution.</u> It is like trying to fix a broken computer with a solid gold hammer because you have a solid gold hammer and have to justify its use.

With most items in the world a higher cost usually means a better product. However - the fact is that during the Vietnam War the US Military was spending $1,000 for every one ($1) the Viet Minh was spending and North Vietnam defeated the US Military. Since the END of the Industrial Age of Warfare in 1960 having **more money $$$ has been proven to NOT equal winning the war.**

One of the reasons for WHY the USA's Generals and Admirals in WW2 were so much better than today's US Officers is because pre-1941 the US Military had extremely limited budgets. **All of the pre-1941 US Army Officers had been forced to learn how to think creatively and manage efficiently with small budgets and very limited equipment.**

The pre-1941 US Officers (rather than only throwing large amounts of money at problems as they do today) had to apply intelligence, imagination, creativity, courage and use every tactic possible to solved problems. <u>They were NOT rich and thus did NOT become lazy.</u>

When General **Douglas MacArthur** was the US Army Chief of Staff (1930-1935) and Congress wanted **MacArthur** to testify before Congress about something - General **MacArthur** would take a Trolley Car (10 Cents per ride) from the War Department to Capitol Hill.

When WW2 finally came to the USA all of the US Army's senior Officers had their heads on straight and their feet on the ground.

(263) The Grand Strategist NEVER Trades Worthless Status Symbols for Real Advantages.

Great Powers in decline (like the USA in the 21st Century) always become embroiled in chasing yesterday's status symbols. In many cases those status symbols have become worthless. **Chasing now worthless status symbols usually comes with a cost of losing real advantages.**

The US Pentagon's old weapons systems designed for yesterday's wars are taking: time, money, planning, work and energy away from the Artificial Intelligence Age of Warfare weapons developments (and the US Space Force and US Communications Force) that are must haves for winning tomorrow's wars.

(264) In Battle (and in Peace Negotiations) the Grand Strategist will Maneuver Towards Positions Where what he Gives Up are Non-Vital Positions/Interests (or best of all only Worthless Status Symbols). While what he Fights For (and Negotiates for) are Real Advantages for his Nation's Vital Core Interests.

The Grand Strategist is willing to compromise on all issues that are not vital to the Strategist's core National interests. **The most skilled Strategist will have the talent and courage to trade worthless status symbols for things that have real value.**

(265) The core Vital Interests of the Great Power are NEVER Negotiable. However the core Vital Interests of the Great Power's Allies Are Negotiable.

If you are a Great Power (like the USA) and you can receive something of long-lasting true value to your Nation by giving up something from an insignificant Ally - then do so.

(267) The Grand Strategist will Never Start a War without Being 100% Sure that He has a Way to End the War Short of Nation Surrender.

A violation of the above Rule was Japan's 1941 bombing of Pearl Harbor. It was a brilliant tactical victory but where was Japan's viable "Exit Strategy" for ending the war with the USA and United Kingdom short of total defeat?

The Japanese were hoping that after Japan seized the Dutch oil fields in Indonesia (and captured the Philippines) the USA would become frightened and ask for peace. However it was President **Roosevelt** and Prime Minister **Winton Churchill** who had tricked the Japanese into attacking the USA and UK. Both **Roosevelt** and **Churchill** never had any intensions of making peace with Japan short of Japan's destruction and unconditional surrender.

Churchill's and **Roosevelt's** main objective with engineering the Japanese attack on the USA and UK was to bring the USA into the war. **Churchill's** secondary objective was to destroy Japan as a colonial rival to England in Asia. **Churchill** wanted England (and NOT Japan) to continue its exploitation of China.

All big wars take twists and turns that no human is smart enough to see at the beginning. **Winston Churchill** succeeded in destroying Japan as a colonial rival for land in Asia. However **Japan then turned to Economic Warfare** and succeeded in destroying major segments of both the UK's and USA's industry. Before WW2 England had a good automobile manufacturing industry. After WW2 Japan decimated English car manufacturing. Today Rolls Royce is owned by German BMW. Bentley is owned by German Volswagen and Jaguar is owned by Tata Motors of India.

(268) When Planning a War the Grand Strategist Remembers that the Most Important Battle is NOT the First Battle but the Last Battle.

It is NOT how a war begins (any military moron can start a war) or how many tactical battles are won during the war. **What is all important in Grand Strategy is how the war ends and what form of peace follows.**

The USA's greatest Generals - **George Washington** (1732-1799) lost more battles than he won but General **Washington** won the battle that counts the most - the last battle. He also won the peace and was able to launch what would be a great Nation.

The grand strategy student should remember that from the end of the American Revolution in 1783 to the start of the Civil War in 1860 there was a 77 year period when the only formal war was the War of 1812. From the end of the Civil War in 1865 to the US going into WW1 in 1917 there was a 52 year period with only one formal war (the 1898 Spanish American War) that lasted only four (4) months. **The USA became a world power because the USA did NOT have endless stupid wars in stupid places for stupid reasons.**

(269) A Top Priority of the American Voters should be Avoiding being Duped into Another Losing War (this time in Iran) by the Same Warmonger Con Artists who Duped the US Voters into the 2003 Iraq War and Before that into the Vietnam War.

There will be future reasons for why Iran is supposedly guilty of this crime or that crime. There is a 98% probability that the "evidence" is a fraud. If Iran officially says "Yes we did it" then believe them. Otherwise take any other "evidence" as a Lie.

One of the "signs" that supposed war crime evidence is really fraud news is that its reporting is made by some low level government official who is a nobody - or worse by the fraud news press. **The supposed "war crimes" should be announced by the President of the United States who says they are true - otherwise we can be 90% sure that the charges are BS.**

(270) The Best General Never Puts his Military Forces into a Position Where he Cannot Retreat Without Significant Loses. Every Major Action Must Have a Viable Exit Strategy.

The strategy that General **George Washington** and North Vietnamese General **Giap** had in common was they always had a wilderness at their backs they could retreat into and could thus never be enveloped.

The warmongers in Washington DC dream of using Afghanistan as a staging area for an invasion of Iran. However - that area is ideal for an envelopment of US Forces. In that same area (in 1842) an entire British Army was enveloped and destroyed with only one survivor.

Afghanistan was ALWAYS a "Bridge Too Far" for any kind of viable USA military actions. In addition the grand strategy in Afghanistan was wrong from day one. It was a repeat of the USSR's failed strategy with the same end results for the USA. The USA's Generals once again confirmed their incompetence by refusing to see what should have been crystal clear at the beginning.

(271) **Only Incompetent Generals have Long Shooting Wars.**

Shooting Wars are by definition the process of losing control of the situation. The more you lose control the more things there are that can go wrong and will go wrong.

It is an Axiom of Math (Murphy's Law) that if a war goes on long enough then by Mathematical Law everything that can go wrong will eventually go wrong. It is like a car - if you drive it for enough years then everything that can go wrong will go wrong. **The US Pentagon's never ending wars will see everything that can go wrong eventually go wrong.**

(272) **In an Active Shooting War (unless there are Compelling Reasons) your Opponent's Strongest Positions are OFF your Major-Attack List Until Your Opponent has been Sufficiently Weakened.**

Ideally you attack your opponent's strong points only after you have attacked all of his weak points first and have greatly degraded his overall position and power.

Shooting wars are more than attacking your opponent's troops, guns and tanks. Vital areas for attack include: your opponent's supply lines, electrical power plants, telephone communications centers, communications satellites, water purification plants, war material manufacturing factories, oil supplies, oil refining, food distribution points, ship and boat ports, shipping channels, bridges, commercial airports, railroad centers and all of the many dozens of indirect targets that will reduce your opponent's military capacity.

(273) **One of the Many Factors that Makes every War Different is National Character.**

A vital factor in good Strategy planning is understanding the National Character of your opponent. If your opponent has a history of giving up after the smallest struggle (and you have great power) then it will be a short war. However - if your opponent is a Nation like Iran (Persia) with a 2,700 year history of NOT giving up easily then an attacking US Military must be prepared for a very long and highly difficult war that will have many dead Americans.

(274) A Maginot Line Defense Mentality can be Highly Dangerous and Sometimes Self Defeating.

One of the dozens of dangerous myths and misconceptions about WW2 is the often heard - "We should have stopped Hitler at Munich." However - **in 1938 there was no - "We."**

When the Munich Agreement was signed (on September 30th 1938) the USA Army had 185,000 troops who were over 4,000 miles away from Germany. The British Army in 1938 had only 220,000 troops and most of the British troops were in Africa, India, Burma, Hong Kong, Singapore and so forth. The German Army in 1938 had over 600,000 troops on the border of Czechoslovakia.

The only Western Army of any size in the area was the French Army. If the French Army had advanced into the Rheinland in 1938 that might have politically toppled **Hitler**. Ironically it was France's Maginot Line defense system (and the false sense of security that the Maginot Line gave to the French) that guaranteed **Hitler's** success at Munich. **The French had put so much money into their Maginot Line they felt compelled to fall back onto its use.**

The US Pentagon is in the same strategic box today. The US Pentagon has put so much money and work into obsolete weapon systems for yesterday's wars it cannot build the vital Space Force that would be decisive in a future war with China or Russia.

If there was a USA-China shooting war. To save all of the Navy's 10 Super Aircraft Carriers the Chinese satellites would have to be taken out in a few hours. However when this Book was published there was NO US Space Force and the Pentagon was clearly resisting the idea.

The ancient Greek Spartans deliberately never built forts or walls. They deliberately forced themselves to keep their Army mobile. If you have a giant fortress then your opponent knows exactly where you are. However - no enemy of the Spartans could be sure where the Spartan Army might be and where it might attack.

In the Great Asia-Pacific War of 1941-1945 the Japanese fortified many Islands. The US Navy thus know where the Japanese where located (and often about how many there were per island) and the US Navy was able to concentrate its power where and when it wanted.

(Trivia Note) The Japanese "followed the book." They built one outhouse toilet per a given number of troops. They often built the outhouses on the beach and over the high tide mark and/or they did not camouflage them. US observation airplanes would count the outhouses and have a good idea of how many Japanese were on an Island.

Sometimes it is the little stuff that you never think about that will kill you. After the cold war started in 1946 the newly formed CIA (that replaced the OSS) sent many agents into Eastern Communist Europe to gather information. The agents had CIA provided Identification Books and the printing and paper was 100% flawless. However the CIA used stainless steel staples and the Russians were using Iron staples that rusted. All the Russians had to do was look at the outside of the Identification Book and if the staples were not rusted then they knew its holder was a spy (who was usually tortured and then executed).

(275) The Key to Finding the Correct Answers to Challenges is Having the Ability to Find and Ask the Precisely Correct Questions.

If you have exactly the correct questions then you are a giant step closer in finding the correct solutions.

Part of the process of finding the correct questions means you have to ask many questions that are often silly and irrelevant. This becomes very difficult (and usually de facto impossible) in old overly formalized and ridged cultures like the US Pentagon.

(276) Every New Major Weapons System will Change the Tactics Used on the Battlefield.

In 1914 the new weapons system of Machine Guns had greatly changed Battlefield Tactics. However all of the Generals refused to change their obsolete Tactics mindsets for the first 3 years of the war.

Try to imagine how battlefield Tactics are going to change in the last half of this Century when the frontline "Troops" are highly Intelligent-machines and not afraid of death.

Picture 40 PT type fast stealth configured Boats (each carrying a nuclear bomb) and each panted flat black and simultaneously attacking a US Navy Super Aircraft Carrier in the middle of the night. If only one of the 40 gets through the US Super Carrier is dead.

During the Great Asian-Pacific War of 1941-1945 the Japanese Kamikazes caused great damage. The Kamikase's great damage to the US Fleet was censored out of the news at the time and has thus stayed out of most of the history books. However - many smaller intelligent devices that are not afraid of death are powerful weapons.

During WW2 Pacific the Kamikazes sank 47 ships and damaged over 300 ships (many so seriously the ships were out of the war). About 7% of all USA combat deaths in the Pacific were caused by Kamikazes.

(277) Retreating from Your Opponent Can Be a very Powerful Tactical and Strategic Tool in the Hands of a Talented Commander.

If you are stalemated in a battle (and you retreat) your opponent will 90% of the time follow you. If your opponent proceeds at a fast pace then at a certain point your opponent will be over extended. If you retreat back to pre-built defenses and supply stations then you are retreating back onto your reserves and supplies. **You will be in a superior position to counter attack an overextended opponent and ideally envelop his advanced forces.**

(278) If the Strategist is Facing an Opponent who has almost Unlimited Territory to Retreat Back into then the Strategist must View that Situation as the Most Disadvantageous for the Attacker.

One of the main reasons for why the British (with a mighty Army and Navy) lost the American Revolution (1775-1783) is because the Americans always had an unlimited deep forest at their backs to retreat into. Unlike most European wars the British could never envelop the American Forces (or capture the American Capital because there was none) and thus the British could never destroy the American Forces.

Another Nation that cannot be enveloped is Iran. Iran is 900 miles by road from South to North and most of that distance is deserts and mountains. Iran is 637,000 square miles with 85 Million people. Texas is 269,000 square miles with 28 million people. There are railroad lines and roads going into Iran that both Russian and Chinese would use to supply Iran with arms and ammunition.

Iran cannot be enveloped without the help of both Russia and China but they would be supplying Iran in a war with the USA. Iran (exactly like North Vietnam) cannot be enveloped and the Iranians (Persians) would never surrender.

In the Iran-Iraq War (from 1980 to 1988) Saddam Hussein of Iraq (with the military backing of Washington DC and Israel) attacked Iran. In 9 years of intense fighting Saddam's Army advanced about 50 miles into Iran. Eventually the Iranian fighters pushed the Iraq Army out of most of that small gain.

The Iraq military suffered 300,000 dead soldiers and 100,000 dead Iraq civilians for a end of the war land gain of a few miles.

US Force could also attack Iran from Afghanistan - but there is 450 miles of hills and desert between Afghanistan and Tehran. Any US Military Thrusting-Attack over 450 miles of desert would be open to flank attacks and would be enveloped and destroyed.

Afghanistan is also landlocked for US supplies. Everything has to be flown in over Nations controlled by Muslims. If Turkmenistan said the US Air Force could no longer use Turkmenistan's air-space the US Forces in Afghanistan would be trapped.

A USA-Iran War would be Vietnam all over again. Washington DC once again would having no way to end the war. The US voters would turn against the Iran War (exactly as they did with the Vietnam War) and Iran would win.

(279) In Grand Tactics you always Try and Cut Your Opponent's Lines of Communication and Supply. Sometimes Cutting Supply Lines can Produce a very Economical Victory.

During WW2 in the Pacific the US Navy 80% cut the supply lines to many Japanese Islands and they never were captured by US Forces because they became useless to the Japanese.

(280) When it Becomes Clear that You are in a Conflict that You Cannot Win then a Much Used "Exit Strategy" is to Declare Victory and Say that What Happened is What You Wanted to Happen - And Then Move On.

During the British Empire's decline (1919-1997) when one of their colonies wanted to break away the British would usually resist with armed force. However **the British were good at knowing when they had lost and if it became clear that the independence movement was not going away the British would then do an about-faced and act like the idea of Independence was theirs.**

The British would celebrate a victory for what they supposedly had wanted. The breakaway former colony usually joined the British Commonwealth. That was NOT as good as outright control of a colony's economy but it retained some British common market economic advantages and some advantages for the British Pound.

The British understood the rule that the fewer people you kill in a conflict the easier it is to make a lasting peace. The British also understood **the Rule that when it is clear you are going to lose then make peace and save what advantages can be save.**

(281) If you Have Strategically Already Either Won the War (or Strategically Already Lost the War) but the Fighting is Continuing then You Must Create an Excuse for You and Your Opponent to End The Fighting.

By 1970 every intellectually honest person could see that the US Military had long ago lost the Vietnam War (1959-1975). Dr. **Henry Kissinger** (then Secretary of State) knew that the war was lost and the USA needed an excuse to end the war. Dr. **Kissinger** (Secretary of State 1973-1977) created the 1973 "Paris Peace Accords" that officially ended the combat in Vietnam and made it possible for US Troops to retreat out of Vietnam in an orderly fashion.

U.S. Secretary of State Dr. **Henry Kissinger** (and the North Vietnamese) knew that the 1973 "peace" was only an excuse for the US Military to declare victory and leave. The peace agreement was a face-saving maneuver for the US Military to remove US Troops without officially admitting that the US Military had been defeated.

The above is **the "British Strategy." When you are in a losing war and have no chance of winning you declare victory (or sign a peace agreement that can be news-spun to look like a victory) and then leave.**

Dr. **Kissinger** and President **Richard Nixon** (President from 1969-1974) and North Vietnam's representatives at the peace talks **all understood that the peace agreement was a way for North Vietnam to take over South Vietnam while giving the US Military a way to say that it was not defeated.**

(282) A Crushing Victory far Beyond what was Needed to Win has Many Times Generated Revenge that Was the Creator of a Future War.

Never deliberately cause massive damage to your opponent (far beyond what is needed) to win unless you are 100% sure that opponent will never recover.

Remember that every war is always followed by peace. In peace you have to live with your former opponent. **The more you stay on the moral high ground during a war the easier (and safer) it becomes to live with a former opponent.**

The USA was exceptionally lucky at every turn in WW2 and very lucky after the war. After WW2 in Europe the Germans were far more concerned and worried about the Russians than any revenge against the USA.

In Asia (after WW2) the Japanese were far more worried about the Russians and the Red Chinese than military revenge against the USA. **The Japanese achieved their revenge against the USA in the form of an Economics War** that successfully colonized a large part of the USA's automobile industry and enveloped 90% of the USA's electronics retail product industry and 99% of its camera industry.

The Japanese attack on the US Auto manufacturers (starting in the 1960's) is an example of classic good strategy. The Japanese analyzed the US Auto industry for Detroit's biggest weakness - and that was the poor and shoddy mechanical quality of Detroit's US cars. The Japanese attacked Detroit's biggest weakness by engineering and selling cars that did NOT break down as much as US made cars.

(283) Every War is Different but at the Same Time Every War has Some Relationship to Previous Wars.

Part of the Art of War and Peace is being able to look at past wars in relation to your current war and (1) correctly choose what ideas and Strategy from previous wars and peace efforts and you want to emulate. (2) What you want to ignore as not applicable and (3) what is now dangerously obsolete and must be avoided.

(284) The Closer you can Cut your Opponent's Communications Lines at their Sources the more Immediate and Significant will be the Positive Benefits for You.

In a shooting war the US Air Force (and US Space Force) must have at the very top of their attack-list the opponent's satellites. The US Space Force must destroy as many opponent satellites and a possible - and as soon as possible. The US Space Force must create ways to defend the USA's satellites. A technology that at this time does not fully exist.

The USA's "Space Force" is the future of decisive combat if there was a shooting war with China or Russia.

In a USA-China Shooting War the USA's Internet, Cell Phones, most TV, Credit Card use, Uber, GPS service, on-line shopping like Amazon, on-line stock market trading - everything connected to Satellites could be **down for several years.** The USA cannot invade China or Russia and China and Russia cannot invade the USA. That means a "state of war" might continue for many years.

(285) In Grand Tactics one of your Top Targets should be your Opponent's Centers of Administration. You should always be Trying to Cut and Sever the Connections Between the Brain and the Limbs.

In Guerrilla warfare (of the type that the USA is mainly fighting in the Middle East) the opponent's "command centers" are the opposition's ideas and persuasion arguments on the Internet.

The only way to effectively attack arguments on the Internet is with better ideas and more persuasive salesmanship for the USA's ideas and objectives.

(286) During a Shooting War the Exact Location of Your Command Centers must be Kept as Secret as is Possible. The Exact Location of Command Centers Down to the Division Level should Only be Told to People on a Need-to-Know Basis.

(287) When Fighting an Opponent who is Over Regulated from the Top Command Center on Down - if you Envelop a Section of your Opponent's Troops and Seal Off their

Lines of Retreat then you can Sometimes Cause "Command Bewilderment" and you could Capture a Substantial Number of the Opponent's Troops.

"Command Bewilderment" does NOT last forever. **You have a very limited window of opportunity and you have to act fast.**

In the 21st Century (during future major shooting wars) if the US Space Force takes down enough of the opponent's communications satellites (and cuts a significant part of the opponent's communications) that might create "Command Bewilderment" in some areas and open a window of opportunity for enveloping and/or destroying a large amount of your opponent's assets. On the other hand the same tactic can be used against the USA by China or Russia. What would the US voters give up to get their Cell Phone service restored?

(288) Ideally Your Plans for Attack Should Be What Your Opponent Would Least Likely Expect.

In all of the ages of warfare the factor of surprise has been the consistent mother of success. **Great effort should always be put into NOT being predictable.** Do things differently (even sometimes illogically) on a regular time period to make yourself as unpredictable as possible.

One of Washington DC's (and the US Pentagon's) many big weaknesses is that they are 97% predictable.

(289) The Strategist must Understand What Has Been the 3 Step Post WW2 "Standard Operating Procedure" in Washington DC "Swamp." (1) Deny any Mistakes or Wrong Doing of Any Type. (2) Never Admit to Any Mistakes or Failures of Any Type No Matter How Obvious the Mistakes. (3) Always Accuse Your Accusers of Doing the Mistakes (or Crimes) You Have Been Guilty of Doing.

Washington DC's fanatical insistence of never admitting any type of mistake is a self-serving con game. **If you never made a mistake then there is nothing to fix** - right?

Never admitting to any type of mistake is one of the biggest reasons for WHY Washington DC keeps making the same old mistakes over and over again decade after decade and lost war after lost war.

Washington DC fanatically refuses to admit that it made any mistakes - but who are the fooling? The "approval rating" for Congress has been at about 12% of decades. The US Military has not clearly won a significant shooting war since 1945. The only people in the world that Washington DC "Swamp" are fooling are themselves.

(290) The US Pentagon and Washington DC must Find some Way to End its Compulsion for Wanting to Refight WW2.

WW2 is over and gone - it is ancient history. The Industrial Age of Warfare that WW2 was part of ended in 1960. Very soon WW2 will be two ages of warfare in the past.

The so-called "terrorists" in the Middle East are not modern day Nazis. Some Middle East so-called "terrorists" are like the Viet Cong fighting against outsider invaders who are trying to reestablish White man colonialism in Asia. It is the US Military that is in somebody else's Nations and murdering the local populations.

The US Military refused to understand the true motivations of their opponents in Vietnam and the US Military is refusing to understand the true motivations of their opponents in the Middle East. It is a debatable question as to who are the real terrorists?

(291) The Strategist is Always Creating Propaganda for His Purposes but Never Becomes a Victim of his Own Propaganda.

The good Commander creates propaganda to motivate his population and troops. However the good Commander must always be crystal clear in his mind as to what is the truth and what are the propaganda spin-lies.

When National Leaders and Commanders start to believe their own propaganda-spin-lies then defeat is usually not far away.

(292) The USA's Economy of Over Consumption (that is Financed with ever Increasing Debt) is NOT the Economy of a Nation that is going to be a Super Great Power Forever.

The USA is already on the downside of its Bell Curve and will be for the remainder of the 21st Century. **Because of momentum the USA will be a great defensive military power for many decades but will continue to decline (relative to the world) both economically and with the USA Militarily's offensive capabilities.**

(293) The Main Foundation of Military Power (in the Long Run) has always been Economic Power.

The US Federal Government (and the USA Private Sector) Cannot Forever keep Increasing the USA's Debts by over a Trillion Dollars Per-Year.

The USA will be able to live on expanding debt for many more decades - but NOT forever. The more the world's economy turns away from the US Dollar and to other currencies (because of US Dollar debt creation that is de facto counterfeiting) the more the USA will lose worldwide military power.

The economics idiots say - "But one person's debt is another person's asset." However - **Debt is worth NOTHING if it cannot be paid back.** In 1866 what was $1,000,000 in Confederate States of America Bonds worth? The answer is $0 money.

The US Federal Government already has many Trillions Dollars of debt. The Federal Government's debt keeps going UP by a Trillion dollars a year. By year 2050 the US Federal Government's Debt will be well over $50 Trillion Dollars. How is the US Government going to every pay that back that amount of money in like value money? It is impossible - that means **US long-term debt** has increasing negative value. **The world is NOT going to play that sucker game forever.**

What is Washington DC's long term economic strategy? The Washington DC Government will continue (for many decades) to under estimate US Inflation by 30% to 60% every year and thus inflate a large amount of "the Debt" away. However the incomes of the bottom 50% of the population (and those on Social Security) are NOT now (and will NOT in the future) keep up with the real world inflation rate. That is going to cause a growing amount of poverty and serious social and political problems.

The USA's future picture is clear. If you are in a position to stay ahead of inflation you will be doing great. However - if you are in a position to NOT keep up with inflation (i.e. those that actually live on Social Security) or have ordinary retail jobs - they will be struggling.

(294) A Nation can Lose only So Many Wars Before it Suffers Serious and Permanent Damage to the Nation's collected Ego and to the Reputation of its Military.

The reputation of the US Military has already been seriously damaged. When this book was published the war in Afghanistan had 2 US citizen mercenaries (making $100,000 to $200,000 per year) for every Regular US Army soldier in Afghanistan. **The US Military is already at a point where it cannot fight shooting wars (even losing wars like Afghanistan) without highly paid mercenaries.**

How many young American men are rushing to join the US Military as a US Army Private making $22,000 a year when the US citizen mercenary next to him in Afghanistan or Iran is making $150,000 a year?

One of the many games is to go somewhere in the Middle East as a US soldier. Acquire a reputation as a good fighter. Leave the US Military and then sign-on as a $100,000 a year special contractor.

The Pentagon's game is saying they have only 9,000 "Troops" in say Afghanistan. They do not count the 15,000 mercenaries because they are NOT officially US Troops.

How many USA college graduates in the top 5% of their class are joining the US Military because that is where cutting edge actions are happening? Not many when this book was published.

(295) Since the End of WW2 Most of the People in the US Pentagon have Sincerely Believed that the Pentagon Has a Mandate from God to Win Every War.

God is always on the same side in every war. God is on the side that can most successfully meet the challenges of change that **God** has created. **God** is on the side that wants to live in the future. **God** is never on the side that wants to die with the past.

(296) The 21st Century USA Grand Strategy Team must have Experts at Managing Limited Wars and Managing Politically Complicated Wars that are often Convoluted Proxy Wars for Something Else.

Warfare is becoming increasingly complicated and that is why the USA must have a greatly expanded Strategy Team. The current Pentagon's organization is obsolete and clearly failing. The USA must have major upgrades in planning and analysis as soon as possible.

(297) The Complexity of military Grand Strategy is Growing at an Increasing Speed and that means Strategy Skills and Organizations Must also be Evolving at an Increasing Speed.

The USA is already dangerously obsolete in many areas of Grand Strategy planning and organization. Most of the USA's Secretaries of Defense knew very little about Grand Strategy. The USA's Generals have not clearly won a significant shooting war since being on the winning side in WW2. **US Generals since WW2 have been proven incompetent at winning significant shooting wars.** There is a desperate need for the USA to create a new Grand Strategy Team as soon as possible.

(298) All Ethical Grand Strategists and National Leaders Reject the Idea of "Preventive Wars."

The Washington DC Warmonger Lobby's idea that some Nation (or group) should be attacked because they might do something wrong in the future is morally and legally wrong by the Laws of all Nations. By the United Nation's Charter and **preventive wars are forbidden by all major religions.**

The world's greatest military grand strategist - General **Sun Tzu** wrote that Generals and National Leaders must always stay on the moral high ground because being on the moral high ground is a vital factor for winning wars.

One of the main reasons for why Washington DC has been losing wars for over 60 years from the Vietnam War to the Afghanistan War is because their wars have in large part NOT been on the moral high ground.

Adolf Hitler's Holocaust against the Jews was "Preventive Law Enforcement." In the minds of the many thousands of people who helped run the Holocaust they were killing the Jews before the Jews could break any laws. **Those who believe in "Preventive Wars" are morally standing next to Adolf Hitler**.

You could point to anyone in the world and say they "Might" do something wrong in the future so that means they should be killed now to prevent the crime. **Once you start killing people for what they might do wrong in the future then you are kill anyone everywhere.**

The warmongers in Washington DC who say the USA should attack Iran now because Iran "Might" do something wrong in the future have **Adolf Hitler's fascist psychopath mindset.**

If China or Russia were going to launch a full 100% of their nuclear capacity in an attack against the total USA mainland (and US Intelligence was 100% absolutely certain it was coming and unpreventable) then there is an argument for the USA to attack first. However anything less should be high profile stated by Washington DC as morally unacceptable.

(299) Sometimes the "Tripwire Tactic" Has a Place in Grand Strategy.

The "tripwire tactic" is when you place your troops in locations where an aggressor cannot attack without killing your troops and thus having an act of war. The Nation that had troops killed then has legal and moral cause to declare war against the other Nation.

US troops in Berlin from 1945 to 1990 were the "tripwire." The USA has had "tripwire" troops on the Korean DMZ since 1953.

The US tripwire troops on the Korean DMZ have been the correct Strategy because South Korea is on the correct side of history and North Korea is not. Time is working against the Communist dictator-monarchy in North Korea.

The USA's correct strategy with North Korea and Iran and Cuba would be to do everything possible to get inside of those Nations. Then using the many non-shooting types of warfare change attitudes from the inside out. US sanctions against North Korea started in 1950. After 70 years of failure it is time to try something different.

(300) The Grand Strategist from a Technologically Advanced Nation like the USA Avoids Putting a Significant Number of his Troops into a Situation where Superior Technology can be either Neutralized and/or has Limited Advantages.

One of the many mistakes that the US Military made in Vietnam was deleveraging the USA's technology. American soldiers were out in the jungle with early model 1960's M-16 rifles and the Viet Cong were in the same jungle with Russian AK47 rifles. The immediate differences between the two groups was the Viet Cong soldiers had better rifles and had a far superior will-to-win.

In future USA shooting wars the "US Troops" on the battlefield must **as soon as possible** be highly intelligent battle machines so the USA can leverage its technology.

(301) In the Last Half Century the US Pentagon's Standard Strategy of "Surging Troops" into Faltering and Failed Wars Has Been the Bankrupt Strategy of WW1's Failed Generals.

The USA's Generals must STOP being 1915's WW1 Generals. The US Pentagon's Strategy of the last 50 years in - Vietnam, Iraq, Afghanistan etc has been proven and reproven to be wrong. There must be major changes in mindsets, tactics and grand strategy as soon as possible.

(302) An Imperfect Peace is many Times Better than Starting a Shooting War you have No Chance of Winning.

When this book was published the Warmonger Lobby in Washington DC were constantly complaining about the Iran Peace Agreement NOT giving the USA 100% of everything the warmongers want. However the US Warmonger Lobby wants Iran to be a West Bank slave state and anything less is "unacceptable."

The proven and reproven historical fact of the last 4,000 years is that **Peace Agreements that are all one sided never last unless one side is massively superior and can easily envelope and destroy the other side.**

Secretary of State **John Kerry** (born 1943, a Vietnam war hero and Secretary of State 2013-2017) negotiated a Peace Agreement (in 2015) that **opened the door to Iran and made it possible for the USA to get inside of Iran and apply many of the non-shooting types of warfare from the inside out.** However President **Trump** slammed that door shut on what was a hard won excellent beginning opportunity to get inside of Iran and hopefully **once again** make Iran an Ally of the USA.

President **Trump** made the mistake of seeing the "Iran deal" as the END of the deal when it was only the BEGINNING of the deal.

In international relations there is no "End" there is only "Next."

Hopefully President **Trump** will learn some history and understand that less than perfect Peace Agreements are often 1,000 times better than a shooting war (in this case with Iran) that the US Military would definitely lose.

The Versailles Treaty of 1919 was vindictive and one-sided in favor of England and France. It lasted until 1935 (16 years) and gave England and France **Adolf Hitler** and eventually WW2.

The axiom is - **"Allies are NOT forever and Enemies are NOT forever."** There are no unsolvable reasons for why the USA and Iran cannot once again be Allies with compromises that both sides can live with and where both sides can benefit.

(303) Nations should NOT aggravate other Nations for Stupid Ego Games and Risk Being Suckered into a War.

Many wars are started with trickery. Especially in modern times when world public opinion is now a major factor in the Grand Strategy equation.

In 1941 **Churchill** and **Roosevelt** tricked Japan into attacking the USA and UK.

In 1960 the reasons given by Washington DC for invading Vietnam were lies and con games.

In 2003 the reasons given for the USA attacking **Saddam Hussein** were lies and con games.

When this book was published the very powerful Warmonger Lobby in Washington DC was trying to **trick Iran into doing something that could be used as an excuse by Washington DC to attack Iran.**

The Iranians should constantly be on-guard against anything the warmongers could spin into a reason for a USA-Iran War. When the Washington DC Warmonger Lobby tries to set up a con game and accuse Iran of a crime that is when Iran should demand a trial in the International Criminal Court at the Hague and demand that the warmongers prove their case.

Iran needs to spend a little money on three or four good public relations agencies in London, Paris, Berlin, Tokyo etc.

(304) If Two Sides in a military Conflict are about Evenly Matched the Winner of the Conflict will be the Side that Makes the Fewest Mistakes.

It is often the same in life and in business. The corporation (or person) that wins is often the one that makes the fewest big mistakes.

(305) **The Concentration of Decisive Force onto a Specific Target of Value is the Correct Strategy in Everything from Grand Strategy to Managing your Personal Life.**

Choose what you want as your life objective (or next big objective) - have something that is very challenging but doable. Concentrate all of your thought, time and energy towards that objective and keep working towards the objective until you are successful.

Most people are depressed at the start of a long hard road towards something significant. This is why most never start. **Mao Zedong** (1893-1976) said to NOT think about the five thousand mile march in front of you. **Only think about the next step you have to take.** If you keep taking steps forward you will arrive at your destination. **Mao** said - "A march of 5,000 miles is made one step at a time."

(306) **There are Always Going to be Black Swans (Unexpected Turns of Events). However An Unexpected Black Swan is Often an Unexpected Opportunity.**

The Grand Strategist must not see Black Swans as ONLY something that is "messing up your plans." Look and think carefully to see if the Black Swan has opened a brand new opportunity that you could turn into an advantage. If so - then move fast, boldly and seize the opportunity.

(307) **The Invention of Railroads Profoundly and Permanently Changed Tactics and Strategy. Railroads greatly Expanded the Ability to Move Large Numbers of Troops Quickly and Decisively.**

Railroads were the leading-edge first great weapons system invention of what would become the Industrial Age of Warfare.

In 1812 the few settlers and soldiers in what would become Chicago USA were defeated by the Indians and either killed or driven off. However - **by 1900 the railroads had turned Chicago into the 2nd biggest city in the USA and the 5th biggest city in the world.** World city sizes in 1900 were - (#1) London 6.4m, (#2) New York 4.2m. (#3) Paris 3.9m, (#4) Berlin 2.4m, (#5) Chicago 1.7m, (#6) Vienna 1.6m, (#7)Tokyo and Saint Petersburg both 1.4m.

Starting in WW2 air transport has greatly changed tactics and strategy. Today all US Troops that go any distance are moved by air. The US forces in Afghanistan are almost 100% supplied by air that has to travel over Muslim governed territory. That makes the US Military in Afghanistan position highly precarious and dangerous in a shooting war with Iran.

(308) In Grand Strategy One of the Best Positions (and a Position to Always Work Towards) is the Balance of Power Position.

In both WW1 and WW2 America was the balance of power. The USA was thus able to secure great long term advantages at a relatively small cost in blood.

In WW1 the USA had 53,000 military personnel killed in battle in a war that had 11 million military personnel killed.

In WW2 the USA had 407,000 military dead in a war where the USA's Allies against **Hitler** and Japan (the UK, Russia, China, France, Poland and so forth) lost over 15,000,000 troops.

In WW1 the USA was the psychological balance of power. In WW2 the USA was both the war-materials balance of power and the troop balance of power.

(Note) In WW1 the USA military suffered 53,000 battle deaths but 117,000 military deaths from all causes. This was a very high non-combat death percentage total (almost 60%) for a Western front Nation in WW1. The USA's non-combat dead ratio in WW2 was also high compared to the British and Germans. **The US Troop's weakness factor continues into the 21st Century with high rates of US Military suicides, sickness and very high rates of US veterans going onto mental disabilities payments for the remainder of their lives.**

The percentage of USA troops that go onto disability payments for stress disorders (payments that continue for the remainder of their lives) is many hundreds of percentage points higher for the USA than for any other Nation in the world.

There are clearly very troubling US Troop cultural weakness factors that need skilled in-depth analysis by the new Grand Strategy Team. **Better that the USA understand the weaknesses of its own Troops and makes adjustments before the Chinese understand what is happening and exploit the weakness.**

(309) You NEVER Deliberately Start a Major War when you Already Have an Ongoing Undecided Significant War Going.

When the first-edition of this book was published the US Military had shooting wars going in: Afghanistan, Iraq and Syria. There was a possible war with North Korea and the Washington DC Warmonger Lobby was pushing for a USA-Iran shooting war.

During 1941 - Germany was in a major war with the British Empire. Germany was not losing that war but Germany was not winning the war. The British Empire in early 1941 had 24% of the world's total population (plus the world's biggest Navy) and was holding its own against **Hitler**. It was something of a stalemate. Germany could not invade England and England could not by itself invade the Continent.

In June of 1941 **Adolf Hitler** made a decisive fatal mistake by greatly expanding his war with an attack against Russia. **Adolf Hitler** deliberately opened a two front war with a major military power. Germany was then fighting a 2 front war when it did NOT have to fight a 2 front war.

Adolf Hitler's deliberate and **arrogant overreach** attack against Russia eventually proved to be a complete disaster for Germany.

Washington DC has been in a USA versus Muslim World shooting war since 1991. **The war is overall going nowhere for the USA.** Like Germany against the UK in early 1941 - the USA is not exactly losing but it is also not winning. However - the very powerful Washington DC Warmonger Lobby keeps calling for a USA attack against Iran. That would **greatly expand the US Military's already shooting war overreach.**

If Washington DC made the horrible mistake of attacking Iran with a shooting war (and things turned bad with North Korea) and there was a shooting war with North Korea. Then the USA would be in a 4 front shooting war: Iraq-Syria, Afghanistan, Iran and North Korea - and all on the other side of the world making supply and maintenance especially costly.

The US Military could not maintain such a war without re-starting the military draft. That would elect an ultra left-wing anti-war government and ALL of the USA's wars everywhere would be lost.

Excellent USA Grand Strategy would be (1) the USA ending all of its current never ending wars and (2) NOT expanding into any new wars anywhere.

Every American born after 1989 cannot remember a time when the USA was not in a shooting war somewhere in the world. For the author of this Book that is a USA National disgrace and disaster.

(310) In Grand Strategy it is Best to be Loved - but if You Cannot be Loved then the Second Best is to be Feared.

A Great Power can often secure what it wants with either love or fear. You always start out nice with love - but if that goes nowhere you change to creating fear.

(311) One of the Most Important Grand Strategy Rules is that if a Serious Systemic Problem Exists You Must Replace the People who Created the Problem.

The current US Pentagon that started the systemic problems of never ending wars in stupid places for stupid reasons cannot solve the organizational and planning problems that it created.

A completely new group (and mindset) is the ONLY solution. This is why there must be (as soon as possible) the new Grand Strategy Team as outlined in Chapter 31.

There are examples of some groups that have created a large systemic problem and then solved that problem but **NOT without first having a major disaster that changed the group's mindset.** The bigger and older the organization the harder it becomes to change its collective mindset. The US Pentagon with 26,000 full time people pushing papers around and losing wars is impossible to change using the system that created the problem and is the problem.

(312) In <u>Super Grand Strategy</u> Your Core Objective is to Envelop your Opponent's Core Grand Strategy.

In the "Cold War" between the USSR and the USA (1946 to 1991) **the Grand Strategy of the USSR was to politically turn most of the world into either Communist or ultra Socialist Nations** that were friendly to the USSR - and thus **envelop the USA in a world full of Communist Nations.** However - because of the fatal core contradictions of the USSR's Communist economic system (and America's superior free market economic system) the USA eventually both economically and organizationally enveloped the USSR.

A second major reason for the USSR collapsing is because the USSR had massive "Diversity." The USSR had many "diversity groups" that did NOT like other groups inside of the USSR and refused to identify with other groups in the USSR. **The "diversity" pulled the USSR apart and destroyed the USSR from the inside out.**

A vital to understand core question is "WHY did Communism in the USSR fail but the Communism in China is succeeding?" The main reason for WHY is because after **Mao Zedong** died in 1976 and **Deng Xiaoping** became Chairman (from 1978 to 1989). Chairman Deng (1904-1997) **kept Communism as the political system but turned the economy into a Capitalist Economy.**

China is 96% Chinese. "Diversity" is actively discouraged in China. In year 2050 China will continue to be one Nation of non-diverse very hard working intelligent people (and the world's biggest economy). **Diversity is NOT going to destroy China as it destroyed the USSR.**

By 2050 there is a high probability that "diversity" will have succeeded in pulling the USA apart in the same way it tore the USSR apart.

When this Book was published Russian President **Vladimir Putin** was bitter about the breakup of the old USSR. Mr. **Putin** has forgotten that the breakup came because the different groups in the old USSR hated other groups. The Leaders in Moscow were no longer willing to use mass murder and government terror to keep everyone "in line" and that was the end of the USSR.

(313) Shooting Wars of Movement (that Keep the Initiative by Constantly Attacking the Opponent with as Many Types of Warfare as Possible) are Historically More Productive than Never Ending Wars of Entrenchment Stagnation.

Marshal of France (and one of Europe's greatest soldiers) **Maurice de Saxe** (1696-1750) wrote in his book ***My Reveries Upon the Art of War***. "The same troops who if they were attacking would have been victorious may be defeated if defending entrenchments. Few Generals have accounted for this happening in a reasonable manner for it lies in human hearts. It proves the instability of the human heart and how little Generals should depend on it."

The US Pentagon's mud-hole never ending wars (like Afghanistan) have a WW1 "entrenchment" mindset. **The US Pentagon believes it is protecting its reputation by never leaving the mud-hole wars. However - in reality the US Pentagon is continuing to lower and destroy its reputation.**

When this Book was published the US Regular Troops in Afghanistan had become so mentally depressed and discouraged the Pentagon was forced to rely on mainly $100,000 to $200,000 per year US citizen mercenaries who were equal in number to the US Army Regular Troops.

The US Pentagon must either fight the "Terrorist Wars" correctly and win (as specifically outline in this book) or stop embarrassing and mentally-depressing the American people (and their own Troops) by constantly repeating their proven failed tactics decade after decade and generation after generation. **America's brave frontline soldiers deserve much better Generals than the shooting war losing mediocrity they have had to suffer under since the end of WW2.**

The USA's military cadets in today's military schools should NOT be happy with the USA's many military failures since the end of WW 2. All military cadets should have as their career priority significantly changing the systems and changing the Pentagon.

(314) Periodically the Leader Should do Something (even if somewhat Illogical) to Make Himself more Unpredictable.

The British have been the most clever in the modern world at military trickery. One of many examples was during WW2 the British were always pretending to be making preparations for an invasion of Nazi held Norway. Especially during the D-Day preparations in early 1944. The British never had any intentions of invading Norway. However - the British tricks successfully kept a large Germany Army in Norway doing nothing for most all of WW2.

In January of 1945 there were 15 German Divisions in Norway waiting for an invasion that was never intended. Those German Troops were effectively in a British prison. Much like the US forces in Afghanistan (310,000 US Troops since 2001) have de facto been in a Muslim prison. **Afghanistan has been the Taliban's Guantanamo for US Troops.**

(315) Once you have Mastered the Application of the 37 Different types of Warfare you Have Also Greatly Expanded Your Ability to Use and Apply Indirect Attacks.

The 20th Century Grand Strategy writer **Liddell Hart** (1895-1970) popularized **the return** to the Strategy of "Indirect Attacks." **Liddell Hart's** philosophy of war was a reaction to the disasters of WW1 **Clausewitz** inspired "Direct Attacks" that slaughtered millions of troops.

Hart's Rule is to have few Direct Attacks and mainly Indirect Attacks. That means NOT directly attacking your opponent's strongest points but attacking his weakest points first. In the 21st Century an "Indirect Attack" against the Radical Muslim Groups means using as many as possible of the non-shooting types of warfare-tools to the greatest extent possible.

What **Liddell Hart** did after WW 2 was to reintroduce what had been **Sun Tzu's** main tactical advice of attacking weakness rather than attacking straight. General **Sun Tzu** was NOT widely known or studied in the West until the 1960's.

China (starting in 1979) has been **Indirectly Attacking** the USA with the **Economics Warfare** tool. China is grand strategy correctly attacking the USA's weakest parts first.

China started its Economic War in 1979 by making toys for the USA. Today China makes the USA's smart phones and computers and thousands of advanced products sold in the USA. China has expanded into the US market to the point where at this time China supplies food to some of the USA's fast-food chains.

(316) There is Never Going to be Anything Close to a Re-Play of WW2 in the Pacific.

The US Navy is insane when it believes the Chinese Navy is going to replay the US Navy's WW2 Pacific-War fantasies while using the US Navy's rulebook.

Eventually (many years from now) China's "main target" in the Pacific will be the value of the US Dollar. The US Navy cannot defend the US Dollar's value with US Aircraft Carriers. However **the Chinese attack against the US Dollar will only happen after the Chinese have "gamed" the US Federal Reserve for many more years** and greatly expanded their worldwide strategic assets bought and paid for with US Dollars.

(317) Speed of Action is One of the Most Decisive Factors in Warfare.

One of **Napoleon's** greatest "secrets" for success was speed of action. The speed of going into battle and especially the speed of exploiting the mistakes of his opponents. Time is always essential in wars and **speed can multiply the value of time and put the power of time in your favor.** Speed is like you moving in fast motion and your opponent moving in slow motion.

Napoleon specifically said - "The strength of an army, like the momentum in mechanics is established by the weight multiplied by the velocity."

The basic equation of (Mass times Velocity = Energy) holds for all of warfare. If you increase your Velocity of actions you increase your Energy towards victory. You could shoot a ping-pong ball through 75mm of steel on the side of a Tank if you could get its Velocity moving fast enough.

(318) The Grand Strategist always Positions his Forces so His Next Target is Never an Absolute Certainty.

Do everything possible to cloud your intentions. Make your next target and actions as hard to discern and discover as possible.

(319) In Warfare God Favors the Side that Has a Superior Ability to Adapt.

Sir Liddell Hart (1896-1970) wrote "Adaptability is the law which governs survival in war as in life - war being but a concentrated form of the human struggle against the environment." Humans have succeeded in taming much of the environment and Humans can succeed in reducing shooting wars to small isolated happenings.

(320) In Active Shooting Warfare one of the Basic Tactics is always the same. If your Forces are: (1) Attacking Forward or (2) if your Forces are Gathered in One Place and Waiting or (3) if your Forces are Retreating. In all three Situations You must Constantly be Throwing as much Steel into the Face of your Opponent as Possible.

In an active shooting combat war **your opponent should constantly be suffering casualties.** You want to make conditions for your opponent as costly and uncomfortable as possible - and for as long as possible.

In 21st Century wars against hostile Radical Muslim Groups the USA's strategy must be primarily based on salesmanship and the many non-shooting tools. The "salesmanship" of the USA's ideas must never stop, never slow down and always be increasing in intensity.

A prerogative for the above salesmanship strategy working to its highest degree is **the USA unquestionable being on the moral high ground fighting for equal freedom and equal justice for everyone everywhere.**

Washington DC has to make a command decision. It has to stop paying for and supporting the West Bank slave colony while fooling itself that nobody in the World (in today's Internet Age) notices the Lies and ethical Schizophrenia. The fact is that everyone in the world (except Washington DC) can see the truth. **The correct Grand Strategy is for Washington DC to be high profile advocating equal justice and equal rights for everyone everywhere in the world.**

(321) If You have the Initiative then You have the Ability to Start the Battle when You Want (and in a Prepared Location of your Choice) and That is a Great Advantage.

During the American Civil War (1861-1865) the tactic of US General **William Sherman** (one of the USA's two best Civil War Generals) was - "Get there first - and with the most." Sherman seized the initiative and kept the initiative and had his battles where he wanted to have his battles.

In the US Civil War the two best Union Generals were **Sherman** and General **Ulysses S Grant.** They were not "Hollywood's central casting" for Generals - but they accomplished what the "Showboat Generals" never could accomplish.

For the last 2,000 years of warfare there have always been Generals and Military-Advisors who could talk a good game of war but who could never deliver. That has been the norm with US Generals for many decades.

(322) If you do Not Seize the Initiative and Drive the Actions of your Opponent then your Opponent has the Opportunity to Seize the Initiative and Drive your Actions.

General Sun Tzu (544-496 BC) wrote "Those who are skilled in conflict will take the initiative first to prevent others from taking the initiative over them."

(1) When you have the initiative then **your opponent must have troops on-guard in many places at once.** Your opponent's forces must be spread out protecting against you possibly attacking in many different areas.

(2) When you have the initiative you can **concentrate your forces on the most advantageous targets** and bring overwhelming force to a specific battle area of your choice and time and be assured of victory.

(323) The Grand Strategist Will at Times Give his Own Staff Disinformation to Cloud his Next Moves and be Unpredictable.

Your staff might be 100% loyal but they talk to people and those people talk to people. The Grand Strategist does everything possible to keep his opponents (and everyone except those that absolutely need to know) guessing as to his immediate objectives and his future actions for securing those objectives.

(324) You must Never Assume that there is NOT a Spy on Your Staff - or Close to Your Staff.

Every spy (and embezzler) was at one point absolutely trusted.

The smart Leader regularly gives out false information (at different times and to only one person at a time) and then sees if it comes back to him from other sources.

The standard historical tactic is to ask somebody for advice - and during the conversation you say something like "This is the area where I really need your opinion." You make it appear that is the area where you intend to act. Then you see what happens.

It is painful and hard to be a successful Leader because you CANNOT trust anyone.

(325) If your Opponent's Forces are Occupying Tactically and Strategically Useless Positions (or have Useless Allies) then Let Your Opponent Keep the Burden of Supplying and Maintaining the Worthless Positions and Worthless Allies.

On June 3rd 1942 Japanese forces invaded and captured the Alaskan Islands of Attu and Kiska. That was a diversion for the Japanese attempted invasion of Midway Island 1,486 miles to the west of Hawaii and 2,500 south of Attu and Kiska. The Japanese thought the Alaskan Islands (USA territory) would be of great psychological-status importance to Washington DC.

The US Navy had broken the Japanese code and knew Attu and Kiska was a diversion-trick.

The Japanese forces held the Alaskan Islands for over a year. The islands were a burden to the Japanese to supply and maintain. The Islands were recaptured by American forces in August 1943.

The USA decision to NOT rush American forces into retaking the Islands was correct. The Islands were strategically and tactically useless to the Japanese. For over a year the Japanese wasted men and supplies defending useless positions.

Today Washington DC's support for the West Bank slave state is a "gift from heaven" for all of the USA's enemies in the world. Every time somebody in Washington DC praises the Zionist's West Bank slave colony the enemies of the USA open up a bottle of Champagne and raise their glasses saying "May the idiots in Washington DC never change."

(326) <u>The Mistake of Misplaced Importance</u> can Sometimes be Used as an Advantage by an Opponent. Nations will many Times put great Psychological Importance onto an Area or Item that has No Strategic Importance.

If you can damage or threaten a psychologically important item of your opponent (at an acceptable cost to you) then you have thrown your opponent mentally off balance at a small cost to your side.

The **Doolittle** Raid on Tokyo (Saturday April 18th 1942 - in time for the Sunday papers) caused almost no physical damage to Japan but was **a great psychological blow to the Japanese leadership.** The **Doolittle** Raid motivated the Japanese to attack Midway Island so they could build an airfield that would prevent another **Doolittle** Raid. The Japanese attempt to take Midway (June 4-7 1942) turned into a decisive disaster for Japan and became the turning point for the USA-Japan Asian-Pacific War of 1941-1945.

(327) General Sun Tzu wrote "When your Enemy's Strategy is Bait - then Never Bite. If the Fighting seems like it is Going Too Easily (or the Opponent is Retreating with only a Token Fight) that is the Time to Watch Out for a Trap."

General Sun Tzu wrote the above advice about 250 years before the historic Battle of Cannae (216 BC) between the Romans and **Hannibal**. When **Hannibal's** battle line started retreating backwards (with **Hannibal** falling back with his retreating troops) **the Roman's should have smelled the trap** and proceeded with extreme caution. (**Hannibal** at the front of his battle line had previously never retreated).

The Romans were over confident (and believed their own propaganda about being invincibility). **Hannibal's** retreat suckered the Romans into a trap and they were destroyed.

Many great Battles in history were won by what looked like the other side retreating. The Battle of Achelous in 917 AD. The Battle of Liegnitz in 1241 AD. The Battle of Russia in 1812.

(328) Sun Tzu wrote that if the Strategist Finds Himself in a "Waiting Attack" he should do Whatever is Possible to Make his Troops More Comfortable than his Opponent's Troops.

The classic "Waiting Attack" was for thousands of years enveloping a Fortress or Walled City and waiting for the people inside to run out of food.

The USA versus USSR Cold War from 1946 to 1991 was a Waiting Attack. The USA was in a much more comfortable situation than the USSR. The USA had an easier time waiting while <u>the forces of history eventually won the struggle</u> for the USA and its allies.

During a Waiting Attack time-period try to make your opponent's troops as uncomfortable as possible while giving your own troops as many benefits as possible.

(329) General Sun Tzu wrote that Once a War has Started the Leaders must Not Have a Rigid Strategy that Inhibits Creative Counter Movements.

Supreme Commanders should be cautious about interfering with the Frontline Commanders ability to take fast advantage of the opponent's disorder, confusion, unpreparedness and miscalculations when the opportunities present themselves.

Germany from 1939 to 1942 accomplished some amazing WW2 wins with a limited population and very limited natural resources compared to Germany's opponents. One of the several reasons for success was the German Army gave their front-line Officers authority to exploit any advantageous opportunities that presented themselves.

(330) If your Opponent Calls for a Peace Meeting but Clearly is Not Defeated either Militarily (and Especially Not Defeated in your Opponent's Mind) that is The Time when You Should be Looking for a Possible Trap.

Sun Tzu wrote "Those who are executing a scheme (as perhaps a surprise attack) might use the tactic of calling for a truce and a meeting while pretending they are seeking peace but are really preparing to attack."

In November 1941 the Japanese called for a "formal peace meeting" for December 7th 1941. It was NOT a peace meeting but the day the Japanese were planning to attack Pearl Harbor.

Churchill and **Roosevelt** had deliberately engineered and maneuvered the Japanese into attacking the USA. If the attack on the USA and UK had NOT been a planned trick on **Churchill's** and **Roosevelt's** side (but a supposed legitimate Peace Meeting with Japan) - **the first question anyone with any Detective talent whatsoever would have asked is "WHY do the Japanese want a meeting on a Sunday?"**

Traditionally the Japanese (and all of Washington DC) love to take Sunday's off. There was no apparent immediate crises and super rush. The Japanese and USA had been talking about the USA and UK oil embargo for 6 months. Why not have the meeting on the Friday before the weekend? Or the next day Monday? **A good Strategist (or Detective) would have seen that something was way out of place and highly suspicious.**

A Sunday was the very wrong time for a formal peace meeting in Washington DC - but it was the perfect day of the week to attack Pearl Harbor.

The "December 7th 1941 Peace Meeting" was definitely a trap. However it was **a double trap**. The Japanese believed they were tricking the Americans and British. However - it was **Roosevelt** and **Churchill** who were tricking the Japanese.

(331) Throughout the History of Warfare one of the Common Mistakes made has been an Underestimation of an Opponent.

Usually both sides in a conflict (at the beginning of a war) tend to think of their opponents as dumb, weak, cowardly, disorganized and with no commitments.

The best situation is to have a completely accurate and intellectually honest understanding of your opponent. Historically it has been safer to slightly overestimate your opponent rather than underestimate your opponent.

In 1914 the Germans at first dismissed the British Army as a "collection of shop keepers."

In early 1941 **Adolf Hitler** greatly underestimated the size of the Russian Army and underestimated the industrial capacity of Russian. **Hitler** in his mind was seeing 1918 Russia and NOT 1941 Russia.

German "Intelligence" had too many "yes boys" telling **Hitler** what he wanted to her. Most decisively **Hitler** did not understand that Russia had reset its National Bell Curve after the disaster of WW1. Russia's will-to-win was enormous compared to WW1.

When **Adolf Hitler** declared war on the USA (December 11th 1941) he dismissed America's capacity to quickly build a military. **Hitler** also greatly over estimated the ability of his U-Boats to stop US troops and supplies from reaching Europe.

Germany's U-Boats were already <u>a proven failed weapons system in WW1.</u> **Hitler** tripled the number of German U-Boats during WW2 and ended up with a bigger failure. Remember the rule - **if you make a failure bigger then what you will get is a bigger failure.**

When the US military entered the Vietnam War (1959-1975) America's Generals dismissed the Viet Cong as "A collection of stupid Gooks in black pajamas."

When the USA-Iraq War first started (in 1991 with Operation Desert Storm) some Iraqi soldiers would run out in front of moving US Tank because **Saddam Hussein** had told his troops the US Tanks were phony and made out of cardboard. Yes - they were phony Tanks at Dover, England in May of 1944 - but NOT in 1991's Operation Desert Storm.

When the USA-Iraq War first started in 1991 some in the US Military said the Iraqis were a collection of "ragheads" and the war would be over in a few weeks. Decades later US Troops continued to be in Iraq battling "the enemy."

Sun Tzu wrote - **"Those who think their opponents are easy without making careful calculations are certain to find trouble."**

(332) There are Times in Warfare when Both Sides Faced Each Other for very Long Periods of Time without Much of anything Happening. It is Essential that the Strategist Maintain constant Extensive-Observations, Constant Intelligence Analysis and Preparedness.

During WW2 there was almost no fighting on the Finnish-Russian front for almost two and a half years. When a Russian Army did attack (in June of 1944) the Finnish troops had grown complacent (and were caught by complete surprise) and were overwhelmed. Finland had to sue for peace and accept considerable land lose and pay large financial reparations.

(333) When the Basic Factors in the Grand Strategy Equation have Turned Against You the Sooner you Make some Form of Peace Agreement the Better.

In the current USA versus the Muslim World war the Grand Strategy equation of history is running against Washington DC. Washington DC needs to be making as many Muslim allies and friends as possible.

(334) To be the Most Effective a Commander Should Be Truthful and Honest with his Troops.

In the Information Age of Warfare everyone knows everything - including your own troops. If the reasons for Washington DC's Muslim-USA Wars in the Middle East are not believed by the US Troops then **the US Troops will go through their movements but their results will be insignificant.** Especially in comparison to the massive costs. This is what happened in the Vietnam War during its last years.

If a Commander has a choice it is better to say nothing specific rather than BS his own troops.

(335) Competent Generals do NOT have Long Wars.

Sun Tzu wrote, "Do not prolong your wars. **Delay during the fighting conflict will sap your strength and undermine the determination of your troops." Sun Tzu** had it figured out 2,500 years ago - today's USA's Generals are struggling to understand what's happening.

One of the US Pentagon's core dreams is re-fighting WW2 with today's weapons. However - one of the most decisive reasons for the USA's great victory in WW2 was because (for the USA) the war was very fast - 3 years and 6 months for the USA in Europe.

For the last half Century **the US Pentagon's wars have been measured in decades and lifetimes.** There has been some type of US shooting conflict in the Middle East since 1991. **Never ending wars are the hallmark of intellectually bankrupt Generals who have little understanding of Strategy's big pictures.**

(336) Never Have a Frontal Attack on a Well Prepared Opponent's Position Unless there are Overwhelming Reasons to Do So.

In frontal direct attacks you must assume that your opponent is prepared and your casualties will be very high.

(337) The Grand Strategist is Always Looking for Past War Examples that Can be Applied to his Current Military Challenges.

The problem is **you need gifted strategists who have the talent and training to recognize the correct examples** from past history that can be applied to current situations. This is a far more difficult strategy analysis job than it first appears to be.

(338) A Nation's Military (or its National Police Force) Should NOT Become a Government Inside of the Government.

Agencies like the FBI, NSA, CIA and so forth are NOT elected. If they become de facto governments inside of the US Government then the USA is no longer a true Democracy.

The FBI, CIA, NSA etc have been **proven and reproven to have had highly flawed judgments. Over the years Washington DC's main "Intelligence Agencies" have been proven to be significantly wrong.** Washington DC's so-called "Intelligence Agencies" have been **proven** to NOT be an improvement over the elected US Congress and elected US President.

If the CIA, FBI, NSA, Pentagon Intelligence etc are so smart then where were they when the **Saddam Hussein's** supposed "Weapons of Mass Destruction" con game fraud was being perpetrated against the American people? Maybe the correct answer is that the "Intelligence Agencies" were part of the fraud?

(339) The 1,000 Year Old Classic Rules of International Diplomacy-Strategy continue to be Valid and Should be the Core of all USA Policy and Action Decisions.

Some of the time-tested classic rules of Diplomacy-Strategy based on a thousand years of experience are:

(A) **Strategy must be completely devoid of any crusading passion.** When a National Leader (like President **Bush** in 2003) starts talking about a Crusade that is the time to panic.

(B) **The objectives of National Strategy must always be defined only in terms of real and valuable National interests.**

Mr. Donald Trump was elected President in 2016 on a campaign promise of "Putting America's interests first."

There should be a National written list of objectives for the USA and the American people. A specific detailed outline-estimate of exactly how the USA is going to make it to the end of this Century.

(C) **The USA's International Strategy must Look at the entire world's political realities from the point of view of ALL other Nations.**

This is the rule of putting your opponent's shoes on and seeing things from your opponents point of view. The above Rule is something that Washington DC almost never considers. **For Washington DC it is Planet Washington DC.**

(D) **The USA must be willing to compromise on everything that is Not vital to the USA's core interests and survival.**

The USA in recent decades has violated this rule by entering into shooting conflicts over issues and areas that have little or no value to the USA's National security.

On the other hand - any vital desire of the USA's insignificant allies (if those desires are NOT vital to the USA) are negotiable. The desires of insignificant allies should be used as bargaining items by the USA (with an opponent force) if the USA receives something of real and long term value to the USA.

(E) **Never fight a shooting-war for worthless rights and worthless status symbols. ONLY fight shooting wars for real advantages that have true value. Never trade worthless statue symbols for real advantages. In your International Relations you should always be looking for ways to trade worthless status symbols for items that have real value to your Nation.**

The world history of warfare shows that a large number of wars were fought for worthless statue symbols.

(F) **Never put (or allow) your Nation to be maneuvered into a position from which your Nation cannot retreat without losing something of value and a position from which you cannot advance without great danger.**

The above is also one of President **Donald Trump's** rules "No matter how good a deal looks do NOT get into it if its failure will take you down."

(G) **Never allow a weak Ally to make your National Policy decisions for you because their vital interests will almost never be the same as your vital interests.**

Violating of the above rule (that has caused great harm to the American people) has been Washington DC's passionate religion since the 1960's. It is long past the time for some sanity in Washington DC. It is a simple rule - "America First."

(H) **The Military (and the National Police Forces like the FBI and CIA) are instruments of National policy and must never become the creators of National policy and a Government inside of the US Government.**

The "USA's Federal Police" are a problem. When this book was published it was painfully clear that many people in Washington DC's key "Intelligence Agencies" had been working hard to be a government inside of the US Government.

The FBI and CIA need some humility about how wrong they have been. The FBI and CIA should not have a "higher loyalty" to their own twisted subjective delusions. They should NOT have a "higher loyalty" towards trying to stop history and returning to yesterday. The FBI's and CIA's "higher loyalty" should NOT be to their own organizations. Their "higher loyalty" must be to the Intent in the US Constitution that they all took an oath to uphold and to what is the guardian of American justice and democracy.

Winston Churchill had it exactly correct when he said, "Democracy is the worst form of government except for all of the other forms of government." Yes - Democracy is highly flawed but it is the least bad of all of the bad choices.

(I) **The Government must be the leader of public opinion and Not the slave of public opinion.**

Enacting government policy and programs based on popularity polls is giving control of the USA's Government to the mainly left wing TV Press. The Left Wing Press knows how to (1) successfully half-truth the facts (and spin the "news") and thus (2) create public opinion for or against specific policies. The fraud-news can (and does) create public opinion that influences Congress and becomes Law. That is a process that makes the TV Press like CNN and MSNBC a big part of the US Government.

During the Obama administration the left-wing TV Press was a government inside of the US Government. The left wing press manipulated the facts (i.e. lied) and created public opinion and then Congress passed laws based on that public opinion. **The left-wing press became a de facto non-elected Congress.**

The #1 reason for the USA's left wing press's intense hostility towards President Trump is because President Trump took control of the Government away from the unelected left wing fraud-news press. President **Trump** returned the control of Government to the elected office of the President of the United States.

SUMMARY.

The objective of attacks and battles in every war should be the eventual destruction of the Strategy of your opponent. **Your attack objectives are to make the wishes and dreams of your opponent impossible to achieve.** Your attack objectives are to destroy any possibility of your opponent achieving his core desires. Your attacks must use as many of the 37 different types of warfare as possible to destroy your opponent's core Strategy for achieving his core desires.

The overall ultimate objective in every war should be the same. A sustainable peace that is the most advantageous for your Nation and for as many people as possible.

Few peace-periods are perfect for everyone. A peace that is "perfect" for one side but bad for the other side has 80% of the time NOT lasted.

A very common historical mistake is to go from a peace that a Nation is not totally happy with to something that turns out to be much worse.

Germany had a disadvantageous and insulting peace from 1919 to 1938. Germany started a war in 1939 in large part to avenge the insulting peace. However - **after WW2 Germany had a peace that was 10 times worse than before the war.**

Never destroy an opponent's government based on the naive assumption that what replaces the destroyed government is automatically guaranteed to be better. **If you destroy a government you must have extensive and detailed realistic doable plans for exactly how to quickly build and establish a new government that will successfully replace the destroyed government.**

The essence of military battle tactics **is the concentration of decisive force onto an achievable objective of big picture value.**

Your objectives must be doable. Difficult is okay - but being something that is doable is essential. The Commander can (and should) demand a 100% effort from his subordinates but **he must never give an order that is impossible to accomplish.**

The standard rule is that <u>indirect attacks</u> will have a higher rate of success than <u>direct attacks</u>. **There are 37 main types of warfare. Many of those types of warfare use <u>indirect attacks</u> that your opponent is probably not going to be expecting. The types of attacks your opponent is not expecting have the best chances for long term success.**

Your plans, policies, programs, actions and equipment must never be about how to win yesterday's war. Everything must always be about how to win tomorrow's wars.

Chapter 11

The Subjective and Objective in War and Peace

The "Subjective" is the way you (or your group or your Nation) see something. The "Objective" is the true real happening and situation. Subjective is opinion that may or may not be true. Objective is reality.

A Subjective belief of individuals and groups can be (1) 100% in agreement with Objective reality or (2) 100% different from reality (and thus 100% wrong). Or (3) usually what happens with almost all complicated beliefs is they are a percentage mix of both truth and falsehoods.

Every individual and group has thousands of beliefs and they range from 100% correct to 100% wrong and every percentage combination between.

Disagreements and many wars are created when Nations sincerely believes that their Subjective wrong views are the Objective correct views. The disagreements between what are Subjective and Objective beliefs is one of the many causes of wars. This is why Nations and National leaders must spend the maximum amount of time talking to each other. Fighting an issue out with words is a billion times better than fighting an issue out with soldiers that die.

It is vitally important for the Grand Strategist to always keep in mind the fact that his Subjective view of reality (or his Nation's subjective view of reality) might NOT be the true Objective reality.

You must always make your plans based on the true Objective reality and never on your desired Subjective wishes that could be false and thus produce a highly flawed and possibly disastrous Strategy.

The core Rule is that just because YOU believe something to be true that by itself does NOT make it true. What is true is what can stand up to intense cross examination by experts in that area of knowledge and can be proven by the verifiable accurate facts.

The reason for WHY radical groups throughout history have always (and are) attacking free speech is because their flawed subjective beliefs cannot stand up under intelligent cross examination. Intellectually honest people (and sane people) do NOT fear free speech - they welcome free speech and the cross examination of ideas.

If your opponent went to war based on false Subjective views then a large part of your Strategy could consist of changing his flawed Subjective views to what are the world's Objective views. For changing false Subjective beliefs the salesmanship tool can be much more effective (and much less costly in money and blood while being more ethical) than guns and bombs and killing people to force them into accepting your ideas.

What is vitally important for Grand Strategy planning is understanding that if large groups of people believe that something is true (but in Objective reality it is false) <u>for Strategy planning the falseness of the belief is NOT an automatic solution.</u> Using all of the tools of salesmanship to convincing your opponent that his belief is flawed is the solution.

National Leaders (and the Grand Strategy Planning Team) must clearly understanding that their opponent's false beliefs are his truth and <u>your opponent will act on his beliefs as if they were true</u> because for him they are true.

Never assume that something is true only because it was at one time true. Every belief must be constantly tested and cross examined. In warfare yesterday's truism is often tomorrow's obsolete and highly dangerous falsehood.

There are 37 kinds of warfare and many of them are based on changing opinions and beliefs. Only a psychopath or war criminal or military moron would choose the shooting war tools as the starting point without first completely exhausting all of the salesmanship tools.

(340) The Grand Strategist constantly keeps Asking Himself if His Views and Plans are Based on the Objective Reality or Based on his Subjective perhaps False Views of the World.

In the 1950's - psychological research proved that it was possible to take 10 individuals and show them the same movie of people interacting and then after the movie was over the researchers might receive 10 different interpretations of what happened in the movie. **Each person saw the same movie through their own subjective mindset filters.**

<u>The fact that the strategy analyst must never forget is that nobody "sees with their eyes."</u> Everyone is actually "seeing" with their mind. A Donald Trump supporter and Donald

Trump hater will look at the same speech from Mr. Trump and both "see" and "hear" to different things.

An example of the dangers of confusing your subjective views with the objective reality is **Adolf Hitler.** In 1942 the Objective reality of the world's power balance against **Adolf Hitler** became an increasing contradiction to the Subjective delusions of **Adolf Hitler**. Like so many mentally flawed leaders in the past - **Hitler** tried to bring the Objective Reality into line with his Subjective Delusions via force of arms. **Hitler** failed and killed Millions of innocent people in the process.

(341) A Flawed Subjective Belief Cannot Change the Objective Reality Unless the Subjective has Overwhelming Power to Alter Reality.

After the Cold War (1946-1991) started the USSR's subjective self-view of its economy became an increasing contradiction to the objective economic reality-facts.

The USSR ignored the contradictions (typical of most governments in history) and tried to "double down" on their subjective delusions (also very typical of most governments). The USSR's subjective-objective contradictions were resolved in 1991 when the USSR fell-apart and was officially dissolved.

The War in Vietnam (1959-1975) was for the USA **a Subjective war for freedom.** For the Viet Cong it was an **Objective war for freedom** from White-man colonial armies that were murdering Vietnamese in their homeland.

Washington DC's wrong subjective view of "freedom" was a contradiction to the objective reality of what side was actually fighting for freedom. Washington DC has continued to make the same mistake up to today by supporting colonial subjugation and calling it freedom.

One of the dozens of questions that the USA's Generals (and war planners) in Vietnam never asked (but should have been asking) is if their subjective views of freedom were the objective true-reality or Washington DC's mistaken subjective falsehoods.

When you cannot change an important subjective false belief to the objective reality (and the issue is viewed as important to all sides) - that is when you often have shooting wars.

The solution is to bring flawed subjective viewpoints into line with the world's objective reality even if that is difficult and painful.

Chapter 12

Anger and Emotion in Strategy and Peace Planning

You must never do any Strategy planning while you are in state of anger. A highly skilled opponent (like Osama bin Laden) will build his Strategy around your anger and leverage your anger for a major victory far beyond the original damage that caused the original anger.

Many Generals and government people think that using profanity and threats behind closed doors makes them smart and tough. However - profanity is the language of thugs. Profanity is empty-words that are useless to help solve complicated problems. The Einsteins and Stephen Hawkings of the world never use profane words as their thinking tools. Showboat profanity words have no creative problem solving value and are often used to cover up the user's ignorance.

The unabridged Dictionary of English has about 470,000 words. The more words you know the better your thinking will become because most problem solving thinking is done mainly with words.

Different languages create different pathways in the brain. Some languages are better than others for some types of thinking and creating. If you want to achieve advanced thinking that is accurate and creative then you must be highly skilled in an advanced language. The advanced Language might be words, music, art, mathematics or visual conceptions.

(342) Anger Alone is Never a Good Reason to Start a War.

General **Sun Tzu** wrote **"Anger and intense feelings are NOT reasons to launch a war. Intense feelings of dislike can cycle back to fondness. Anger can cycle back to satisfaction. But those killed in a war will never cycle back to life."**

At this time relations between America and Vietnam are cordial. The USA and Vietnam are doing business together. American tourists go to Vietnam and some Americans have moved there to live. All is calm and friendly. However the 58,220 USA military killed (11,465 of the USA killed were 18 and 19 year olds) are dead forever. The over 100,000

American soldiers that lost eyes, hands, arms and legs - those young Americans were crippled forever.

If the USA-Vietnam War had never happened (if the USA had never invaded Vietnam) the world's political power situation today would be 98% of what it is currently.

The world's historical record has proven that **if you are not crystal clear about a situation it is better long term to err on the side of peace rather than on the side of a shooting war.**

During the Vietnam War over 58,000 young Americans died to supposedly stop the spread of Communism. However when this Book was published a large part of the US Democrat Party had turned into the de fact Communist Party of the USA.

(343) Genghis Khan (1162-1227 AD) said "An Action Committed in Anger is an Action Doomed to Failure."

The Iraq War's second US Military invasion of 2003 was an "action committed in anger." Or more accurately it was a con game exploiting the anger and insecurity of President **W Bush.** All of the reasons that Washington DC said were the justifications for the second invasion of Iraq in 2003 were proven to be con games and lies.

The end results of Iraq War 2 (for both the USA and Iraq) have been horrible. Washington DC's muddled, confused and on-going conflicts in Iraq are proof that **Genghis Khan** was correct. "**Do NOT start wars based on anger alone."**

With every proposed future war the American voters must be asking "Exactly and specifically what are the real provable and doable advantages for the average USA worker that will come from this war?" Without very specific and exact in-writing answers on record from Leaders and Generals the US Voters should NOT support war. **If the Generals and Politicians cannot put down in writing the exact and specific meaningful reasons for the war then why are you going to war?**

The usual "Glittering Generalities" must NEVER be accepted as reasons for starting a shooting war. Propaganda slogans like "defeating terrorism forever" are con games. **There must be specifics of exactly what valuable advantages the average USA worker will be getting that justifies the money costs and the dead young USA Troops.**

(344) With every New Challenge the Grand Strategist Starts by Asking "Is this Challenge Important to the Core Vital Interests of my Nation? Or is this Only a Short Term Ego and Status Aggravation that in the Big Picture Means Nothing?"

(345) The Grand Strategist Never Forgets that Homo sapiens have Brains that are Hard-Wired to at Times be Highly Irrational and Needlessly Aggressive.

The Grand Strategist never forgets that he is human and capable of highly irrational actions. The good Grand Strategist also keeps asking himself, "Are my plans sane?"

One place where "mind management" becomes especially tricky is the fact that **few people are universally smart 100% of the time and for 100% of their lives.** Some overall smart people (like President **Trump)** can nevertheless be capable of being very wrong.

(346) In many Conflicts the Politicians will Say "This War is Different" and/or "This Enemy is Different." The Wars and Enemies are Never Fundamentally "Different." Presenting Wars and Opponents as Fundamentally "Different" is a Planning and Analysis Mistake and/or a Con Game.

Calling your opponent "Different" and your proposed war "Unique in history" is a propaganda con game. Usually those who believe a war is "different" are the victims of their own propaganda lies.

Understanding the fundamental thinking and behavior of Home sapiens is the basis for understanding military Grand Strategy. The very best Grand Strategist will know what his opponent is going to do before his opponent knows. **Osama bin Laden** knew that President **W. Bush** would falsely blame **Saddam Hussein** for the 9/11 attack before the first airplane crashed into the World Trade Center.

As technology progresses (and the world transitions from one age of warfare into another) the tools and tactics of warfare will change. However **the basic emotions and mind-mechanics of Home sapiens will remain the same.**

Homo sapiens' brains are in large part hard wired and their fundamental behavior has been consistent for thousands of years. This is why we can today watch ancient Greek plays like *Oedipus Rex* (429 BC) and relate to all of the people (and their emotions) in a play that is 25 Centuries old.

The enduring patterns of human nature is why some of the military Grand Strategy advice from: **Aristotle, Sun Tzu, Polybius, Vergil, Vegetius, Julius Caesar** and others from over 2,000 years ago continues to be valid today.

(347) One of the many Reasons for having a USA Grand Strategy Team with Diversity of Talent and Backgrounds is because a Diversified Team Will Have More Control Over their Collected Emotions.

If at all possible - major decisions must not be made when in either (1) anger or (2) deep fear.

(348) Osama bin Laden Used One of Sun Tzu's Rules Against the USA with Spectacular Success.

Sun Tzu wrote "If your opponent becomes angry easily and is of bad temper then irritate him and use his anger against him."

The above is exactly what **Osama bin Laden** successfully achieved against the USA. The 9/11 attack (September 11th 2001) was a tiny cost to the attackers. However - the attacker (**Osama**) very successfully stampeded Strategy-ignorant President **George W. Bush** into believing the con game of an Iraq connection to **Osama bin Laden**. A connection that was a con game lie and never existed.

The anger of President **George W. Bush** multiplied-up the original 9/11 damage by over a Trillion US Dollars (and continuing to count) and multiplied-up the American dead by over 130% (also continuing to go up).

Osama bin Laden was smart enough to know that his con game would have much help from some "close Allies" of the USA and from "Intelligence Agencies" inside of the USA. **Osama** know what "Allies" would double cross the USA because it was advantages to them. **The Nations and groups that helped start the USA-Iraq War2 all had their own reasons for promoting Iraq War2 - but the end result was a help to Osama's master plan.**

The lesson that Washington DC must learn is that **anger can reinforce stupid Strategy actions that can then turn into multi-Trillion Dollars mistakes - or worse.**

The current challenge for the American people is to NOT permit the same "Allies" and same "Intelligence Agencies" that sucked the American people into Iraq War 2 to now successfully sucker the American people into a shooting war with Iran.

Chapter 13

The Decisive Importance of Intelligence

Intelligence has always been important. During the Greek-Persian Battle at Thermopylae (480 BC) the Intelligence of the Persians finding the goat path around the Greeks was decisive. Intelligence systems in the 21st Century are the top most vitally-important weapon system.

The good Grand Strategist understands that Intelligence is much more than data. It is more than facts and figures. <u>The ultimate Intelligence is the correct analysis of what the massive amounts of data mean and then the correct programs and actions for the maximum effective use of the Intelligence.</u>

During the Vietnam War the US Military (and various Washington DC agencies) collected enormous amounts of data and numbers. However <u>the analysis of the meaning of the data was disastrously incompetent.</u> The "missing link" in the USA's Intelligence was the courage to come to the correct conclusions. Apparently the junior USA Officers who might have know that something was very wrong were too cowardly (or too smart) to tell the Generals what the Generals did not want to hear. The junior Officers who said nothing probably ended up as Generals. The junior Officers who said the system was flawed were probably never promoted because they were "defeatists."

If you are a competent analyst and tell a US General something they do not want to hear you are NOT going to be seen as "intellectually honest" - you will be "a defeatist." The US Military promotes "yes boys" and that is why the US Pentagon has turned into a collection of yes-boy mediocrity that only knows how to lose wars.

This Book is "intellectually honest." Do you think it will ever on sale in the Pentagon's Book Store or West Point's Book Store? Not in my lifetime.

(349) In the 21st Century Good Intelligence is the Crown Jewel for Winning Wars.

Intelligence gathering has always been vital to successful Strategy. In the 21st Century we are well into the "Information Age of Warfare" (1960 to ?). The correct use of good

Intelligence in all of its forms (and all of its uses) has become the most overall important weapons system for winning conflicts and ideally winning without shooting wars.

Correct and complete intelligence can prevent wars, shorten wars and win wars. Twenty-five centuries ago **Sun Tzu** wrote,

"Intelligence gathering is vital and critical for winning conflicts and winning the peace. Nothing should be more favorably regarded and cultivated than your intelligence operations and nothing should be as generously rewarded as intelligence. Nothing should be kept as confidential and secret as your intelligence system. Do it subtly, very subtly but nowhere neglect the gathering of intelligence."

Twenty-five centuries ago **Sun Tzu** in his classic book ***The Art of Strategy*** had a specific outline of the five groups of intelligence that the Grand Strategist should be using.

1 Local Intelligence. This comes from natives in the area and from captured local opponent troops. The smart commander is also constantly gathering feedback from the local people about who they support (and do not support) and why. What the local people think of your opponent. What the locals think of your troops. How would the locals fight your opponent - and all other vital local-issue questions for that particular conflict.

2 Insider Intelligence. This comes from the opponent's politicians and administrators and what they are talking about and doing. Today this would include a careful reading and analysis of the opponent's TV news programs, newspapers, magazines and email communications within and between government agencies. **In modern times gathering the private emails and phone calls of your opponents (even if they are only minor bureaucrats and Officers) are excellent for analyzing the (1) true attitudes and (2) mental sets and (3) the current will-to-win of your opponent.** It also helps to build a correct profile of your opponent's Government Culture (its flaws and blind spots) and its collected mindset decision making process. You need to be looking for the mental/emotional weaknesses and exploiting them.

3 Counter Intelligence. This comes from double agents and spies who have access to your opponent's intelligence gathering organizations. Counter Intelligence is understanding your opponent's intelligence and finding the weaknesses in your opponent's systems. Counter Intelligence is learning about your opponent's plans in advance. It is acquiring a true and accurate understanding of your opponent's desires, motivations and Strategy for securing those desires.

4 Deadly Intelligence. Is intelligence that is leaked to your opponent in such a way that your opponent does not know that you know. Double agents are Deadly Intelligence. The intelligence is false but your opponent believes it is true and also believes that you do not know that your opponent knows the intelligence.

During WW2 the British put a dead body dressed in a British Officers uniform ashore where they knew German agents would find the body. It was made to look like a British airplane had crashed in the Sea. The body had a copy of the "secret plans" for the Allies (who were in North Africa) to invade Greece.

Adolf Hitler was fooled and believed that the Allies did NOT know that he knew of their secret plans. **Hitler** sent a German Army to Greece under the command of General **Erwin Rommel**. Sicily was always the Allies target and on July 9th 1943 General **George Patton** and General **Bernard Montgomery** invaded the Island. The dead body that tricked **Adolf Hitler** was Deadly Intelligence.

5 Secure Intelligence. Is information that is made public about what you are doing or planning or not doing. **The objective of made public Secure Intelligence is misinformation that disorders your opponent's priorities** to areas of no threat to you. A modern name for the process is "disinformation" and **Winston Churchill's** "bodyguard of lies."

In April, May and early June of 1944 there were British and US Newspaper reports and pictures of US General **George Patton** (1885-1945) inspecting his "newly forming Army" in Dover England across the English Channel from Calais France. The British and Americans at Dover built wooden airplanes, inflatable Tanks, had hundreds of empty troop tents, stock piles full of empty boxes and so forth. The British and USA successfully used "Secure Intelligence" and tricked **Adolf Hitler** into believing that the D-Day invasion would be at Calais, France rather than 200 miles away (by road) in Normandy.

General **Sun Tzu** wrote **"When all five types of intelligence are working together and connected together at the same time, and your opponent is not aware of them. This is called the divine golden web and is the treasure of the ruler."**

Sun Tzu also wrote that: spies, double agents, paid informants, traitors to the other side, disgruntled and disloyal officials in the opponent's camp - **anyone who can be bought with money or turned with promises are the Gold Standard for Intelligence.**

Intelligence in the 21st Century is mainly technology based. However despite all of the technology the very best situation (if possible) is to have a human spy on the inside of your opponent's organization.

A spy satellite will tell you where your opponent has 200 jet bombers parked but only an insider informant can tell you what your opponent's Generals are planning to do with the airplanes and when.

Nicolo Machiavelli (1469-1527) who lived 2,000 years <u>after</u> General **Sun Tzu** - **Machiavelli** gave the same advice in this classic book ***The Prince***. **Machiavelli** wrote "In your enemy's state you can always find malcontents and such as desire a change. Such men if used correctly can open the way into your enemies' state and render the victory easy."

(350) If your Opponent is Competent he will Constantly be trying to Analyze Your Strategy. This is why You Must Do Everything Possible to Disguise, Cloud and Obscure your True Objectives and Strategy for Securing Your Objectives.

Winston Churchill said "The leader must surround himself with a bodyguard of lies." Every Grand Strategist must surround himself with lies.

With several 24/7 TV News Networks one of Washington DC's many Strategy challenges is keeping its intended future actions secret from the world. **The best defense for Washington DC is to <u>return to the moral high ground</u> and then Washington DC will have fewer secrets to keep from the world.**

(351) Understanding Your Opponent's Culture and Your Opponent's National Character (in All of its Parts) is Vital Must-Have Knowledge to Correctly See the Complete Intelligence Grand Strategy Picture.

All older Nations have a National Character and look at problems and situations sometimes decisively differently from other cultures. It is vital that the Grand Strategy Team have a clear and precise understanding of the (1) different ways that different cultures make decisions and (2) the different value and priority systems that are in operation.

(352) Sometimes You can "Turn" a Nation the Same Way an Individual can be "Turned." However - <u>the Prerequisite is You Must Have an Operating Presence in that Nation.</u>

Within most Nations there are usually different "groups" of "diversity" with different ways of seeing the world and different political objectives and political philosophies.

Sometimes it is possible for operatives inside of a Nation to turn one or more groups into allies. Then you have friendly forces with political power inside of your opponent's Nation. Many "tools" can be used to make those groups (who are friendly to you) the majority ruling power and/or a significant political power that can influence events from the inside out.

Nations can be "turned" the same as individuals can be "turned." However - to "turn" a Nation you need an effective operating organization INSIDE of that Nation. **Getting inside of a Nation has to be the #1 first priority and <u>Top Priority</u> in your Strategy.**

Washington DC's standard tactic of trying to seal-off Cuba with sanctions and seal-off North Korea and seal-off Iran (etc) with sanctions is a proven WRONG strategy and stupid. **When Washington DC seals off a Nation with sanctions it seals the USA OUT of that Nation.**

Washington DC's sanction against places like Cuba, North Korea and Iran make it possible for the Dictators INSIDE of those Nations to operate in complete safety. **It gives the Dictators a golden excuse to continue being Dictators** so they can protect their Nation from Washington DC's sanctions-aggression that inflicts great harm and suffering onto ordinary people. **When you threaten to destroy a people 98% of the time they fight harder.**

President **Trump's** dumping of the 2015 Iran-deal was a serious grand strategy mistake. The "deal" opened Iran to many American influences and companies. Those influences and companies could have been used to turn the minority of Iranians who want more co-operation with the USA into a bigger and stronger group. The warmongers (rather than the peace-builders) won when President **Trump** returned to the proven failed dumb sanctions.

Mr. **Trump** should have asked himself "How did President **Ronald Reagan** win the peace? By putting many sanctions onto Russia? Or by working at being a true friend to **Mikhail Gorbachev**? How did West Germany envelope East Germany? By putting as many sanctions as possible onto East Germany? What worked was the exact opposite.

News flash to the sanction-dead-heads in Washington DC - it was East Germany's Dictators who built the Wall to keep West Germany out and to keep the East German Dictators safe.

After the fall of the USSR (in 1991) planeloads of US cash were sent to Russia. President **Reagan** also give **Gorbachev** ownership of the best house in San Francisco. Apparently the Germans also gave **Gorbachev** a large amount of money and a nice home in Germany. A small price (for everyone in the world including Russia) to pay for NOT having a nuclear war.

Good strategy would be President **Trump** getting rid of the military morons in the White House and have private donors give **Kim** of North Korea three adjoining big homes in Hollywood California.

(353) The Reason for Why Wars must be Started with Great Caution is Because (especially in Highly Complicated 21st Century) NO Person has Enough Intelligence and Brain Power to Know How a Significant Shooting War will End.

President **George W. Bush** had his "Surge of Troops" into Iraq in 2003 - promising the American people that the war would be over in a few weeks. How many 2003 US Generals at the Pentagon knew that 16 years later US Troops would continue to be in the Iraq war fighting shooting battles with "the enemy"? The answer is - **None**.

In 1962 Vietnam how many US Military Officers knew that in 1972 the shooting war would be continuing and the US Military would have already been de facto defeated? The answer is - **None.**

What is the current Pentagon's "assessment" of a future significant shooting war's outcome worth? Based on Post WW2 history the assessments of the Pentagon's Generals as to how a significant shooting war will end have been **worth less than nothing.**

How would a USA-Iran shooting war end? Based on the Post WW2 history we can be 99% sure it would NOT end in a way that was anything close to the way the US Pentagon's Generals say it would end. There is a 95% probability that a shooting war would end like the Vietnam War with another US Military defeat - but a far worse defeat than Vietnam.

(354) At the End of many of History's Wars it Was Clear that the Justifications and Reasons for Beginning the Wars were Mainly False Reasons of Little (if Any) Value to Those Who Started the Wars.

The US voters should be exceptionally carefully and on-guard about being con-gamed into another dead-end losing war - specifically this time with Iran.

In 2003 the US Military attacked Iraq. The justification for the aggressive attack was that **Saddam Hussein** had "Weapons of Mass Destruction" and **Saddam Hussein** along with his "good friend" **Osama bin Laden** were going to use the nuclear weapons against America etc etc...

All of Washington DC's reasons for the aggressive start of Iraq War 2 in 2003 were eventually proven to be fraud-lies, fabricated disinformation and a collection of con games.

In 1959 the US Military started entering the Vietnam Civil War in force. The propaganda justification for invading Vietnam in force was that if the Communists won there would be a "Domino Effect and Communism would spread into all of Southeast Asia."

The US voters were told that Cambodia, Laos, Thailand, Malaysia, Indonesia and the Philippines would all become Communist. **The US Military invading Vietnam in force was "sold" as "Stopping Hitler at Munich."**

At the end of the American-Vietnam War (1959-1975) 58,220 American troops were dead - forever and the US Military had lost the war. The end results were all of Southeast Asia did NOT turn Communist as the US voters (for almost 18 years) had been told would happen.

The majority of the people in Asia believe the US Military attacking Vietnam was a war crime. They are correct. The US Military in Vietnam murdered 650,000 civilians by dropping FIVE TIMES more tons of bombs on Vietnam than on Nazi Germany - and by having the biggest Chemical Warfare weapons attack (20 million gallons of poisons) since WW1.

Some of the poisons the US Military used were "forever poisons" that never break down and are to this day killing people and animals.

If the USA had never sent one soldier into Vietnam the overall situation in Asia today would be 98% of what it is now. The 2 biggest differences the Vietnam War made for the USA is that (1) the war was paid for with borrowed money and every taxpayer today continues to pay the interest costs on the Vietnam War. And (2) - the 58,220 US young men murdered in Vietnam are dead forever. The 650,000 Vietnamese men, women and children killer are dead forever.

The **US voters must never permit the Washington DC warmongers to con-game the American voters into another criminal war crime.** Any reasons for the US Military invading another Nation must be double verified and triple checked and **there must be specific and detailed reasons for exactly and specifically what benefits are in the war for the average American worker - and for the world.**

The government of Iran has to be exceptionally careful and mentally controlled and on-guard all of the time because the Washington DC Warmonger Lobby is definitely trying to trick Iran into doing something that can be spun into looking like an act of war.

(355) The best Grand Strategist will have Himself Psychoanalyzed Every Five Years so He is Aware of his Own Mental Weaknesses and Contradictions.

During the last 4,000 years of history there were many "Leaders" who's most dangerous opponent was the opponent between their own ears.

Good, professional psychoanalysis at this point is only a hundred years old. However the US Military is far behind the Bell Curve for using internal psychoanalysis to improve its own planning, reasoning and strategy analysis.

The US Military needs many new 21st Century standards. One of them is that anyone entering the General Grade Officer's ranks should be psychoanalyzed in depth. The conclusions must be presented to the person being psychoanalyzed and that person can discuss the points with the doctors (in an ongoing process if needed) towards the aim of improvement on all sides. The idea is NOT to "get" anyone but to improve everyone.

No US Military Officer (active or retired) has been part of a significant US Military shooting war that was an eventual clear success. **US Military Officers know how to win all of the battles but de facto lose all of the wars.** US Generals of the Post WW2 years are like Germany's Generals in the 20th Century. Germany's Generals won many battles brilliantly but lost all of the wars. The bottom line is that there is a proven great need for improvements at the US Pentagon.

(356) The 21st Century "Leader" Should Have Himself Psychoanalyzed. Plus Everyone on his Grand Strategy Team, his Nation's Major Institutions, world Leaders, Allies and Opponents.

The world is into the Information Age of Warfare and the more information that the USA's leadership has about everything (including information about themselves) the better they can organize successful peace strategies.

(357) The core of good Grand Strategy Analysis is to Always be Questioning Everything. There are NO Irrelevant or Silly Questions because Questioning Everything is the Process. There Must Be NO Forbidden Questions.

During the analysis process it makes no difference if 90% of the questions are silly, trivial or irrelevant. **Nobody should be ridiculed for asking an irrelevant question because asking irrelevant (and perhaps silly questions) outside of all of the usual mental boxes is a Key Part of the Creative Problem Solving Process.**

Unfortunately the "culture" of the US Military (and almost all militaries) is that ONLY "culturally correct" questions are tolerated. If a question inside of the professional military culture is NOT also inside one of the approved steel boxes in the professional military's mindset then you will end up like **Billy Mitchell** - drummed out of the military for NOT "fitting in."

The conundrum with the US Pentagon's culture is that "fitting in" with the obsolete group-thing culture IS the problem.

Think of Grand Strategy questions and scenario playing as gold mining. In gold mining you have to go through tons of mud to find a gold nugget. Likewise - **it will usually take many unimportant and trivial (and often dumb and silly) questions to arrive at an important question.** If you ask 4,000 questions and 3,995 are unimportant and/or silly but 5 are the correct questions with the correct follow up answers - then your process is a brilliant success.

The most difficult part of the above process (and where it becomes ***The Art of War and Peace***) is you need a Strategist with the **God** given talent to recognize the correct questions when they are asked.

(358) Never Say that Something Cannot Happen Because it Makes No Sense and Is Insane.

One of the "Arts of Grand Strategy" is being able to understand the logic of illogic and knowing that sometimes opponents can (and will) do insane things.

In the Spring of 1941 (before Germany attacked Russia) the **Dora Soviet Spy Ring** in Switzerland and Germany headed by Mr. **Sandor Rado** told **Joseph Stalin** that the German Army was massing in Poland and East Prussia (on Russia's frontlines) and **Adolf Hitler** was going to attack Russia.

Joseph Stalin did NOT believe **Sandor Rado** and said that **Hitler** would have to be insane to attack Russia while engaged in a major war with the British Empire. A war with England/UK that **Hitler** was not losing but was also not winning. **Joseph Stalin's mistake was believing that Hitler was sane and logical.**

Nazism for the Nazis was a semi-religion and in religions people are especially capable of doing insane acts. Ironically Stalinism was also a semi-religion and it is own way just as fanatical and ruthless as Nazism.

YES - **Hitler** was insane to attack Russia when the British Empire remained undefeated and the USA was not yet in the war. **YOU must never assume that your opponent is sane and rational.**

Every professional strategy analyst should read **Adolf Hitler's** declaration of war against the USA. The high quality of its organization and salesmanship reveals why so many people trusted and followed **Hitler.**

(359) Periodically Most Nations and Political Parties Go Through Periods of Insanity. It is Vital and Decisive that the Sane People Recognize the Insanity and Protect Against its Often Nation Destroying Dangers.

Often brilliant people with past excellent records can be incredibly dumb at recognizing National insanity. When this Book was published the Democrat Party of the USA was going through a period of insanity but the insanity was not fully recognized.

Joseph Stalin was not alone in greatly underestimating **Hitler's** insanity. Two of the best known Grand Strategy writers and analysts of the 20th Century: **Liddell Hart** (1895-1970) and British General **John F. C. Fuller** (1879-1966) both (well into WW2) thought that **Hitler** was a sane and rational person. During the 1937 to 1941 time period both General **Fuller** and **Liddell Hart** destroyed their reputations by writing that **Adolf Hitler** was a reasonable person and England could (and should) negotiate with him.

The end result was that the British Military during WW2 had nothing whatsoever to do with General **Fuller** who was the brilliant WW1 inventor of Tank warfare. General **Fuller** was one of the most original military thinkers of WW1 (and the 1920's) who wrote several books and papers about armored warfare. General **Fuller** was the true father of Tank Warfare.

During WW2 the British Military also had almost nothing to do with **Liddell Hart**. In 1944 the British Military had **Liddell Hart** update a few of his pre-war training manuals - and that was all.

Both the British and US Military had almost nothing to do with **Fuller** and **Hart** for the remainder of their lives. However - it was the brand new Israeli Army of the 1950's and 1960's that totally embraced the teachings of **Fuller** and especially **Liddell Hart**. Israel's Generals spent much time talking directly with Sir **Hart**. The result was that Israel's Generals spectacularly won the 2 most brilliantly fought wars of the last half of the 20th Century and 2 of the most brilliantly fought wars in the history of warfare - the Six Day War in 1967 and the Yom Kipper War of 1973.

The historical fact is that **those with brilliant past track records can nevertheless (at times) be completely 100% wrong.** That is why all plans from anyone and everyone have to be cross examined intently and never automatically accepted only because of past success and past performance.

All US Generals alive today have NO past USA shooting war clear wins. All of their war strategy plans should be taken as wrong until they can pass an extensive strategy cross examination conducted by proven strategy experts.

Individuals and political parties and Nations can (and do) go from being rational and sane to being dangerously insane (Germany under **Hitler** and Japan under **Tojo** were insane Nations). It is absolutely vital and decisive that the newly insane groups be recognized as insane that the **sane and rational people make the mindset changes needed (and action changes needed) to meet the new dangers.**

During the 1930's there were many prominent and mainly brilliant Americans (like **Henry Ford**) who were big supporters of **Adolf Hitler**. In 1937 **Hitler** gave **Henry Ford** the Grand Cross of the German Eagle medal. Mr. **Ford** during the 1930's said that **Adolf Hitler** was a sane and rational political leader.

In 1940 there was the German America Bund (i.e. the USA's Nazi Party) that supported USA political candidates. Over 10,000 of the Bund's members were German citizens (and **Hitler** suppers) who were permanently living in the USA.

When insanity arrives (and it arrives in every Nation at one time or another including the USA) the sane and rational people must acknowledge the insanity in their mindsets and actively defend their Nation.

(360) In the 21st Century a Talented International Macro Economist has the basic Understanding for being a Better Grand Strategist than most of the USA's Military's Generals.

A good macro economist would immediately understand the decisive military implications of what the Chinese are doing with their Silk Road Project.

The Chinese are already much bigger manufacturers of hard-goods than the USA. One example - **China builds over 100% more cars and trucks per year than the USA.** 70% of the products that are sold in the USA (and use electricity) are made in China. For hard goods products every Super Walmart in the USA is already a de facto "Super Chinamart."

The USA has been in an Economic War with China since 1980 and for the first 30 years **90% of the people in Washington DC (and the US Pentagon) never knew the war was happening. They never knew that the USA has lost the war.**

When the Roman Empire in the West came to an end in 476 AD there were no bells. There was no grand happening. 99% of the people in Italy did not realize that "the Roman Empire" had finally and forever ended in Italy and Western Europe.

Today most Americans do NOT realize that the USA has been on the Downside of its Bell Curve (relative to the world) since the 1975. The USA cannot afford endless stupid wars in stupid places for stupid reasons that have zero benefit to the average American. **Washington DC must stop acting like the USA continues to be on the upside of its Bell Curve and must stop the endless stupid wars that are speeding up the USA's decline.**

(361) Sun Tzu Wrote that for a Sure Victory You Must Have a Complete Understanding of Both Yourself and Your Opponent.

The Grand Master of Military Strategy - General **Sun Tzu** wrote his Three Core Rules for war planning victory:

(1). If you (1) know yourself and (2) you know your Grand Strategy and (3) you understand the mental process and mental-set of your opponent and (4) you understand the objectives of your opponent's Strategy - **then you will WIN ALL of your wars.**

(2). If you (1) know yourself and (2) you know your Grand Strategy but (3) you do **NOT** understand your opponent's mental-set and (4) you do **NOT** understand your opponent's Strategy objectives then **you will WIN HALF of your wars.**

(3). If you (1) do **NOT** understand yourself, and if (2) you do **NOT** understand your own objectives, and (3) you do **NOT** understand your opponent's mental-set and Strategy for obtaining his objectives then **you will LOSE ALL of your wars.**

Since the end of WW2 the US Military has NOT understood itself and has NOT understood its opponents. This is one of the major reasons for why the US Military has NOT clearly won a significant shooting war since 1945.

The **"solution"** for ending this great USA weakness is the new Grand Strategy Team as outlined in detail at the end of this book.

(362) General Sun Tzu wrote "If the National Leader does Not have Skills and Talent in the Art of War then the National Leader must Find a Skilled Commander who does have Talent and Give the Management of the War to that Skilled Commander."

Franklin Roosevelt (1882-1945) was Assistant Secretary of the US Navy from 1913 to 1920 (during all of WW1) and helped manage the safe shipping of two Million US troops to Europe. **Franklin Roosevelt** had a substantial understanding of military matters and was qualified to make military decisions. However President **Roosevelt** (President from January 1933 to April 1945) did not make the mistake of trying to personally micro manage the USA's part in WW2.

During WW2 President **Roosevelt** set the overall Grand Strategy objectives. Then President **Roosevelt** gave his commanding Generals and Admirals a free hand to win the battles that were needed to achieve the overall objectives.

(363) An Active Shooting War (were Your Troops are being Killed) is NOT a Good Place for the On-The-Job War Training of National Leaders.

The political Leaders who do NOT have an extensive proven background in Military Strategy must build a Grand Strategy Team of Experts that will hopefully (at the very least) avoid the usual disasters. **Especially in this 21st Century when there are many nuclear weapons in the world.**

Chapter 14

Warfare Trickery Used for Moral High Ground Ends

Military trickery and espionage are the active parts of military intelligence. Military trickery at its highest levels is skilled Judo using the weight and force of your opponent against your opponent.

Military trickery is making your opponent believe that something is real when it is an illusion. It is disinformation that your opponent believes is true. The British and American successful trickery of WW2 saved many thousands of British and American lives and shortened the war.

(364) The Best Grand Strategist is a Skilled Magician who has the Talent to Create Illusions and Trick the Mind of his Opponent.

Some of the **military magic tricks** are:

(1) The trick of appearing vulnerable in an area where you are prepared and want your opponent to attack.

(2) The trick of being confused and disorganized in an area where you are highly organized and prepared against an attack.

(3) The trick of retreating when you are setting a trap.

(4) The trick of appearing to be ready to attack in one area while you are preparing to attack in a different area.

(5) The trick of talking peace while you are preparing for war.

(6) The trick of talking surrender while you are preparing to attack.

(7) The trick of threatening war when you want to negotiate.

(8) The trick of bluff - appearing to be ready to fight when you do not want to fight.

(9) The trick of giving up ground and assets for peace when what you are doing is buying time to rebuild your forces for another attack and/or another war at a later time.

(10) The trick of being reconciled to losing a war (and not being a future threat) when you are planning revenge.

(11) The trick of using some of the 37 different types of warfare on your opponent in ways your opponent's mindset will NOT see and thus NOT comprehend the danger until it is too late.

The modern western world's masters at military trickery have been the British. The Chinese in the 21st Century are also going to be excellent at trickery because they are the direct genetic and cultural inheritors of **Sun Tzu's** brilliant mindset.

Two successful WW2 battle tricks (on the Allie's side) were (1) the Battle of Midway when the US Fleet out-tricked the Japanese Fleet. (2) the D-Day Invasion NOT being where the Germans thought it would be.

The most successful WW2 grand strategy trick was **Winston Churchill** and President **Roosevelt** maneuvering and tricking the Japanese into attacking the USA and UK.

The Germans on D-Day (June 6th) had the bulk of their forces at Calais about 200 miles away by road from the actual D-Day landing areas in Normandy. The Germans had 15 Army Divisions waiting at Calais for an attack that was never intended for that area. By the time **Adolf Hitler** was convinced that Normandy was the real invasion area the German Tank Forces at Calais were pinned down by US and British overwhelming air superiority.

The D-Day trick worked brilliantly and saved many allied lives. On D-Day - at Utah Beach - the allies landed 21,000 troops with only 197 casualties. At Juno Beach it was 21,000 troops landed with only 359 allied troops killed the first day. Other areas had higher casualties but on D-Day only 2,499 total Americans were killed. A very small number by WW2 major battle standards.

Total allied troops (USA, UK, Canada, etc) killed on D-Day were 8,443 (2,499 were Americans) and 47% of the 8,443 killed on the first day were USA and UK airborne troops that were killed far inland from the beach landing areas.

For battle-dead comparison on the first day of the Battle of the Somme (July 1st 1916) the British first attacking wave of 100,000 men suffered 57,470 casualties with 19,240 of that number being fatalities. 19,240 British were killed on the first day of the battle of the Somme compared to 2,499 US dead on the first day of the D-Day attack.

The final British total for the whole Battle of the Somme was 158,786 dead and 263,000 wounded.

Frederick the Great (1712-1786) the King of Prussia in his 1747 book of Military Instructions for his Generals wrote "The more you employ stratagems and ruses the more advantages you will enjoy over the enemy. You must deceive him and induce him to make mistakes in order to take advantage of his faults."

The British have always had a vicious sense of humor about their Military Trickery. In 1865 the Chinese Government was talked into buying 7 British warships (paid for in advance) for their desired new Navy. However - when the British warships arrived in China they all had British crews and Officers who said there was nothing in the written purchase contract about the ships NOT having and keeping British crews. The Chinese were outraged but could do nothing except send the ships away.

Remember the Rule - **good strategy is positioning yourself so no matter what happens you win something.** If the Chinese sent the ships away and the British kept the money - the British win. If the Chinese tried to take the ships by force there would have been another war the British would have won and they would have taken more Chinese territory - the British win.

(365) The Grand Strategist Must Periodically Stop Looking at the Trees and Step Way Back and Try to See the Forest.

The Strategist must be periodically asking himself - "does what appears to be happening make sense in the big picture? Or is it a con game on me?"

The FACT is the US Pentagon cannot "see" the Forest of China's highly successful Economic War because the Pentagon is looking at the Trees of WW2 ancient history. The delusion loving Pentagon is a failure everywhere - including (when this book was written) a failure at defending the USA's southern border.

(366) Always Question the "Secret Information" You Have Received for Possibly Being a Con Game Trap Against You.

"The weapons of mass destruction" that in 2003 were supposedly in Iraq - that was all a con game trap against the American people. The added twist was the con game was greatly aided and advanced by some of America's "Allies" and also advanced by the FBI and CIA.

The US Congress should have stepped back (and looked at the forest) and asked "Even if **Saddam Hussein** had a few nuclear weapons would using them against the USA (that had 5,000 nuclear weapons) make any sense whatsoever?" The answer is "No." **Saddam** and all of Iraq would have been wiped off the face of the world.

Dr. **Sigmund Freud** (1856-1939) **said that the most insane people nevertheless understand pain.** Even if **Saddam** was crazy he would have understood pain and dying. A person might be crazy but they understand that pain is bad when applied to them.

(367) If Your Opponent Does Something Significantly Wrong (and You can Exploit the Mistake) that Could Be Decisive for Your Side.

Sometimes it is not so much what you did correct as what your opponent did wrong. For many decades Sears and Montgomery Ward were America's 2 biggest retailers and were about even as competitors. In late 1945 the management of Montgomery thought the Depression would return after WW2 (a widely held belief at the time) and pulled back in many areas. However - Sears' management thought the economy would boom and aggressively expanded their stores and operation. Sears jumped way ahead of Montgomery in the 1945 to 1960 period and was clearly dominate and booming by 1960's.

(368) The USA's Never Ending Wars in Iraq and Afghanistan are Examples of how Trickery can Cause Major Damage to a Great Power (the USA).

Osama bin Laden is arguably the best Grand Strategist of the 21st Century (**up to this point**). That is NOT an approval of what **bin Laden** did it is only a factual statement of his success.

Osama used his 9/11 attack to trick then President **George W Bush** into expanding the Iraq War. The expanded Iraq War was (and is) a slow motion disaster for the USA. **Osama** brilliantly leveraged his 9/11 attack Judo style and caused his opponent (**W Bush**) to inflict on the USA additional (ongoing to this day) damage that is many times greater than the original 9/11 attack damage and costs.

In 2001 **Osama bin Laden also** tricked Washington DC into invading Afghanistan. Osama made himself the "bait" in Afghanistan knowing that once the US Army was in Afghanistan it would never leave.

There never was anything of any value to the USA in Afghanistan and never will be. **For the USA the area of Afghanistan is WW1's Salonika - a strategically useless dead end.** Since 2001 the US Troops in Afghanistan have de facto been in **Osama bin Laden's** prisoner of war camp.

The warmongers in Washington DC dream of Afghanistan as a someday jumping off base for an invasion of Iran - but that is logistic-supply insanity. The only way into Afghanistan is by going over the airspace of: Turkey, Armenia, Azerbaijan and Turkmenistan. If anyone say's "No" then there is no way into Afghanistan except through Pakistan. Currently USA-

Pakistan relations are NOT reliable. Aside from that the idea of a USA-Iran shooting war is insane.

(369) The good Grand Tactician and Grand Strategist will Try to Trick and Bait his Opponent into Attacking Non-Vital Points while Trying to Disguise and Draw Attention Away from Vital Points.

In 1940 the British tricked **Hitler** into bombing London by bombing Berlin. **Bombing civilians has been proven to be almost useless for any decisive military advantage.**

If the Germans in 1940 had continued to concentrate their air attacks on England's (1) airfields (2) radar stations (3) electrical power plants (4) Ports (5) railroad hubs and main bridges and (6) armament factories - all of that combined with a German offer of Peace would have probably persuaded the English to negotiating a peace agreement.

If **Hitler** had ended the war with the British Empire then **Hitler** would have been free to attack Russia with 100% of his military forces. The Germans would have defeated the Russians in the first 7 months of fighting and would then have access to all of Russia's oil and vast natural resources and industry. **Hitler** would also have had a bomber-proof direct land connection to Japan. WW2 would have ended with German in control of almost all of Europe.

Hitler's biggest mistake was the insane passion for total complete victory by using ONLY the shooting war tools. Basically the same mistake that Washington DC has continued to making for decades.

(370) Sometimes You Have to Use Trickery Against Your Own Citizens.

In 1939, 1940 and 1941 up to December 7th 1941 the American voting population was 75% strongly anti-war. Only 5% of the USA voters were strongly for the USA joining the European war. However - it was vital for the USA to enter the war on England's side. England could NOT defeat Hitler and in 1941 it looked like Russia would be defeated by Hitler. That would give Hitler all of Russia's vast resources and Hitler could battle England forever. To solve the "decisive problem" of getting the American voters into the war **Winston Churchill** and President **Roosevelt** conspired to trick the Japanese into attacking America and the UK. In July of 1941 the USA and UK put an oil embargo on Japan and told the Japanese Government it had to leave China. **Churchill** and **Roosevelt** know that was both psychologically and politically impossible for the Japanese. If any Japanese politician had proposed doing so he would have been assassinated. **Churchill** and **Roosevelt** know that the Japan would attack and take the Southeast Asia oil fields by force.

The trick worked brilliantly. On December 6th 1941 - 75% of the USA's voters were adamant about NOT going into any war anywhere. On December 8th 1941 - 95% of the American voters were demanding a war with Japan.

At 4:10 PM on December 8th President Roosevelt signed the Declaration of War against Japan. **Winston Churchill** had declared war against Japan nine hours earlier.

The cheery on the cake for **Churchill** was 3 days after Pearl Harbor (on December 11th 1941) **Adolf Hitler** declared war on the USA. **Churchill** know he had won the war.

The USA needed to join the war on the side of England and China. **Hitler** and **Tojo's Japan** were enslaving and murdering people by the millions and were clearly evil. It was a trick on the American voters but something that in the big picture was morally correct. Sometimes you have to drag people kicking and screaming into doing what is clearly needed.

(371) Thinking in Simplistic Dumb Terms of 100% Good Versus 100% Bad Destroys your Ability to have an Accurate Analysis of Your Opponent's Motivations and True Strategy.

Automatically thinking of yourself (as the US Pentagon does) as the 100% good-guys in the White hates and your opponents as the 100% bad-guys in the Black hates is stupid and counterproductive. **Many of the US Pentagon's opponents have valid and ethical reasons for opposing the US Military** in their Nation. The US Military supports Apartheid and racism and many of its "enemies" in the Middle East are fighting against a foreign colonial power (the USA) exactly like what happened in Vietnam.

(372) In the 20th Century Most Wars were Started for Reasons that were Deceits and Frauds. It Looks like the 21st Century will Be Worse for Starting Shooting Wars with Deceits and Frauds.

WW1 was started by the Austrian Government falsely saying that the Serbian Government was behind the assassination of Archduke **Franz Ferdinand**. The Serbian Government was not behind the attack. The Austrian Government used the assassination as an excuse to invade and occupy Serbia.

WW2 in Europe was started with a supposed attack against a German position by Polish soldiers - it was all a staged set up by the Germans and was a complete fraud.

The reasons for the USA going into Vietnam in 1960 were lies, frauds and con games against the American people and the world.

The reasons for Iraq War 2 "the weapons of mass destruction" was a con game fraud.

The reasons for US Troops staying in Afghanistan (after 2005 when **Osama** left Afghanistan) were con game lies.

The Japanese attacking Pearl Harbor in 1941 was **Roosevelt's** and **Churchill's** successful manipulation and tricking the Japanese into attacking the USA and UK.

If the warmongers cannot sell their war to the voters (or keep their wars going) with honest truthful logical reasoning and facts then they use lies and fraud. As the "Pentagon Papers" proved towards the end of the Vietnam War.

The leaders who sell their wars with lies and frauds tell themselves that their dishonesty is justified by the "eventual greater good." However - often their "Higher Loyalty" is to something that is in reality a "Higher Insanity."

USA Secretary of Defense **Robert McNamara** kept the Vietnam War going with lies and fabrications and dishonesty. **McNamara's** "Higher Loyalty" lies murdered 58,220 Americans and 650,000 Vietnamese including children, women and old people.

The phony "Saddam Hussein Weapons of Mass Destruction" fraud that was used to start Iraq War 2 turned into another highly expensive US Military debacle and world public relations disaster for the USA.

For over 60 years the USA Warmonger Lobby's lies and frauds have turned into disasters for the USA and the American people and the world.

Currently the US Warmonger Lobby wants the USA to attack Iran with a shooting war. The reasons they use for demanding a shooting war are lies, half-truths and puffed up spin. It is the **Saddam Hussein** "weapons of mass destruction" con game all over again.

When this Book was published the Washington DC Warmonger Lobby was trying to reproduce **Churchill's** and **Roosevelt's** 1941 trick of getting the Japanese to attack the USA. However this time the trick is to force Iran into attacking. The Washington DC Warmongers are trying to forced and con game Iran into attacking because that is essential for changing the US voter's now very anti-war attitude into a shooting war with Iran attitude.

The sane people of the USA should demand that the Warmonger Lobby (who are trying to get Iran to attack the USA) provide a written detailed explanation of exactly and specifically what they want from Iran. **Up to this point there have been NO specific demands from the USA to Iran. The US Warmongers do NOT want specific demands that Iran might meet. The Warmongers want a shooting war with Iran.**

Chapter 15

Propaganda - a Decisive Tool if Used Ethically

For thousands of years propaganda was NOT significant in warfare because of two major factors. (1) The lack of mass communications. (2) Wars were mainly wars between Kings and what the "public" believed was not important.

As late as WW2 most of the world's population only knew what their governments wanted them to know. However today in the 21st Century (and with non-standard Guerrilla types of wars) successful propaganda and idea communications have become a decisive weapons system for the long term outcome of many conflicts. Public opinion in 21st Century Democracies (in the current Information Age of Warfare) can now be more decisive for the outcome of most wars than what happens on the shooting war battlefields.

(373) The #1 Most Important Thing with Propaganda Lies is to NEVER Become a Victim of Your Own Propaganda Lies.

Nations, Leaders and Groups often become the biggest victims of their own Propaganda and Lies. If your thinking is dominated by lies and falsehoods then your Strategy planning will at the best be highly flawed - and will often be disastrously wrong.

(374) The Basic Mechanics of Destructive Propaganda (and TV Fraud News) is (1) the Highly Skilled Use of Half-Truths. (2) The Omission of All Facts that Counter the Fraud Narrative. (3) Spinning the Half-Truths and Making it Look Like what is Only Half of the Story is the Complete Story.

Much of what is on the USA's TV News is malicious propaganda. The fraud-news Networks taking selected parts of the truth and combining that with "puff" and "spin" to produce a conclusion that is false.

Most of the time there are few direct 100% lies in Propaganda. The core of Propaganda consists of using only **part of the facts that are then presented as ALL of the facts.**

Most Propaganda contains "lying by omission." This is what a large part of the TV Nightly News does every day. This is WHY one of the most valuable skills the Strategist can be born with (and cultivate with training) is the skill of (1) hearing what is NOT being said and (2) seeing what is NOT being presented.

Propaganda in all of its many forms is a tool to influence the thinking of populations and troops. It is a tool to plant disinformation for both your public's consumption and the consumption of your opponent.

(375) God is Often Used as a Propaganda Tool. However the Grand Strategist Knows that God is Always on the Same Side. God is Always on the Side that Can most Efficiently and Effectively Meet and Resolve the Challenges that God Has Created and Puts in Front of Everyone.

The belt buckles of German uniforms in both WW1 and WW2 said "**God** is with us." Germany lost both wars.

(376) One of the Lowest Moral Propaganda Ploys is Saying that a Proposed War is Going to be a War "to Prevent the Next War."

Suppose you were a White American male living in Chicago. In the month of July you went into the bad part of West Chicago at 3:00 AM and saw five young black thugs on a street corner. You stopped your car and pull out your gun. You shoot and kill all five of the Blacks. In court your defense was "Those Black thugs were criminals. I prevented them from committing future crimes." Would the Judge say "Oh, preventing crime is wonderful and what you did was wonderful." Or, would the Judge charge you with murder? The Law would correctly charge you with murder. **The Leaders and groups that advocate "preventive wars" are insane war criminals advocating murder.**

In 1949 and through the early 1950's there were MANY calls inside of the USA (from some well known Americans) for a "preventive war" with Russia before Russia could build several nuclear bombs. The arguments for having a "preventive" nuclear attack (during that time period) against Russia were relatively commonplace. Thank **God** the sane and moral individuals prevailed. The USA was enormously lucky to have **Harry Truman** and **Dwight Eisenhower** as Presidents.

Today we have calls in Washington DC for a "preventive war" against Iran. The Washington DC Warmongers are standing next to **Adolf Hitler.** They are the type of psychopaths who would murder young African-American males to supposedly "prevent crime."

(377) Nations Being Con-Gamed into Fighting Other People's Wars is a Trick as Old as Warfare.

Both the British and French were conned and tricked (by the Turks) into supporting the Crimean War (1853-1856). England and France had as many vital interests in the Russian Crimean of the 1850's as the USA has vital interests in Iraq, Iran and Afghanistan today - and that is **zero**.

In the Crimean War the British Army suffered 40,462 dead and the French Army suffered 135,485 dead for a war that was of absolutely zero value to the average British or French citizen.

The Turks were the brilliant ones who tricked the British and French into fighting their war with Russia. The Russians had been threatening the Turks because of the Turk's brutal mistreatment of Armenian Christians inside of Turkish territory. **It was Russia that was on the moral high ground** and it was England and France that were protecting the Turk's brutal Colonialism against Armenian Christians.

The British and French being con gamed into the Crimean War to protect Colonialism is like today's US Pentagon that has been con-gamed into protecting Apartheid and brutal racist Colonialism against West Bank Palestinians.

A prominent Member of the British Parliament - **John Bright** (1811-1889) gave a speech in the House of Commons after the Crimean War. He said "It is a painful and terrible thing to think how easy it is to stir up a Nation to war and you will find that wars are always supported by a class of arguments which, after the war is over, the people find were arguments they should NOT have listened to."

All of the arguments for the USA invading Iraq in 2003 (and for greatly expanding the Vietnam War in 1960) were con games that the American people should NOT have listened to. Today the arguments and lies trying to peddle a shooting war with Iran are con games the American voters should not be listing to.

(378) Throughout History Ruthless Leaders (and Fanatical Groups) have tried to Prove that Their Propaganda, Lies and Flawed Belief-Systems were Correct by Jailing and/or Murdering Anyone who Disagreed with Them.

The sick illogic (for thousands of years) has been that if there is nobody around to say that the Fanatical Group's (or Leader's) ideas and actions are wrong then that proves their ideas and actions are correct.

(379) The Grand Strategist Knows that 2,000 Years of Warfare has Proven that Victory Almost Never Fulfills all of the Pre-War Propaganda Promises.

The more unrealistic your promises for what a war will accomplish the deeper the pessimism, cynicism and future apathy will be when the promises do NOT come true.

WW1 was a horrible and costly war (10,000,000 dead from all causes). The Allied propaganda machines during WW1 kept saying over and over that "the Great War" was the "War to End Wars." 21 years later in 1939 - WW2 started in Europe and killed 30,000,000 people.

The usual bloated and false promises of most shooting war propaganda ends up producing deep anti-war sentiment in the long run. The more the US Pentagon mucks around in never-ending-wars decade after decade the (1) deeper the USA's anti-war voter sentiment will become and (2) the more the US Military's reputation will be discredited and lowered.

(380) Starting with the Industrial Age of Warfare (1885-1960) Control of the Media has been Increasingly Decisive.

Now that the world is into the Information Age of Warfare (1960 to ?) the world's internet media like Facebook, Google, Twitter etc have become major weapons systems.

The Internet and salesmanship have become decisive for the USA resolving idea conflicts with the Radical Muslim Groups - plus decisive in many other idea-conflicts.

The core objective is to NOT have disagreements turn into shooting war disagreements. Over a Century ago Winston Churchill said - "Talk talk is always better than shoot shoot."

(381) The USA's Laws Regarding the Government's Restricted Active Use of the Internet Must be Reworked and Changed.

The US Pentagon's obsolete WW2 mindset (and the USA's obsolete laws about Government communications) are decisive and fatal flaws that **guarantees continued USA worldwide defeats** in the current Information Age of Warfare.

(382) The Key to Countering Propaganda is YOU must NOT Put Your Mind into Neutral When You are Watching the TV News.

The TV News is NOT a movie where you suspend reality and enjoy the ride. YOU must be constantly "fact-checking" in your own mind as the "News" is presented to you.

Chapter 16

The Strategy of Managing Allies for Winning Long Lasting Peace

Allies (or the lack of Allies and/or the wrong Allies) is often decisive in wars.

Throughout history Allies have many times been "fair weather friends." If you are winning and/or your Allies are receiving something of value from you then your Alliance will continue. However - when you are no longer useful to your Allies then many will stop giving you any real help and/or they could go over to your opponent's side.

The core rule with both Allies and Enemies is that Allies are NOT Forever and Enemies are NOT Forever. You must constantly be evaluating the cost/benefits ratio of your Allies. You must constantly be working towards getting closer to your Enemies and working to turn your Enemies into Allies.

With the USA - the only 2 Nations in the world that can cause massive damage to the USA are Russia and China. The USA's maximum effort and work most go into maintaining an effective working relationship with Russia and China. A working relationship with Russia and China must be the USA's highest priority for the foreseeable future.

(383) **Great Powers (i.e. the USA) Use their Insignificant Allies for the Benefit of the Great Power. The Great Power NEVER Lets an Insignificant Ally Use the Great Power for the Insignificant Allies' Purposes that are Often Counter to the Great Power's Important Interests.**

Remember the axiom - "**The USA must be willing to compromise on everything that is Not vital to the USA's core interests and survival and that includes the USA's insignificant Allies."**

Washington DC in the Middle East has violated this rule for decades by entering into shooting conflicts over issues and areas that were valuable to the USA's insignificant Allies but had little or no vital value for the USA's National security.

The USA's core vital interests should NEVER be open to negotiations. However the vital desires of the USA's insignificant allies should always be open to negotiations if the USA receives something of real value to the USA in return.

The USA has ZERO vital allies in the Middle East who are "must have" strategic important allies. Ethically and morally everyone in the Middle East is to a high degree anti-freedom and anti-equal justice for some other groups. There are no "good guys" in the Middle East. The interests of every US ally in the Middle East should be open to negotiations if the US People can receive significant long term advantages in return.

(D) **The USA must be willing to compromise on everything that is Not vital to the USA's core interests and survival.**

The USA in recent decades has extensively violated this rule by entering into shooting conflicts over issues and areas that had little (or no) vital interests to the USA's National security. For example - there was NEVER anything vital to the USA interests in Vietnam. There was never anything vital in Afghanistan.

(384) A Great Power (i.e. the USA) must Never Let Weak, Flawed, Fascist, Low Quality Allies be a Detriment and Negative Factor for the Moral and Ethical World Public Image of a Great Power and the USA in Particular.

The United States of America should ONLY be supporting equal justice and equal rights under the law everywhere in the world. **The USA Government should high profile condemn racism, apartheid and injustice everywhere in the world without exceptions.** Washington DC's practice of disapproving of apartheid in some areas (as in N.W China) but approving of apartheid in other areas (as in the West Bank) is schizophrenic. Washington DC's moral schizophrenia and has been (and continues to be) a worldwide public relations disaster for the USA.

(385) In the Current Information Age of Warfare Nations (i.e. the USA) Can No longer Hide the Truth of About Their Allies. Nations Can No longer Hide their Ally's Hypocrisy Behind Schizophrenic Double Standards Because with the Worldwide Internet all the World Is Seeing the Truth.

When the vital interests of the great power - and the vital interests of an insignificant moral low ground ally clash - the Grand Strategist for the great power has three choices:

(1) Tell the insignificant ally that the great power is the boss. The great power tells the ally what is going to be done and what is going to happen - **not the other way around.**

(2) The great power becomes an equally phony ally and tells the insignificant ally the great power loves the ally but then does nothing for the ally while putting as much distance as possible between the great power and the useless ally.

(3) The great power dumps the negative-value ally on the side of the road and drives away.

Remember the Rule - if you want to always be Mr. Nice Guy then become the manager of a flower shop - NOT the Leader of a Great Power.

(386) Taking On An Ally Also Means Taking On All of the Ally's Enemies. The Grand Strategist Must Constantly be Analyzing and Updating the Benefits versus Disadvantages Ratio Trade-Offs.

For Allies you want Nations that have valuable strategic friends and assets that will help you (and ideally help you in combat with large numbers of troops). What you do NOT want are Allies that have little or no strategic value to you but have many very powerful enemies. Then **the powerful enemies of your Ally also become your enemies.**

If a significant power group in the world (like the Muslims) was not the USA's enemy - but the USA takes on a worthless ally who is an enemy of a significant power group (like the Muslims). Then the significant power group becomes the USA's enemy. That is a bad trade-off mistake that a competent Grand Strategist would never make.

In the long Cold War between the USSR and the USA (1946 to 1991) the USA took on many Allies that were Fascist and Totalitarian moral low ground Leaders and Governments. **The moral and ethical trade-off during the Cold War was Grand Strategy expedient. The USA resolved the Cold War** without a major nuclear-shooting war. However **when the Grand Strategic paradigm changes then your basic Strategy must also be up-dated and change in an ethical upwards direction.**

The USSR ended in 1991. **Washington DC no longer needs the scumbags of the world.** However Washington DC continues to support Fascist and anti-freedom Governments **following a Cold-War Strategy that was already obsolete for 30 years when this Book was published. Washington DC's Cold War mindset is currently highly counterproductive to the USA's worldwide public relations and image.** The enormous worldwide "goodwill" that the USA had after WW2 has been 85% destroyed by Washington DC supporting Fascism, racism and colonial oppression.

In this Century the US Government (and the American people) MUST be super high profile advocates for (1) equal freedoms (2) equal justice and (3) equal rights for everyone EVERYWHERE in the world.

The USA cannot correct every one of the dozens of wrongs in the world and must not destroy itself trying. However - the USA can STOP supporting Fascism and Apartheid with massive amounts of money and armaments. If the USA cannot be part of the solution then at the very least it can **STOP the world public-relations disaster of high-profile financing injustice and anti-freedom and thus being part of the world's problems.**

(387) Washington DC's Obsolete Mindset is that the USA is the World's Most Profitable Nation and will Forever be the World's Richest and most Profitable Nation. Therefore the USA Can Support any Worthless Ally and Anti-Freedom and Anti-Equal Justice Ally and it Means Nothing

The American Grand Strategist should always be asking himself "Exactly and specifically what is this Ally doing for the USA that is a real advantage to the USA's citizens here and now?"

When an Ally becomes more of an overall problem than it is worth at that point Washington DC has to distance itself from the problem Ally. A Great Power (like the USA) should high profile (and very publically) make it crystal clear that the USA does NOT approve of its Ally's moral low ground anti-freedom actions.

(388) China and the USA are Natural Allies. In Addition China Winning the Economic War with the USA Has a Big Upside. China is Going to Be Exceptionally Reluctant to Ever"Nuke" the USA When China Owns so Much of the USA.

(389) The Grand Strategist will Use Any Nation He can to Achieve his Core Vital Objectives.

The great power should use other Nations for what it wants but should NOT become "married" to anyone who is not absolutely vital.

(390) The Two Most Important Nations in the World for the USA are China and Russia.

There has been an Economic War (starting in the 1980's) between the USA and China. China won the first round. However - the USA and China can cooperate on issues that are vital to the world. In the 21st Century if part of the world is damaged than all of the world is damaged. We are ALL on the same spaceship.

China and Russia are the top two Nations that Washington DC MUST constantly and vigorously be working towards building and expanding better relations. WHY? Because they are the only two Nations in the world that have the power to destroy the USA. **The interests of other Allies must be secondary for building closer USA-Russia and USA-China partnerships.**

The best long term situation for the American people (and for China and Russia) would be a USA-Russia-China Common Market Alliance.

A USA-Russia-China Alliance would control 90% of "Space" and that means they would control 90% of the World.

(391) For the Grand Strategist the Words: "Friend, Ally, Enemy and Neutral" ALL have the word "TEMPORARY" in Front of Them.

The above is one of the more important rules for having a superior Grand Strategy mindset. **The Strategist is always working towards (1) turning enemies into allies and (2) upgrading his collection of allies and (3) tuning what were former allies into being current allies.**

The World's mix of Allies and Enemies has always been (and always will be) in a state of constant flux and change. During WW2 - China, Russia and Iran were all allies of the USA and today that list is reversed. However there is no absolute forever reason for why China, Russia and Iran cannot once again be allies of the USA.

Sometimes Nations (like the USA) cannot acquire new and more valuable Allies without first discarding some old and no longer valuable Allies. China had that problem when this book was written - is North Korea more valuable as an Ally to China than good relations with the USA?

(392) Deadly Overreach Comes in Many Forms. One Form of Overreach is to Support Allies that are more of a Detriment Than an Asset.

Sometimes Great Powers have to choose sides and sometimes it is emotionally painful.

The USA will be declining (relative to the world) during all of this Century. The USA cannot afford to pay for and carry worthless Allies. **One of the main factors that destroyed the USSR's Empire was Deadly Overreach in the form of trying to support too many worthless Allies while the USSR was in relative to the world Bell Curve decline.**

Currently Washington DC is making the same overreach mistake that the USSR made in the 1970's and 1980's - that is trying to control too much in the world with an economy that is powered mainly by massive debt creation at every level.

(393) The Great Power Grand Strategist is never Naive About his Allies having some Objectives that are very Different from the Objectives of the Great Power.

After the (December 1941) Pearl Harbor Attack (and the USA came into WW2) everyone in Washington DC said that America's and England's "special relationship" put the USA and England on the same sheet of paper for all objectives. However the USA's and England's Grand Strategies had a few completely different objectives.

In Asia during WW2 the American people (and USA Government) wanted Asia (after WW2 ended) to have as many independent Democracies as possible. However the post WW2 objective of England was the reestablishment (and expansion) of the British Colonial Empire in Asia. England did NOT want independent Nations in England's Asian Empire.

A sub-objective of **Winston Churchill** in tricking the Japanese into attacking the USA and UK in 1941 was the elimination of Japan as a colonial rival in Asia. However - immediately after WW2 ended the Socialists took control of the British government and England and did not pursue significant colonial expansions.

Other WW2 "Allies" of the USA like France and the Netherlands also had post-WW2 objectives of reestablishing White Man colonial rule in Asia.

In 1960 the USA "went over to the Dark Side" and tried to reestablish White Man de facto colonial rule starting in Vietnam. Then proceeding to Iraq and so forth.

When is Washington DC going to understand that White Man colonial rule has been the very wrong side of history since the end of WW2?

(394) Many Ally "Friendships" between Nations are Similar To Rich and Powerful Hollywood USA Movie Industry "Friendships."

In Hollywood if you have something that somebody else wants (or needs) then you are their "best friend" and they "love you." However - if you do not have what the typical Hollywood "big shot" operative needs then you do NOT exist.

(395) The Grand Strategist Never 100% Trusts Any of His Allies.

The probability of your vital interests and your Ally's vital interests both being 100% the same all of the time is ZERO. If your Allies see a turn of events as being vital to their interests (or to their subjective attitudes) but not good for you then your Allies will often take actions to bring about a result **favorable to what they want to happen** and **NOT** necessarily favorable to you.

The USA's Top Secrets are valuable to many Nations. In the past there are incidents of where an "Ally" knew the USA's Secrets and the Ally traded the USA's Secrets to Russia for items the Ally wanted. **Trust Nobody.**

(396) Because your Allies are (1) Often Temporary and (2) Often have your Opponent's Spies in their Organizations and (3) they have their Own Agendas that are NOT your Agendas. You (4) Never Tell Your Allies More of Your Top Secret Information than your Allies Need to Know at the Time.

Spies are everywhere. They are inside of the CIA, FBI, NSA, the US Congress, State Department, Pentagon and everywhere. The National Leader should NEVER tell his Allies (or everyone on his personal staff) more than what any individual needs to know to do their job.

History has proven that (1) Allies often have your opponent's spies in their organization. (2) What is a vital Top Secret to you might NOT be than important secret to an Ally who will treat it nonchalantly. (3) **Allies will sometimes trade your top secrets** (that are of little relevance to them) **to your biggest opponents for items that are of value to your Ally**. (4) **Always assume that your allies might sell your Top Secrets to your opponent for items that are valuable to them.**

All of the many USA Spy organizations plus the US President, State Department, Pentagon etc **must be planting different pieces of false information (on a rotating regular schedule) into various select parts of all of their systems.** They must do this on a continuous regular schedule and to specific people on a rotating timetable. Then carefully look from what directions (and in what forms of action) the "secrets" come back to them.

(397) Because your Allies are often (in the long run) Temporary you should Never Give them more <u>Weapons</u> than what is Needed to Help You Win the Shooting Battles You Need to Win at that Specific Time.

If a great power gives its Allies too many weapons the Great Power is giving its Ally (1) more opportunities to stumble into trouble and then drag the Great Power into the Ally's problems and/or (2) the weapons might someday be used against their source.

Starting in the 1890's the British built the Japanese Navy. All of Japan's Battleships up to 1906 were built by the British and the Japanese Navy was trained by the British Navy.

The 1905 Sea Battle of Tsushima (between the Japanese Fleet and Russian Fleet) changed world history. Japanese Admiral **Togo's** Flagship (the Mikasa - now a museum in Japan) had been built by Armstrong-Elswick in England. **All of the Japanese Battleships at the history changing Battle of Tsushima had been built in England** mainly by the Thames Iron Works and Armstrong-Elswick.

In 1941-1942 the Japanese Navy captured a large part of England's Asian Colonial Empire using the Japanese Navy that England had trained and originally built. **The Japanese lost WW2 but in the process the Japanese very successfully destroyed the idea of White man colonialism in Asia.** After WW2 White man Colonialism in Asia was the walking dead. **England had built and trained the Navy that was the catalyst for eventually destroy England's Asian Empire.**

During the Algerian Revolution against French rule (after WW2) many of the "rebels" had been trained in the French Army.

One of the heroes of the US Revolution (who won the battle of Saratoga) - General **Horatio Gates -** was a former British Army Officer.

Robert E Lee was the Superintendent of the US West Point Military Academy - and so forth. Today's Ally can be tomorrow's opponent so be aware of what you are giving other Nations.

(398) NATO is a Large and Diverse Military Alliance. The Question is if it Would Work if Put to the Test? Washington DC must NOT Assume that the Answer is "Yes."

With the US Taxpayers paying many of everyone else's NATO costs and expenses the Alliance continues - but what would happen in a significant shooting war?

The USA's best Strategy is to do whatever is needed to NEVER test the NATO alliance with a nuclear issue. **Washington DC should NOT start a significant shooting war with the belief that it will automatically have the support of NATO.**

The "answer" to what amount of support the US Military would receive from its Allies was given in Iraq War 2 (2003 to ?). The British sent troops to police the southern quite area of Iraq but the British did very little combat fighting. Aside from England - the USA's Allies sent only what were de facto "military observers" that were of no significance as meaningful fighting unites. In Syria the Kurds, Iran and Turkey provided valuable troops in the fight against ISIS. However the USA's most expensive Middle East Ally (that Washington DC supports with massive amount of free military hardware and money) sent Zero Troops to fight to help fight ISIS or to help in Iraq or to help in Afghanistan.

In a USA-Iran shooting war 99.9% of the shooting combat against Iran would be only US forces.

The overall Rule for Washington DC must be to NOT start a significant shooting war anywhere in the world unless the USA's mainland (or the USA's truly vital interests) are being attacked in a sustained assault.

(399) After a Victory (either Bloodless or with a Shooting War) the Grand Strategist Tries to Absorb his Former Opponent into His Organization of Allies.

How many "experts" and Generals in 1942 Washington DC predicted that Germany and Japan would be long term good Allies of the USA? The answer is "None." **There are no insurmountable reasons for why Iran cannot return to being an Ally of the USA.**

One of the many reasons for why you should do as little damage to your opponent as possible (to win a war) is because the ultimate ideal long term objective is to turn your opponent into an Ally.

For turning enemies into Allies the non-shooting types of warfare are vastly superior to the shooting war tools.

(400) All of the Numbers for You and Your Allies Versus the same Numbers for your Opponent and Their Allies are highly Decisive Forces in the Grand Strategy Equation.

In early 1941 Germany was fighting the British Empire (England, Canada, Australia, New Zealand, South Africa, India, Kenya, Nigeria, Rhodesia, Egypt and so forth). The British Empire in 1939 was 24% of the world's 2.3 Billion total population. In terms of gross population Germany (and its Allies) were 9 to 1 outnumbered compared to the UK. Then in June of 1941 Germany (with a population of 66 million) attacked Russia with a population of 160 million. The population advantage of the UK and Russia over Germany was enormous even before the USA (1941 population of 134 Million) entered the war.

In Asia the Japanese had been attacking China since 1931. Japan had a population of 73 million in 1940 and China a population of 520 million. In December of 1941 Japan attacked the USA that had a population of 134 million and most decisively **the USA was the largest and strongest manufacturing industrial Nation in the world.**

Japan also attacked the United Kingdom that was 24% of the world's population. At the start of 1942 Japan's opponent's (China, UK, USA etc) had a combined gross population of over one billion people versus Japan's 73 million.

The current war between the Muslim World and the USA started in 1991. In 2019 the "Muslim World" had 1,850,000,000 people compared to the USA's 325,000,000. By the year 2050 the world Muslim population is projected to be 2,800,000,000. The USA population in year 2050 is expected to be 450,000,000. **The massive difference in gross population is only one of the many decisive factors for why the USA will never win a shooting war victory against the Muslim world.**

The one and ONLY way for the USA to win in the Muslim World is by using the many non-shooting types of warfare. Washington DC's Warmonger Lobby keeps pushing for a USA-

Iran shooting war. However **a USA-Iran shooting war is a guaranteed lost war for the USA** and should NOT for one second be considered.

The Japanese during the Great Asian-Pacific War (and the Germans in WW2) relearned three ancient grand strategy axioms the hard way.

(1) An Alligator cannot swallow an Elephant.

(2) Wars often mutate into much bigger and deadlier conflicts than the original attacker had planned.

(3) Wars are very easy to start - but once started many times can only be ended with (A) the complete destruction of one side - or (B) the unconditional surrender of one side.

For the USA there are ZERO doable and workable shooting war military options that would succeed in the Middle East. Any USA significant shooting war in the Middle East is going to be one more guaranteed USA muddle and de facto defeat.

However - **if the USA effectively uses and applies the non-shooting tools of warfare then the USA would have an 85% probability of building an advantageous peace for all of the USA's citizens.**

(401) The "Alliance Lessons" for the USA from WW1's and WW2's Successes (and from the failures in Vietnam, Iraq and Afghanistan) is that if the USA wants to Win Shooting Wars it needs (1) very Powerful Allies who are (2) War/Battle Experienced and (3) Strategy/Tactics Skilled and (4) Willing to do the at least 80% of the Ground War Shooting Combat.

The USA was on the winning side in WW2. However the USA's Allies did 85% of the combat fighting in Europe. The USA's Allies did 97% of the military personnel dying in Europe and 98% of the military dying in Asia.

In WW2 Russia alone lost 10 Million soldiers and 10 million civilians fighting **Hitler**. The USA lost 280,000 soldiers (and Merchant Marines) fighting **Hitler** and a few hundred civilians. During WW2 the USSR lost 14% of its total population. In WW2 the USA lost 1/3 of 1% of its total population.

The USA's biggest contributions to WW2 were industrial war products , financial backing and psychological-morale factors.

The USA's massive production capacity overwhelmed both Germany's and Japan's industrial capacity. However in the 21st Century most of the USA's heavy industry (the industry that was decisive in winning WW2 for the USA and its Allies) has been moved to China and Japan and dozens of other parts of the world.

(402) Great powers have to be Careful about Promising too Much to Insignificant Allies that are in Unsolvable Geopolitical Positions. The Problems of the Insignificant Allies are Their Problems to Solve and NOT the Great Power's Problems.

Many times Nations (usually smaller Nations) find themselves in very difficult geopolitical positions that are basically un-resolvable and/or they desire things that are impossible. **Great powers (and especially the USA) must guard against being suckered into the often self-created problems of their insignificant allies.**

In the early 20th Century the European Nation of Serbia was in an un-resolvable geopolitical bad position. Serbia was surrounded by Nations mainly loyal to Germany and Austria. Russia was an Ally of Serbia but Russia had no easy way to directly reach or help Serbia in a shooting war. **When Austria attacked Serbia (July 28th 1914) the Russians in one of the most insane and disastrous decisions in modern history came to the defense of Serbia.** Russia had no way to directly help Serbia and never did directly help Serbia.

Germany was an Ally of Austria and Kaiser **Wilhelm II** of Germany had for years been wanting a war with Russia. Kaiser **Wilhelm II** used the Russian's threats against Austria (Germany's Ally) as the excuse to attack Russia. The war that followed destroyed the Tsar's family (most were murdered) and Imperial Russia was destroyed and replaced with Communism and **Joseph Stalin**.

In 1914 the Nation of Serbia was less than worthless to Russia. When Austria attacked Serbia in 1914 the Serbian population was 4.5 million. The Austrian Empire's population in 1914 was 51 Million. When Austria attacked Serbia **Tsar Nicholas II** should have said "Too bad - but if Serbia had never existed it would have meant nothing whatsoever to Russia."

One of the dumbest Grand Strategy moves is for a great power to risk a shooting war with another great power to save a worthless Ally.

Today the USA's relationship to Zionist controlled Israel is very much like Imperial Russia's per WW1 relationship to Serbia. The Zionist's dream of ONLY a Zionist Nation and ONLY for Zionists is unworkable and unachievable. It is a racist idea that was unworkable 2,000 years ago and is completely unworkable today. The USA must NOT with the Zionists repeat the same mistake that **Tsar Nicholas II** made with Serbia.

Most of the USA's Generals say that the Zionists are one of the USA's most valuable allies. That is more proof that most of the USA's living Generals are Grand Strategy morons.

(403) Often making Alliances with Groups and Nations that You do Not Especially Like is exactly the Correct Grand Strategy.

There are only two types of groups in international power politics. There are (1) those on the winning side and (2) those on the losing side. The objective is for YOUR Nation (the USA) to do what is needed to be on the winning side.

In 1941 there were no two Nations in the world more apart in political philosophy than the USA and **Stalinist** Russia. However the USA and USSR formed a highly successful Alliance to terminate the greater evil of **Adolf Hitler**.

(404) The Grand Strategist is always Looking for Ways to Exploit the Desires of his Opponent's Allies Towards Ends that are Useful for the Grand Strategist.

Often there are ways to give your opponent's Ally something the Ally deeply wants and turn that Ally against your opponent.

(405) If your Opponent is In a very Strong Position then Attack His Allies Starting with the Weakest Allies First. Turn your Opponent's Allies Into Problems for your Opponent and that will Throw your Opponent Off Balance Mentally and Organizationally.

The German Army by November of 1914 was in strong defensive positions in France - and the German troops were in good spirits. The morale of the Allies was NOT good. However the Allies (England and France) had world naval superiority and proceeded to attack Germany's overseas colonies. **The German Navy was big enough to aggravate the English (and bring England into WW1 on France's side) but the Germany Navy was NOT big enough to defend Germany's overseas colonies.**

The Allies attacked all of the German colonies in Asia and the Pacific: Tsingtao in China and the Pacific German colonies of: New Guinea, the Solomon Islands, the Mariana, Samoa, Palau Islands, the Caroline Islands and German New Guinea.

In Africa the Allies attacked the German colonies of: Nambia, Togo, Cameroon and Tanzania. Most of Germany's colonies were conquered by the end of 1914 - and at a low cost to the British, French and Japanese Allied forces. (Japan was an Ally of Britain, France, Russia and the USA in WW1).

The capture of Germany's colonies was a great morale boaster for the Allied troops in France - and a deep discouragement for the German troops and people who had spend much money and effort acquiring their colonies.

The Rule is that if you can produce a defeat for your opponent early on in the conflict that is an excellent morale booster for your troops and National voters.

If there ever was (God forbid) a shooting war between the USA and China all of the USA's Super Aircraft Carriers (big fat targets) would be sunk in the first 2 weeks of the war. It would be a great psychological blow for both the US voters and US Military and a great morale booster for the Chinese Military.

(406) The Grand Strategist must NOT Fall in Love with His Allies. The Only Love the Grand Strategist Should Have is a Love for His Own Nation.

There are too many people in Washington DC who's first loyalty is NOT to the USA. That is an issue that Washington DC needs to face and make the needed adjustments. The core Rule for the USA's operatives must be "The USA first and ONLY." **The CIA, NSA, US Pentagon, State Department etc need to be constantly reminded that they are NOT being paid to save somebody else's Nation.**

(407) Military Incompetents usually want to Attack the Strongest Part of their Opponent's Alliance. You must Resist that Temptation and Attack the Weakest Parts First.

If you are strong enough to quickly defeat your opponent's most important Ally then do so. Otherwise start with the weakest Allies first.

In today's 21st Century wars between the USA and the Radical Muslim Groups (RMGs) the RMGs are having more attacks in France, England, Spain, Europe etc than in the USA because the Guerrilla's correctly see Europe's voters as more susceptible to voter change than homeland USA voters. The RMGs are attacking the USA's weaker Allies first.

(408) If there is a Nation that Looks like it Might Join Your Opponent as an Ally then the Strategist will Apply a Combination of (1)Bluffs (2) Threats and (3) Rewards to Keep the Nation in Question from Joining your Opponent's Side.

In every war (if you start to win the war) always let your opponent's Allies know (in secret) that you are ready to talk about a separate peace deal that would have advantages for them.

The classic British Tactic from the 18th, 19th and early 20th Centuries was if a Nation was thinking of becoming an Ally of a Nation not particularly friendly to the British then the British would send several of their biggest Man-of-War ships on a "friendship tour" of the subject's area. The British would also send representatives to the target Nation to talk about all the British could do for that Nation. How the British could bring that Nation into England's de facto worldwide common market. How the British Navy (the world's biggest) would protect the Nation in question. **The protection of the world's biggest Navy (England's) was a powerful incentive when Sea Power really did rule the world.**

The British were highly skilled and talented at the process of BOTH (1) **threats** with a big stick - and (2) **carrot rewards** with the British worldwide common market and Naval protection.

During the last 50 years Washington DC's international negotiations (with less than friendly Nations) have been very big on the "stick" part but usually lacking in the "carrot" part. Too often Washington DC's attitude has been that "You are going to do what we want because that is what we want."

(409) When You Start to Lose Your Wars Then You Will Also Start to Lose Your Allies.

Historically the Nations that were winning pulled Allies to their Nation's side because **everyone wants to be on the winning side.** However - if a Nation starts to lose it will also start to lose Allies and/or (most commonly) the Allies will become "half hearted Allies" who say they love you but start to "drag their feet" and give you as little as possible.

Italy in WW2 changed sides after Germany's major defeat in the decisive Battle at Kursk in Russia (July 5-16 1943). Italian Field Marshal **Pietro Badoglio** (1871-1956) could see that Germany had lost the war. **(General Badoglio understood the core Grand Strategy Rule - You have to know when you have lost).** On July 25th 1943 the Italian Government arrested **Benito Mussolini** (1883-1945) and changed sides. The Italian Government joined the Allies.

The above strategy move by Field Marshal **Badoglio** was both brilliant and **exceptionally courageous**. There was a big German Army in Italy.

The war in Europe ended 1 year and 9 months after Italy had changed sides. **Italy was thus on the winning side at the end of the war and NOT on the losing side occupied by foreign (probably Russian) troops.**

Romania was an Ally of **Hitler** but changed sides on August 23rd 1944. The Romania Army joined the Russian Army in their fight against **Hitler** and the German Army.

The Hungarian Government (an Ally of Germany) after the Battle of Kursk (July 1943) also knew the Germans had lost the war. The Hungarian Government tried to make an Alliance with the Russians. However the Russians told the Germans that the Hungarian Government wanted to change sides. This was a maneuver to tie up German troops in Hungary. The German Army in March of 1944 occupied Hungary.

Stalin telling the Germans that Hungary wanted to change sides condemned Hungary's Jews. Hungary in pre-war 1939 had a Jewish population of over 800,000. They were relatively safe until the 1944 German Army occupation of Hungary. Then the German SS (under Adolf Eichmann 1906-1962) murdered about 500,000 Jews.

The US Military in the current USA versus the Muslim World War (1991 to ?) has been in a shooting-war somewhere in the Middle East since 1991. **Washington DC's former excellent support from the World's Nations has suffered greatly since 1991.** In Iraq War 2 (2003 to ?) NO "ally" sent significant numbers of ground troops to fight in Western and Northern Iraq. England sent some troops to garrison the relatively peaceful South and lost about 180 dead over 4 years time.

Most Nations today are "cool to the idea" of being part of the US Pentagon's Team that has not clearly and decisively won a significant shooting war since 1945. For as long as Washington DC's massive money payments continue to the USA's "Allies" those "Allies" will continue to say they love the USA. (Pay most people big amounts of money and most people will say they love you). However - when the US money stops flowing Washington DC will probably be lucky to get the time of day from its "friends."

(410) If Two Nations Cannot Invade Each Other Then They Are Natural Allies.

There have always been many significant cultural differences between Russia and the USA. However the key military facts are the Russian Army cannot march to Washington DC and the American Navy cannot sail to Moscow. **Neither the USA or Russia could occupy the other Nation's land area.** Neither Nation has to fear a direct invasion from the other Nation.

The USA and Russia were Allies in the both WW1 and WW2 in Europe. The USA and Russia were also Allies in the Great Asian-Pacific War against Japan. **The Russian attack against the Japanese Army in China (ending Japan's core strategy objective of a Chinese colonial empire) was the main factor that caused the Japanese Leadership to surrender.**

For the Japanese their core grand strategy objective was always building a colonial empire in China. **When Japan's core grand strategy dream was destroyed (via the Russian invasion of China) the war was over in the minds of Japan's Leaders** and they surrendered. Victory in every war is always in exactly the same place. It is in the minds of your opponent.

It was NOT the Atomic Bombs that caused the Japanese leadership to surrender. It was Russian's invasion of China and the destruction of the Japanese Army in China. Japan's dream of a colonial empire in China was dead and the Japanese leadership (very correctly) feared a Russian occupation of half of Japan far more than they feared a USA occupation of all of Japan. **If the Japanese had not surrendered to the USA then today there would be a North and South Japan like there is a North and South Korea.**

The records of the Japanese Leadership conversations during that time period show that it was NOT the Atomic Bombs that ended the war. **The Atomic Bomb ending the war is a USA self serving (and dangerous) propaganda myth** that is a detriment to US Military correct grand strategy planning.

AFTER the Japanese Leadership had decided to surrender they used the Atomic Bomb (and saving lives) as the excuse for surrendering but that was NOT the true reason. The Japanese Leadership knew that if the war continued the Russians would invade Hokkaido and Northern Honshu and stay forever. **Today there would be a communist North Japan.**

Because of the mistaken belief that it was the Atomic Bombs that caused Japan to surrender **the popular wrong assumption (in the USA) is that a nuclear war between the USA and Russia (or China) would be over in a few days.** However - the USA has no way to invade either Russia or China and thus force a surrender. The Russians and Chinese have no way to invade the USA and force a surrender. So - **neither side would surrender.**

In a USA-Russian (and/or China) nuclear war **a "state of war" might exist between the USA and Russia (and/or China) for many years.** Much like a state of war has existed between North Korea and the USA since 1950. The world's biggest winners would be those Nations NOT in the war.

Chapter 17

Aggressive Shooting Wars are Usually Failures

Military morons believe that if you are the one to start the war then you are going to win the war. However - starting Aggressive Wars in the Industrial Age of Warfare (1885 to 1960) and in the current Information Age of Warfare (1960 to ?) has proven to be far more risky and dangerous for the aggressors than for the defenders.

Any military moron can start a war but winning wars has become a difficult task demanding highly advanced planning talents and exceptional skills in military and Grand Strategy management. Most of the aggressive large wars started in the last Century (and up to now) were eventually lost by the Nation that started the war.

(1) In Europe the Austrians/Germans started WW1 (1914) and the Germans started WW2 (1939) and they lost both wars.

(2) The Japanese started the Great Asian-Pacific War of 1931-1945 and lost.

(3) North Korea started the Korean war (1950) and lost.

(4) The Indian-Pakistan Wars that were mutually started in 1948-1949, 1965 and 1971 killed many people and went nowhere.

(5) China attacked Indian in 1961 but withdrew its forces.

(6) The Muslim Nations started the Six-Day War (1967) and the Yom Kippur War (1973) and lost.

(7) The USA invaded Vietnam (1959-1975) and lost.

(8) Russia invaded Afghanistan (1979-1989) and lost.

(9) Argentina started the Falkland Islands War (1982) and lost.

(10) Saddam Hussein started both the Iraq-Iran War (1980-1988) and the Gulf War (1990 to 1991) and lost.

(11) In 2001 the USA invaded Afghanistan and has de facto lost that war.

(12) The costliest in lives war in Western World history (as a percentage of the European population killed) were the Crusades (1096-1204 AD). The wars were started by the Christians and Christians were defeated by the Muslims.

(13) If Washington DC attacks Iran with a shooting war the US Military will definitely lose.

Starting with the Industrial Age of Warfare (in 1885) most of the significant inter Nation wars were lost by the side that started the war. On the other hand (starting in the 20th Century) <u>most of the Revolutions and wars against colonial domination have been won by the side that started them.</u> The US Military trying to re-establish White man colonialism first in Vietnam (and now in the Middle East) puts the US Military is on the very wrong side of history.

The reason for WHY the aggressor Nations in the last Century have lost the big wars they started is because the world's Generals and Admirals do NOT realize how complicated and interconnected modern warfare has become. Most of the world's Generals (and clearly the US Pentagon's Generals) do NOT comprehend the vast number of highly specialized skills (and exceptional talents) that are now an absolute necessity for long term success in modern aggressive warfare.

Modern warfare is 8X8X8 three-dimensional-chess but the USA's Generals only know how to play checkers.

The most important Rule that the US Pentagon can learn is the Rule that says "Winning Battles is NOT by itself winning wars." If you have the biggest Military Budget in the World (as the Pentagon has) then winning battles is "a given." However - the US Military won all of the battles in Vietnam and lost the war. Since 1950 the US Military has been winning all of the battles and losing all of the wars.

When future US Presidents talk to the Pentagon's Generals the future US Presidents should be asking - "Don't tell me about how you are going to win the battle - tell me about how you are going to win the war."

(411) US Military Schools Should Use the Ideas of Clausewitz as Examples of What Not to Do.

The obsolete **Clausewitz** philosophy of warfare says that the attacking Nation usually wins the war. **The Clausewitz axiom has been proven to be almost completely wrong from WW1 to now.**

During WW1 German General **Paul von Hindenburg** (1847-1934) and General **Erich Ludendorff** (1865-1937) brilliantly won the great battle of Tannenberg (August 26-30

1914) in East Prussia and permanently crippled the Russian Army. However **Hindenburg** and **Ludendorff** lost the war.

During WW2 both the Germans and Japanese started the wars (and attacked first) and both brilliantly won some of the greatest battles in modern history - but they lost their wars.

Napoleon Bonaparte (1769-1821) attacked many Nations and brilliantly won many of the most famous battles in history but **Napoleon** lost the last battle - at Waterloo. **Napoleon** died a prisoner on a remote island far from Europe and the Russian Army was in Paris.

Hannibal (247-183 BC) attacked Rome and brilliantly won the most studied and admired battle in military history - the Battle of Cannae (August 2nd 216 BC) in southern Italy. However **Hannibal** lost the war and Carthage was forever destroyed. **Hannibal** was hunted down like a dog and committed suicide before being captured.

If Washington DC makes the great mistake of a shooting-war attack against Iran (Persia) the USA would easily win all of the opening battles. However **the USA would definitely lose the shooting war and it would be a five-times bigger defeat than Vietnam.**

Persia (Iran) in 2,700 years surrendered once to **Alexander the Great** (in 334 BC) because many Persians wanted the Greeks to rule them. Persia somewhat-surrendered once to the Arabs who were spreading Islam in 633 AD because many Persians wanted to adopt Islam. The Persian "surrenders" to both **Alexander the Great** and the Muslim Religion were NOT classic military defeats for the Persians but a wanted acceptance of the invaders. In 2,700 years no other invader of core Persia lasted for more than a short time.

The warmongers in Washington DC keep saying that Persia (Iran) would be an easy and fast shooting war win for the USA. "Just drop a few bombs on the Persians and they would surrender unconditionally." Sure. In 2,700 years of warfare history the Persians were never big on surrendering. The way to "secure Iran" is to follow **Alexander the Greats** example and make the Persians want the USA to be part of their government, culture and business matrix.

The best USA Strategy for Iran is the USA doing what is needed to get inside of Iran to the greatest degree possible and then work at changing Iran from the inside out.

(412) Stone Castles are a Classic Picture of Warfare (and Stone Walls were a Major Weapons System for Thousands of Years) but the Correct Mental-Set Picture for All Modern Wars Should Be WATER.

Stones can be hammered and broken. However you can hammer water all day and nothing happens. If water meets an obstacle it flows around the obstacle. If water cannot flow around it then water will flow under the obstacle. If water cannot flow under something then water builds up and flows over the obstacle. **Water is the essence of modern Tactics.**

As the environment changes water will fit itself to the environment. Water will fill every crack and every hidden place. Water can be very hot or very cold. Water can be a liquid and a solid and a gas. Water can be 80,000 feet up in the sky and in the deepest part of the Ocean.

Water is the symbol of both successful Tactics and Strategy. You cannot cut water with a sword. You cannot strangle water to death. You cannot bomb water. If you bury water in the ground it is unhurt. If you burn water it becomes a cloud and returns as water. **Ideally your military should be like water.**

The only way to successfully defeat water is to contain it. You must be able to envelop the water. (The main reasons for why the French and US Military were both defeated in Vietnam is because neither could envelop the Viet Cong).

If you cannot envelop water and contain it then you cannot defeat water. This is why **Envelopment is the king of Tactics and the Gold Standard of Strategy.**

The ONLY way for the USA to long term win in the Middle East is to envelope the ideas of the Radical Muslim Groups in a container made out of USA better-ideas than the Radical Muslim Group's ideas. However - this is impossible without a major USA realignment of Allies in the Middle East. **The USA absolutely must be for equal freedoms and equal justice for everyone everywhere.** At the very least (the very minimum least) the USA cannot be supporting Apartheid against Muslims.

Yes - You can envelope ideas. That is how the USA (and its Allies) won the Cold War. The USA winning the Cold War was one of the greatest victories in world history. Plus a victory were almost everyone won something. However - you can ONLY envelope ideas with superior ideas that are aggressively "sold" in every way that modern technology can communicate.

Washington DC and the US Pentagon in the Middle East have to stop being whores who think they can play both sides of the street at the same time and nobody in the world will notice.

(413) In 4,000 Years of Warfare two Popular Types of Attack Have Been: (1) The Deep Thrust Attack and (2) the Even Front Expanding like Water Attack.

Deep thrust attacks can have the most dramatic and quick results but they are also the most dangerous type of attack.

The deep thrust attack is punching your way to a vital objective and/or to a position that envelops your opponent's forces. A classic example is German General **Heinz Guderian** (1888-1954) who in May of 1940 made a very deep thrust of his Panzers around the backside of the French and British forces in Belgium and successful enveloped the British and French Armies.

The advantage of a deep thrust attack is that if you attack deep inside of your opponent's territory and envelop a vital objective then that will usually create a temporary mental paralysis of your opponent.

The great disadvantage of a deep thrust attack is its narrow width. This makes flank-attacks by your opponent easy and effective. As British Field Marshal **Bernard Montgomery** (1887-1976) found out the hard way with his failed deep thrust attack (in September of 1944) to take the Bridge at Arnhem. (A movie was made of the attack "A Bridge Too Far").

The US Pentagon has been losing wars for a lifetime and they have all been "Bridges Too Far" like: Korea, Vietnam, Afghanistan, Iraq etc. They were (are) all (1) too far away from the USA both physically and mentally. (2) It is the US Military that was (is) de facto **enveloped both physically and mentally** in faraway lands fighting dead end wars for reasons that have zero value to the average American.

The US Military in the Middle East is surrounded by the Muslim world. It is the US Military that is de facto surrounded and enveloped in a Middle East. For over a 1,000 years Christian outside soldiers have lost shooting wars in the Middle East exactly like the US Military has been losing for decades.

One of the many things that the new Grand Strategy Team must reform is the Pentagon's suicidal belief that in some magical way world military history (and all of the axioms in this Book) do NOT apply to the US Military.

(414) To Counter a Thrusting Attack you need to Counter Attack very Quickly (with Everything you can spare from your Frontlines) and Attack the sides (Flanks) of the Force that is trying to Thrust Forwards.

In 1940 (during the battle for France) the British and French Generals had their thinking frozen in a strategy for re-fighting WW1. The French and British Armies were on the Belgium-German border - that would have been the exactly correct strategy for 1914.

When General **Guderian** came through the Ardennes Forest and drove his forces in back of the British and French frontlines (that were in Belgium) the British and French Generals should have broken away significant forces and attacked directly into France and into General **Guderian's** Tank force. The Allies turning some of their forces 180 degrees and attacking to their rear would have meant giving up ground on the Belgium-German frontlines. However - General **Guderian** (who was greatly overextended) would have been the one who was enveloped and **Guderian** would have been the one who was destroyed.

The great advantage that a thrusting-envelopment attack (like General **Guderian's)** has is the opposing Generals (who are already in combat on one front) are thrown into a dilemma and 90% of the time they find it extremely difficult (both mentally and in the physical maneuvering) to break away troops from the current active battlefront and send those troops to their rear area to fight another battle.

In 1940 the British and French Generals could NOT turn their forces to counter an attack on their rear exactly like 2,200 years earlier the Roman Generals could NOT turn their forces to counter an attack on their rear at the 216 BC Battle of Cannae.

(415) There was 6,000 Years of Middle East History Before the first US Soldier Arrive in the Middle East and There are Going To Be Thousands of Yeas of Middle East History After the USA is Gone.

Washington DC has to STOP being control freaks who want to control everything according to Washington DC's usually highly flawed and wrong delusions. The world is going to be around for a very long time after the USA is gone. There never has been and never will be a "settled." Washington DC has to relax more and enjoy the ride.

After machines take over the world and humans are gone probably one group of machines will not like what another group of machines are doing and there will be wars?

(416) If you Start a War your Objectives must be (1) Crystal Clear and (2) Valuable to your Nation and (3) Doable to Achieve.

The above sounds like an oxymoron but hundreds of wars have been started with no clear or doable objectives and no specific goals. **If your war has no specific objectives then how do you know when you have either won or lost?**

Many wars in history had no objective and ended only because both sides became tired of killing each other and/or realized there never was a rational purpose to the war.

Does the US Pentagon's War in Afghanistan (2001 to ?) have a doable objective? If so - exactly what is it?

For the first 150 years of the USA's history 99% of all Americans never knew Afghanistan existed. Today if you gave USA Public High School kids a blank map of the world 75% of them could NOT correctly point out Afghanistan. However - for the Washington DC Warmonger Lobby places like Afghanistan are "vital" for their continuing profits $$$ so we can be sure the wars will continue.

(417) The Three Legs of Successful Tactics are (1) the Systematic Concentration of Force onto your Opponent's most Decisive Weak Points. (2) The constant Dislocating and Harassing of your Opponent in as many Ways as Possible and in as many Areas as Possible. (3) The Destruction of Your Opponent's Strategy for Achieving his Desires.

(418) With every War and every Planned Attack the Good Grand Strategist will Position Himself so he will Win Something Regardless of what Eventually Happens.

The master Strategist and Tactician will position his assets and actions so that no matter what happens he will win something. **The sign of a bad Strategist (and bad battle plan) is when only one outcome will produce something of value.**

On Saturday April 18th 1942 General **Jimmy Doolittle** raided Tokyo (in time for the Japanese Sunday Newspapers). **Once the USA Naval forces were in Japanese waters no matter what happened (even if the force was turned back) the USA would have shaken the Japanese** and (1) forced a re-arrangement of Japanese forces away from distant battle areas and back to defending the homeland. (2) Motivated the Japanese to attack and capture Midway Island (to prevent future Doolittle raids) and thus overextend their Japanese Naval forces - exactly what eventually did happen. **No matter what happened the Doolittle raid would have achieved its goal of tricking the Japanese into attacking Midway.**

(419) In classical Military Strategy (going back to Sun Tzu) Starting Wars Requires 7 Types of Surpluses. If You do Not have the 7 Surpluses you should be Exceptionally Cautious about Starting Wars.

The Surpluses you need to win wars are:

1. Surpluses of men and troops compared to your opponent.

2. Surpluses of land and/or sea ports in your battle area (to fall back on) so you cannot be enveloped and destroyed. The best situation is to have an almost unlimited area at your back to retreat into.

3. Surpluses of money (and the ability to acquire or borrow or fabricate money) to finance a long war.

4. Surpluses of raw materials (and a large manufacturing and production industrial capacity) compared to your opponent.

5. A surplus of Allies that will help you with more than words of "we love you" - but with **combat fighting troops**, money and supplies on your battlefield frontlines.

6. A surplus of National determination and a National will-to-win above your opponent's National determination and will-to-win.

7. A surplus of support for the war from your population and especially the voters in a Democracy.

Surpluses are like links in a chain - if most of the links are strong but **one link is weak** then the entire chain is open to failure.

Planning and winning wars is <u>much more complicated</u> than planning and winning an attack. In planning the overall war you must know YOUR weakest factors because if your opponent knows your weakest points then <u>your war effort is only as strong as your weakest decisive factor.</u>

(420) For War Planning do NOT Look at Only Where You are Strong - You Must Look at Where You Are Weak Because Where You are Weak is What Will Defeat You.

In the Vietnam War the USA Military had a 1,000 to 1 advantage in both (1) gross firepower and (2) money spent on the war. However there were two surpluses where the US Military was decisively weak - Surplus #6 (a superior will-to-win at the fighting troop level). Surplus #7 (support of the voters at home).

As the Vietnam war dragged on and on the USA's conscript army's will-to-win kept deteriorating. The fighting ability of the US Troops kept going down. Also - the USA voters came to correctly view the Vietnam War as being on the moral low ground and unethical and illegal - and it was.

The home front voter-link in the 7 Surplus chain broke and the USA's Generals lost the Vietnam War despite having won every battle and despite having a 1,000 to 1 advantage in gross firepower and money.

<u>The USA's Generals in the Vietnam War had their minds frozen in WW2 assumptions</u>. In particular **the wrong assumption that any and every US War will automatically and forever have 95% of the USA's voter support.**

<u>In WW2 the US Military had the voter's support because Japan and Germany had attacked the USA and declared war on the USA.</u> Japan and Germany were clearly an aggressive highly evil force in the world. The Japanese and German Fascist forces were killing millions of people and trying to enslave as many people as possible. However - today it is the US Pentagon's Allies who are the ones trying to enslave people. Today it is US forces that are aggressively invading other Nations.

70% of the US voters are NOT buying the warmonger's and US Pentagon's excuses for the never ending wars. <u>The current left-leaning socialist movement in the USA is in large part</u>

powered by the growing hatred of the Pentagon's never ending stupid wars in stupid places for meaningless reasons.

(421) If You do Not have Large Surpluses Over your Opponent then you need to be 95% Sure that you can Win your Shooting Wars Very Fast.

(1) Never start a war if you have a lack of surpluses in vital areas.

(2) Do not become a victim of wishful thinking that somehow the war will be short.

(3) **Do NOT believe that a war will "bring everyone in your Nation together." Many times wars "tear everyone apart"** and cause a Revolution.

(4) If you cannot end a war (and you are deficient in a vital surplus area) then you have a very high probability of eventually losing the war.

The US Warmonger Lobby will try and trick Iran into attacking the USA. However the US voter support for the war will be very short lived. If the US Military cannot win a war with Iran very fast then the US Military will lose the US voters and lose the war.

(422) Nations at War should NOT be Naive about Warfare being a Two-Way Street - What Goes Around Often Comes Back Around to You.

The Nations that are involved in wars in the Middle East (like Washington DC) are simple minded to NOT expect reprisals at home in the form of terrorist attacks. **What goes around comes around.**

If the US Warmonger Lobby succeeds in tricking Iran into attacking the USA we can expect attacks inside of the USA. The Warmonger Lobby will use those attacks as the reason to expand the war in Iran and beyond.

In WW2 during the London Blitz (September 1940 to May 1941) **Hitler** thought he could force England into surrendering by bombing London. Ultimately it was England and the USA that bombed and destroyed: Hamburg, Cologne, Dresden, Berlin and several other German cities. **What went around came around.**

The US Military in the Middle East is an invading foreign Army in somebody else's Nation. A foreign invading Army that is killing Muslims - men, women and children. However **when a Muslim kills somebody in the USA it is a major shock.** Why is it a shock? Why is it a surprise? **What goes around comes around.**

(423) The Competent General is always Looking for Opportunities to Dislocate his Opponent. Chinese General Sun Tzu wrote that there are Five Classic ways to Achieve Dislocations:

(1) **Changing your front.** You quickly move (hopefully secretly) your main force from one area to another and then attack in force in the new area where your opponent is weak. You thus force your opponent to move his forces in a way that (ideally for you) is difficult and highly disruptive for your opponent.

(2) **You force (or trick) your opponent into separating his forces.** This is done by threatening two separate objectives without your opponent knowing which one you will concentrate your forces against for an attack. **This is the classic maneuver of putting your opponent on the Horns of a Dilemma.** It forces your opponent to only have half of his force at either location or only 25% of his forces if you are threatening four locations. That is when you concentrate 80% of your force at one of the locations and (because you have superiority of force) you win. Then you quickly re-disperse your forces to keep your opponent guessing as to your next target. **The above Strategy depends on you seizing the initiative and keeping the initiative.** Exactly what General McArthur did in the 1941-1945 great Asian Pacific War.

(3) **You dislocate your opponent by endangering his supply lines.** This forces your opponent to allocate troops and resources to defend his supply lines and thus weakens his frontline forces. The Russian Guerilla forces did this to the Germans in WW2.

(4) **You threaten your opponent's line of retreat.** This forces your opponent to allocate troops and resources to make sure he is not enveloped and has a retreat route that is open.

(5) **You have deep thrusting attacks into your opponent's territory**. Even if the attacks are mainly symbolic (like Colonel **Jimmy Doolittle's** raid on Tokyo in April of 1942) **deep thrusting attacks usually produce significant psychological dislocations in the mind of your opponent.** Almost always your opponent will bring back some of his frontline troops for homeland defense - thus weakening his frontline troop strength, and/or will over extend his frontlines and dilute his troop concentrations. **You want to force (or trick) your opponent into stationing as many of his troops as possible in strategically useless positions.**

The best battlefield Strategist would have the talent to use all five tactics in a fast, integrated "rolling" overall plan of attack that kept his opponent constantly mentally and physically unbalanced.

(424) The Laws of Physics have Always (and will always) Apply to Warfare. The Grand Strategist Believes the Numbers and Believes in the Laws of Physics.

In the current USA-Muslim World war (1991 to ?) most of the population and land area numbers are overwhelmingly in favor of the Muslim World. The Middle East is also on the other side of the world from the USA.

The Rule is that an Alligator cannot swallow an Elephant. If the Alligator tries to swallow an Elephant (as the US Military is trying to do in the Muslim World) the Alligator will choke to death. Exactly like the Crusaders choked to death a 1,000 years earlier.

Washington DC keeps telling itself that it is "fighting terrorism." However **it is the US Military that is in somebody else's Nation murdering the local people.** Washington DC has become a victim of its own twisted-words propaganda and is **accusing other groups of doing what it is doing.**

The US Military has been in Afghanistan since 2001. If there was an honest democratic election in Afghanistan about the US Military (1) being asked to stay or (2) being asked to leave Afghanistan - what would the Afghan people vote?

If the US Military cannot win that type of an election wherever it is fighting then the US Military will NEVER win a Guerrilla War wherever it is fighting.

The Middle East Grand Strategy solution for Washington DC is to do what is needed to become a desirable and wanted Ally and valuable friend to as many Muslim people as possible.

(425) The Great Economist Adam Smith (1723-1790) had an Excellent Idea for Peace. He said there should be Laws that all Aggressive Wars be 100% Paid For From Immediate Taxes.

Adam Smith believed that paying for wars as they happened would reduce both the number of wars and shorten the duration of wars.

The Vietnam War and the Iraq War were financed with debt that was never repaid and never will be repaid. **Every taxpayer in the USA today is paying interest on the Vietnam War debt from the 1960's.** When the Grandchildren of Americans are very old they will be paying interest on the Vietnam War loans.

A USA-Iran War (that the USA would lose) would be financed with debt. The US taxpayers would be paying interest on a war with Iran well into the next Century - and for what?

The average American has been tricked and lied to. **It is a Washington DC con game to pretend that the costs of a war stop when the shooting stops.** All of the US Troops that are somewhere in the world fighting some stupid war in a stupid place - those US Troops when they return to the USA (based on the proven records) will 35% to 40% go on some type of monthly disability payment for the remainder of their lives.

Much of the money to finance Washington DC's wars is borrowed and the US taxpayers will be paying interest on that debt far into the next Century.

(426) As We Go Deeper into the 21st Century <u>Flexibility</u> will Continue to Increase in Value for Overall Success.

Mao Tse-Tung (1893-1976) said, "Flexibility is a concrete manifestation of initiative. Flexible employment of forces is indispensable."

In the 21st Century there are 37 types of warfare. Many of them are new to the last half of the 20th Century and there are more new types in the 21st Century. The new non-shooting types of warfare are far superior to the ancient shooting warfare tools for achieving an advantageous long term peace. However - one of the many serious problems is that **the US Pentagon's cultural mindset has almost ZERO flexibility with either wanting to understand the non-shooting types of warfare and/or wanting to apply them.**

(427) The Grand Strategist puts a Maximum Effort into Making Sure that Small Wars do Not Develop into Big Wars.

Keeping small wars small is usually done with hard and constant diplomatic work. You have to be sure that all of the significant powers are in agreement that the war must NOT expand. **If a small war starts to expand it must be ended by whatever means are needed as soon as possible.**

WW1 started with what looked to be a small local war between Serbia and Austria. The Vietnam War started with one US "advisor" being killed.

The maximum effort must be put into NOT starting shooting wars because nobody knows what they might turn into.

(428) History has several Examples of Nations Building Large Effective Armies and Navies in Short Time Periods.

Calculating the outcome of a war ONLY on the relative size of other Nation's militaries at the time the comparative analysis is made (or when the war starts) is a proven major strategy flaw.

Nations can (and have) built great armies (and sometimes great navies) very quickly. This is why **the Grand Strategy equation absolutely must include everything that goes into the analysis of National Power.**

The core questions are - (1) what is the size of a Nation's (or group's) fighting ability at that current time. (2) Given all of the other factors in play what could their fighting ability be in a few years? In 10 years? In 25 years? In 50 years?

The Japanese attacked the USA on December 7th 1941 and **Adolf Hitler** declared war on the USA December 11th 1941. **At the time neither the Japanese or Hitler thought very much of the US Military** - and for good reasons. In early 1940 the US Army had 269,023 troops, the US Navy 160,997 personnel and the US Marine Corps 28,345 men for a total of 458,365. For comparison **in June of 1940 the German Military was over 5,000,000 strong. The Japanese Military in 1940 was 1,600,000.** The Japanese Army in 1940 was over 200% bigger than the US Army. However - 4 years later in 1945 the numbers were: US Army 8,266,373 - US Navy 3,319,586 - US Marine Corps 469,925 for a total of 12,055,884.

If a Nation has a large production and manufacturing economic base (as China has today) then it can become an enormous military power in a relatively short time.

In 1970 the Chinese produced 87,166 car and truck vehicles. In 2016 the Chinese produced 28,118,794 vehicles. In 2016 the USA "produced in North America" 18,200,000 vehicles. However many in that number were actually made in Mexico and Canada. The Chinese are clearly #1 in the world with what used to be the USA's Crown Jewel of auto-manufacturing.

In 2019 the Chinese were major players in expanding the cell phone 5G system worldwide. **The Chinese G5 system is the first big <u>Chinese Space-Force</u> victory for China.**

Good Grand Strategy analysis is being able to correctly analyze the world's power equations 10 years, 25 and 50 years into the future.

The US Pentagon refuses to acknowledge the fact that the USA is on the down side of its Bell Curve relative to the world. That means most of the US Pentagon's long term planning is based on false assumptions and is obsolete.

(429) Wars are like Gunfights at the OK Corral. There are No Prizes for Second Place Winners.

General **Douglas MacArthur** wrote - "In war there is no intermediate measure of success. Second best is to be defeated."

In 1914 the Germans had a large High Seas Navy fleet but it was **second best** to the British and that made the German fleet strategically less than useless.

The German Navy in both WW1 and WW2 won no prizes for being second best. **If you cannot build a world class weapons system equal to your main opponents then you are often better to NOT build anything.**

Having the second-best weapons system (or a big weapons system that is obsolete) means it is big enough to get you into trouble but NOT big enough to get you out of trouble.

The US Military having the biggest obsolete weapons systems (like the 10 US Super Air Craft Carriers) is worse than having the second-best system.

(430) All Attacks should have the Highest Degree of Secrecy Possible.

There are 37 different main types of warfare and they should all be kept as either secret (or low key) as possible. How would you keep Cultural Warfare or Economic Warfare a secret? **By NOT revealing your intent.** By NOT clearly stating your purpose. Example - the Chinese won their Economic War with the USA and **96% of the senior US Officers at the US Pentagon never knew there was a war.** The Chinese never publically said "We are going to turn Walmart - America's biggest retailer into a Chinese colony."

The war that the very best Grand Strategist wants to fight is the war that the opponent is not expecting and will thus probably not see until it is too late.

(431) One of the Core Objectives of Strategy is to Keep the Mind of Your Opponent Off-Balance and In as much Confusion as Possible.

Listed below are 14 of some of history's most used tricks for confusing your intended victims:

(1) A classic battlefield Tactic (for dampening your opponent's attacks) is to have counter-attacks whenever you are attacked. You counter-attack unless the counter-attack's ratio of risk-benefits factors are overwhelmingly NOT in your favor.

When your opponent's troops know that you always counter-attack (the Germans 80% of the time counter-attacked during WW1) the opponent's troops will be highly reluctant to advance forward at a great speed (or great distance) because they know a counter-attack is coming. The frontline troops will know that the deeper they go into enemy territory the higher the probability of them being killed (or captured) in the counter-attack. They will thus advance at a slower rate and a more causes rate.

(2) The Strategist keeps his opponent confused by appearing to be unprepared for an attack when he is actually very prepared. If an opponent attacks what he thinks will be an easy win (but is rebuffed with big loses) he will be off-balance and open to attack in other areas.

(3) You appear to have your reserve forces distant when they are close and ready for a counter attack.

(4) You appear to be at rest when you are actually working hard to prepare your positions for defense and your troops for future attacks.

(5) You appear to be low on supplies and ammunition when you are adequately provisioned.

(6) You appear to have adequate supplies when you are actually low on supplies. Doing things like having "supply dumps" full of what are empty boxes. (The British did this in early 1915 when they had almost no artillery ammunition).

(7) The Strategist keeps his opponent confused by appearing to be retreating when he is actually preparing to attack. (As Hannibal did to the Romans at the Battle of Cannae - and as the Russians did to Napoleon in 1812).

(8) You ask for peace talks when you are actually preparing to attack. (As **Adolf Hitler** did at Munich in 1938).

(9) Using "peace talks" as a time stalling device to harden your defensive positions and build up your supplies. As North Korea and Red China did in 1953.

(10) You appearing to be preparing to attack in one area when you are really preparing to attack another area. As the Allies did before D-Day 1944.

(11) At the critical time the Strategist has his agents feed his victims false information. (In 2003 the US Warmonger Lobby successfully con-gamed the US voters into believing that **Saddam Hussein** had weapons of mass destruction that he and **Osama bin Laden** were going to use against the USA).

(12) At a critical time the Strategist will confuse his opponent by telling people (that he knows will leak information to the press) that the Strategist is planning to do one thing when he is really planning to do the opposite.

(13) The tactic of issuing official statements to the media "news" that are selective half-truth disinformation to obscure what is actually going to happen.

(14) At a critical time the Strategist will confused his opponent by feeding false information to the people that he knows (or suspects) are spies for his opponent and/or have his double agents give his opponent flawed information.

(432) The Grand Strategist in a Shooting War will Cause as Little Damage as Possible to the Opponent - Only What is Needed to Win. The Grand Strategist Thinks Long Term about Using as many of his Opponent's Resources as possible and Ideally Turning his Opponent into an Ally.

Immediately after US forces took control in Iraq War 2 (2003 to ?) the USA made the stupid mistake of disbanding the Iraq Army. The US forces made the stupid mistake of disbanding and outlawing the Iraqi Ba'ath Party that had been running the Iraq infrastructure. The US forces fired most of Iraq's bureaucracy. On USA TV news there were pictures of American Tanks running over and crushing piles of Iraq Army AK-47 Rifles. Why? Because the US Pentagon believed their own lies and propaganda and sincerely believed that conflict in the Middle East had ended forever and ever.

If any of the USA's Pentagon Generals had understood basic Grand Strategy and basic world history they would have:

(1) Kept the Iraq Army (and Iraq Officers Corps) - and kept the Iraq Police Force.

(2) Put the Iraq Army (and Iraq Police Force) on a USA payroll so the Iraq Army and Police would NOT have drifted way and turned into: al Qaeda, ISIS, anti-American terrorists etc.

(3) Washington DC should have kept the Iraq Bureaucracy because they were the people who knew how to run the infrastructure and keep everything going.

(4) Washington DC should have kept the Ba'ath Party because the Ba'ath Party politically knew how to keep the peace between Sunnis and Shiites. Before the US invasion of Iraq (in 2003) the Sunnis and Shiites were living together. After the US Generals arrive and destroyed the Ba'ath Party the Sunnis and Shiites were killing each other in a never ending low grade civil war.

(5) The many Washington DC warmongers who constantly lobby for a war with Iran should have remembered that the Iraqi Ba'ath Party was anti-Iran. **The USA's invasion of Iraq destroyed Iraq's anti-Iran party and created a pro-Iran Baghdad government.**

In 2019 when this Book was published the Iraq Government was close to being a de facto Ally of Iran.

When the Germans won the Battle of France in 1940 the Germans did NOT disband the French Government administration. The Germans kept the French Police and kept part of the French Army and Navy under the command of French Field Marshal **Petain.** The German's kept the French Government Bureaucracy and did NOT create an administrative wasteland and power-vacuum like the US Military did in Iraq.

For the Germans - the French economy and Nation were valuable assets from late-1940 to mid-1944. The Germans extracted valuable services from the French Nation. On the other hand - for Washington DC Iraq has ONLY been a military burden and expense to the American tax payers.

Washington DC's warmongers have learned nothing from their decades of repeated failures. Now they want to turn Iran into the same kinds of problems while promising "everything will be wonderful."

(433) Every War has Mistakes but the Grand Strategist Never Tries to Justify his Past Mistakes by Doing the Same Things Over again (as the US Pentagon does) in an Attempt to Prove the Originally Failed Plan was Not a Mistake.

Miscalculations are part of the learning process and only become "mistakes" when you repeat the same wrong action in the same old wrong way a second time - and third time - and forth time etc etc as Washington DC does constantly.

Washington DC since the end of WW2 has been a collection of 1914-1917 WW1 Generals. The solution for every failure is to repeat the failure but do it bigger. The results are almost always a bigger failure.

In standard Washington DC swamp double-think doing the failure again (but only bigger) "proves" that the original failure was not a failure.

The good Strategist learns from his mistakes and works mightily to avoid repeating the same mistakes in the same way.

(434) If Something has Been Tried and Failed (Either in the USA or the World) then NEVER Do the Same Thing in the Same Way.

How smart do have to be to understand this axiom? Apparently smarter than most of the people in Washington DC who keep doing the same old failed things in the same old failed ways.

(435) In 1917 (during WW1) Infantry Tactics finally Started Entering into the Industrial Age of Warfare (1885 -1960).

For 3,000 years (and well into WW1) a large part of infantry Strategy and Tactics centered on massed troops charging head-on into the opponent's frontlines.

In the Spring of 1918 German General **Eric Ludendorff** (1865-1837) was the de facto commander of the German Army. In 1918 **Ludendorff** brought infantry tactics into the Industrial Age of Warfare. Up to that point WW1 infantry tactics had been mainly the 4,000+ year old tactic of massed numbers of infantry directly attacking head-on into the opponent's frontlines.

WW1 German Army Captain **Willy Rohr** (1877-1930) had been working on more effective infantry tactics. Captain **Rohr** changed the WW1 massed charges of infantry into infiltration (water) tactics of many small groups infiltrating the enemy lines and finding the points of least resistance.

The core of the new German tactics was NOT attacking strong points but flowing around the strong points and either enveloping them or bypassed them. It was a tactic of always looking for the weakest point that your troops can flow through.

Captain **Rohr's** tactics were the opposite of the **Clausewitz** rule that says you always seek out your opponent's army (and attack the strongest parts head-on) with the objective of destroying your opponent's army on the battlefield.

German Captain **Willy Rohr** said "Where does a water pipe under pressure break and explode? It does not explode at its strongest parts but breaks at its weakest point. **You enemy is only as strong as his weakest point."**

Captain **Rohr's** new tactics made finding the weakest points the infantry's objective - and then exploding outwards and occupying the surrounding area - then moving forward and doing it all over again.

Captain Rohr's platoon and battalion tactics became part of all modern infantry tactics and are used in every army up to this day.

Captain **Willy Rohr** did NOT invent anything new. What Captain **Rohr** rediscovered were the tactics that had been used for thousands of years but **had been completely replaced during the 17th to early 19th Century by European formal "set piece" Warfare.**

The Persians (Iranians) at the battle of Thermopylae (480 BC) repeatedly failed with their head-on direct attacks against the Greeks. Then the Persians used a goat path and (like water) flowed around the Greeks and enveloped them.

The American Indians in their warfare (especially the US Southern great plains Comanche) - and the Viet Minh during the Vietnam War - they all used the same tactics. Always looking for the weakest points and then attacking the weakness.

(436) The Current USA Military-Industrial-Complex is Failing to Create the Weapons Needed in Future Wars.

The current USA Military-Industrial-Complex is a weakness. It is a system that is mainly about some people getting tons of money by producing what are in many cases only updates of yesterday's weapons for yesterday's wars.

For the US Military Industrial Complex if it is NOT the 5-B's (Bullets, Bayonets, Battleships, Bombs and Blood) then in their minds it is NOT part of their world and not part of their job.

The system is retired US Military Officers selling overpriced stuff to current US Military Officers who are the buyers now - but who will retire and take over the jobs of the former Officers they were buying from. Yes - there are laws against doing the above but the war-industry is so complex and interconnected those laws can be easily sidestepped. In addition almost nobody is prosecuted.

The proof that the system is highly flawed is the fact that the USA's Military Industrial Complex (that has existed since the end of WW2 and now costs $7 Trillion Dollars every ten years) has NOT clearly won a significant shooting war since 1945.

The fact is that history is moving faster and faster and **yesterday's military experience has become a detriment.** There should be 25 year olds at Apple and Google planning weapons systems for tomorrow's wars and NOT 55 year old retired US Officers (who never won a shooting war) designing weapons for yesterday's wars.

The USA does NOT need weapons to win battles the USA desperately needs weapons to win wars.

(437) In the 21st Century's Information Age of Warfare the Internet (and social media) and "Salesmanship" are far More Important Weapons Systems than America's old B1 Stealth Bombers that cost $750 Million per airplane or the Pentagon's Super Aircraft Carriers at $5-6 Billion Each.

When the B1 Stealth Bombers were first built in 1989 they were already obsolete. There were zero valid reasons to send a manned bomber to drop an atomic bomb on something when an ICBM could have done the same job many times better and safer. The B1 was yesterday's weapon for yesterday's wars built by a US Pentagon full of yesterday's mindset Generals.

In the 21st Century many wars are Guerrilla types of wars where the opposition soldiers live and make war in a Sea of civilian popular support. **The only way to decisively win a Guerrilla war is to bring the overall population over to your side and thus dry up the Sea the Guerrillas swim in.**

Winning Guerrilla wars is a salesmanship job. Long term you can win Guerrilla wars only by sincerely helping the majority of the local population to the point where the local population wants your side to win.

(438) Many of the most Disastrous Mistakes of the Modern Age of Warfare (1885 to now) were Leaders and Generals Using what had been Yesterday's Correct Strategy to Fight What Was Then a Current War.

When **Hitler** invaded France on May 10th 1940 the bulk of the French and British forces were in Belgium on the German border facing the German forces. That should have been the French and British Strategy in 1914 to counter the German's **Schlieffen** Plan. **The**

French and British Generals of 1940 were making the classic mistake of being one war behind in their Strategy plans.

The German Forces facing the French and British on the German-Belgium border attacked and held the French and British in place. Then German General **Heinz Guderian** used his Tank force (like **Hannibal** had used his cavalry at the Battle of Cannae) and swung around in back of the main French and British forces in Belgium. The result was the main French and British Armies had German forces attacking them in the front and German forces in back of them. They were enveloped.

(439) **If what your Shooting War Opponent Wants is NOT that Valuable to You but IS highly Valuable to Your Opponent a Way to End the War is to Give Your Opponent what He Wants.**

Suppose in a Guerrilla war (or any war) the "desires" of your opponent are NOT that important to you - maybe not desirable for you but also not impossible to live with. For example, after WW2 the British were faced with many Guerrilla wars in Kenya, Palestine, Indonesia, Malaya, Cyprus, Muscat, Oman, Jordan, Brunei, Uganda, Dhofar, Aden and Egypt. The main "desire" of the Guerrilla forces was to have the British gone as the Direct Colonial Rulers. **The British successfully "attacked" the desires of their opponents by giving them what they wanted.**

The end result (in 80% of the cases) was the "rebel Nations" became members of the British Commonwealth (at least for a while). Certainly Not as profitable as direct colonial control but **the British saved something**. Unlike Washington DC the British understood that a 50% win is better than a 100% loss.

The ultimate objective in all wars is the best possible advantageous long lasting peace. The British retained a semi-colonial system (with some economic advantages for England) - and "some" is always better than None.

The British during the post WW2 colonial breakup of the Empire did NOT immediately "cave in." The British had some significant shooting conflicts in their post WW2 colonial rebellions. However - when it became clear to the British that their opponents were NOT going to stop their rebellion then **the British did NOT make the mistake that Washington DC constantly makes. The mistake of never leaving a lost war because you have the demented delusion that if you never leave then you never lost.**

The British (in their 20th Century colonial wars) understood three of Grand Strategy's most important rules **(1)You have to know when you have lost. (2) You do NOT fight wars that are not worth fighting. (3) You do NOT fight wars only to be fighting wars.**

The core desire of the Middle East Muslim "insurgents" (who are shooting at US Troops) is to have the US Military OUT of the Middle East. US Troops were NOT in the Middle East for the first 170 years of the USA's existence. **There are ZERO absolutely vital grand strategy factors for the USA in the Middle East.** There is an excellent case for saying that the USA Military should have never been in the Middle East in force. **There are excellent grand strategy reasons for US Military forces leaving the Middle East.**

(440) National Leaders should be on Guard and Cautious about War Plans coming from Generals who have in the Past (1) only Won Easy Battles in Easy Wars. (2) In the Case of All US Generals Alive NEVER Decisively Won a Significant Shooting War Anywhere.

General **Sun Tzu** wrote, "To lift an autumn leaf is not an act of great strength." What **Sun Tzu** meant is that if a General has never won a difficult war then both he and his plans are untested until proven otherwise.

The inconvenient truth is that all US Military Officers alive today only have experience in (1) losing significant shooting wars or (2) experience in muddled never ending stalemates.

What is absolutely certain is that there would be ZERO war winning experience for US Generals in a USA-Iran shooting war.

(441) Maneuvering your Opponent's Troops into Tactically and Strategically Useless Positions is Better than Killing Them.

If your opponent has his troops in positions of little strategic threat to you. Then it is best to leave them alone and have the isolated troops of your opponent be a logistics burden and psychological/morale burden for your opponent.

During most of WW1 the Allies had significant troops in Salonika, Greece. In 1918 the Allies had over 700,000 troops with over 2,600 Artillery pieces and 2,700 heavy machine guns in Salonika. The troops and equipment were strategically less than useless. **The Germans called the Allies at Solonika "our greatest prisoner of war camp."**

Over 3,000,000 US Troops have been in Afghanistan and Iraq since 2001. Many of them for 2 or 3 "tours." The real and true costs to the America people include (1) the interest on the borrowed money to pay for the wars - and (2) the fact that 30%-35% of the veterans go onto some kind of disability for the remainder of their lives. (3) The real long term costs of Iraq and Afghanistan are way over 2 Trillion dollars - and for what?

When this Book was published the US Military had already sent over 3,000,000 US Troops into Afghanistan and Iraq since 2001. The Taliban in 2019 controlled 50% of Afghanistan - and the Iraq Government was working with Iran.

Iraq War 2 and Afghanistan have been disasters for the American people. Now the Washington DC Warmonger Lobby want to do the same thing over again with a shooting war in Iran. This is insanity that is destroying the USA.

(442) When Two Opposing Armies are almost Equal in Size and Power you Must always Keep your Army in One Group and as a Whole.

If part of your army is broken away from you then your opponent might be able to concentrate the majority of his forces against one of your separated parts and (having superiority of force against your detached part) your opponent has a high probability of destroying your detached part. Then your opponent would have overall superiority of force against what remained of your army. **This is the classic "dived and conquer."**

(443) In the 21st Century the Weapon of Salesmanship is Often More Effective than Guns and Bombs. The Internet is now a Major Weapons System.

The old axiom of "He who controls the Seas controls the world" is now obsolete. It was true for 500 years but everything has a Bell Curve including "Sea Power." Today the axiom is **"He who controls Space controls the world."** Controlling Space includes controlling the world's communications that use Space satellites.

Effective Grand Strategy is the process of attacking your opponent's motivations, desires and his Strategy for achieving his desires. **In the 21st Century ideas are far more effective than bullets for attacking your opponent's desires.**

In the Cold War (1946-1991) the USA defeated the USSR with salesmanship, information and the successful spread of pro-capitalism ideas.

The Information Age of Warfare started around 1960 and the USA (without firing a shot at Russia) won the Cold War 30 years into the Information Age of Warfare.

Winning the Cold War was the USA's greatest military victory and **arguably the #1 overall greatest victory in world history.** The USSR's massive military machine (52,000 Tanks, 70,000 Armored Personnel Carriers, 44,000 Artillery etc) was <u>defeated without a single US soldier seeing combat in the USSR.</u>

For size reference - When WW2 Europe started in September of 1939 Germany had 3,531 Tanks. The USSR in 1989 had 52,000.

The USSR's Generals in 1989 had the same flaw as the USA's Generals today. The USSR's Generals were planning to fight yesterday's war with yesterday's weapons and they were completely blindsided by the flow of history.

Even though the USA's greatest victory (changing the USSR) was a victory of ideas (and NOT a shooting war) **the US Pentagon keeps doing more of what continues to be proven failures while greatly resisting doing more of what has been the USA's greatest success.**

For those in the US Pentagon (and Washington DC) who hysterically insist that only bombs and bayonets can win wars here is a question. **"WHY did the US Military lose in Vietnam?"** 2,709,918 US Troops were in the Vietnam War. During the Vietnam War the US Military dropped Five-Times the tonnage of bombs on Vietnam than the US Military dropped on Nazi Germany in WW2. In Vietnam the US Military had the biggest Chemical Warfare Attack since WW1. US Forces eventually murdered 650,000 Vietnamese women, children old people and men from all forms of attack. So - **if the 5-B's of "boots on the ground," bombs, bullets, bayonets and blood are the only way to win wars then why did the US Military lose in Vietnam?**

Why would the same US Pentagon proven failed strategy win a shooting war with Iran? The US Military lost in Vietnam for the same reasons they would also be defeated in a USA-Iran Shooting War.

(444) A 2000 Year Old Standard Tactic is to Attack at the "Junctions" Where Possible.

A "Junction" can be any number of things - a material connection of many types, a mental-psychological connection and so forth.

A classic rule of tactics is you try to attack at "junctions." For example if you were a German General in 1915 facing a frontline with French troops in one part and British troops in the other part you would (other factors being equal) attack at the junction where the French and British forces joined. If the entire frontline was the British Army you would attack at the junction between two British Divisions - and so forth.

(445) There are Only Five Types of Basic Attacks but their Combinations Gives us Every Type of Attack.

The Five Attacks are:

1 The Surprise Direct Attack.

2 The Not Surprise Direct Attack.

3 The Surprise Indirect Attack.

4 The Not Surprise Indirect Attack.

5 The Walting Attack.

There are only the five basic types of shooting war attacks but their different combinations run into the thousands. There are only five basic colors (black, white, red, yellow and blue) but those five colors give us every color imaginable. There are only five basic flavors (sweet, sour, bitter, salty, pungent) but those 5 tastes give us every food taste imaginable. There are only 5 types of attack and **the Strategist must know what percentage combination he is using with every attack he plans.**

A classic warfare example of "Waiting Attacks" are when an aggressor would surround a castle (or walled city) and wait for those inside to run out of food and surrender.

America's greatest military victory was a "Waiting Attack." It was the Cold War between the USA (and its allies) and the USSR and its allies (1946-1991). The USA waited for the economic contradictions of Marxist Communism to eventually destroy the USSR's economy.

In the 21st Century the Chinese are using a Waiting Attack against the USA. The Chinese are waiting for the massive contradictions of the USA's debt driven economy (and the US Federal Reserve's funny-money driven economy) to eventually create significant worldwide US Dollar deterioration.

The Chinese are playing the same game the USA played during the Cold War. The Chinese loan "developing" Nations in Asia and Africa money to build Sea Ports, Airports, Railroad lines etc but it is more money than the Nations can pay back. Then the Chinese take over the Sea Ports, Airports, Railroads etc.

For student's of 21st Century military strategy a must read book is the 2005 ***Confessions of an Economic Hit Man*** by **John Perkins** (and/or the 2016 update of the book). The Chinese are using the US Federal Reserve's flood of cash (and using the USA's money) to **play the same games the USA played during the Cold War that defeated the USSR.**

The Chinese are using the US FED's flood of Madoff funny-money to buy hard assets (that have military value) all over the Eurasian Continent, Africa and the World. **The Chinese are using the ancient Anaconda Strategy and building a new Chinese worldwide Empire.**

The Anaconda Strategy is the process of slowly circling your opponent with strong-points and then when you have enough strong points in place - you squeeze your opponent to death.

Today's Washington DC is squandering its Madoff funny-money on investments in: welfare, diversity, yesterday's weapons for yesterday's wars, useless luxuries, supporting less than worthless immigrants, time wasting sloth and stupid wars in stupid places to give the Military Industrial Complex more hundreds of Billions and to give the US Pentagon live-action playgrounds.

The USA in the 21st Century has ALL of the historical classic vices of past great powers in decline. Especially the vice of living off of borrowed money and ever increasing debt.

Over 4,000 years of history has proven and re-proven that Washington DC's insane Strategy of never ending wars in stupid places for stupid reasons is EXACTLY THE WRONG MILITARY STRATEGY for a great power in decline.

(446) The Strategist always Works Towards Containing and Changing his Opponent's Adversarial Subjective Views but is Very Careful about NOT Driving His Opponent Completely Insane because Insane Nations are the Most Dangerous.

The last situation a sane Leader and Nation should what is to be facing an opponent Nation that has gone insane. Nations that in the past have gone insane have done every type of insane irrational action. An insane opponent is the most dangerous type of opponent and the most difficult to second guess.

(447) It is Possible for a Weaker Opponent to Defeat a Numerically far Superior Opponent by the Strategy of making Deep Psychological Penetrations into the Other Side's Mindset.

A prime example is the Vietnam War. The US Military had a 1,000 to 1 "gross firepower" advantage over North Vietnam and a 1,000 to 1 advantage in money spent on the war. However after ten years of intensive fighting the North Vietnamese forces proved that they were never going to give up and were never going to stop fighting. With "modern TV News coverage" (part of the Information Age of Warfare) the North Vietnamese were able to use America's TV News coverage to make "deep psychological penetrations" into the mental-set of the American voters. **The Viet Cong successfully invaded the USA through the USA's Nightly TV News.** That was something that none of the US Pentagon's Generals saw coming.

It is close to a half Century since the Vietnam War but it is an open question as to if the Pentagon's Generals understand why they lost the Vietnam War? And why they would lose a shooting war with Iran.

In Democracies (like the USA) when you lose the voters you have lost the war. In the age of the Internet all of the old government BS lies will no longer work because every common sense thinking person in the world can find the truth. **The Internet has significantly altered the Rules of Warfare and the Rules of military Grand Strategy.** A profound change that the US Pentagon has both failed to understand and resists understanding.

(448) In Today's Modern Age it is Possible for the American People to Vote on Proposed Wars.

The voting machines can be easily brought out - and today all you need to vote is a US drivers license. Why not have the American people vote on a shooting war with Iran? Why not have them vote on staying in Afghanistan - or wherever the next US Military mistake might be happening.

(449) You Must Always Assume that your Opponent is Aware of the Weakest Parts of Your System and has Given Them a High Priority for Manipulation and Attack.

Washington DC (and most of the US Press) ignore the weakest parts of the USA's system. However - if your opponent is skilled (like the Chinese) they will be giving the USA's weaknesses their greatest amount of study and planning and actions.

(450) The Grand Strategist knows that the Attrition-Strategy can Work Both Ways and Guards Against Being its Victim.

The Attrition Strategy is as old as warfare. It is the Strategy of attacking an opponent until the opponent is defeated. The classic Attrition Strategy is to kill many more of your opponent's troops than your opponent is killing your troops.

The common mistake made over the millennium is that the users of the Attrition Strategy often ended up wearing themselves down and either losing the war - or ending up so weak after the war it was a de facto defeat for both sides.

During the Vietnam War (1959-1975) the Pentagon's Strategy very quickly turned into an Attrition Strategy. The US Pentagon's Strategy was to use their superior firepower and keep killing the North Vietnamese and Viet Cong forces in a war of Attrition until the Viet Cong was defeated. However - **the US Pentagon rather than succeeding in wearing down the Viet Cong succeeded in wearing down the USA's voters who eventually demanded an end to the war.**

The brilliant and tenacious North Vietnam General **Vo Nguyen Giap** (1911-2013) said about the war - "The US lost the war because it did not understand Vietnam. We knew right from the outset that we would triumph. The words 'fear' and 'surrender' never existed in our military philosophy."

General Giap successfully invaded the USA by way of the USA's TV nightly news. Nobody at the US Pentagon saw that one coming. Likewise nobody at the US Pentagon can see what will defeat the US Military in a USA-Iran shooting war.

(451) In Grand Strategy the "Direct Attack" and the "Indirect Attack" are In Computer-Terms the "1" and "0" that are the Core Parts of the Basic War Winning Strategy.

Direct attacks (with limited specific objectives and time frames) should be used to **maneuver your opponent into positions that are less advantageous for him** and more advantageous for you in the next stage of actions.

Indirect attacks are used to keep reducing your opponents assets. Your opponent becomes weaker in small degrees until it all adds up to a significant reduction in overall power.

Destroying your opponent with 100 knife wounds is just as effective as destroying your opponent in one titanic battle. However **the 100 knife wound Strategy is much safer because your risk factor in any one battle is insignificant.**

During the USA's part (1941-1945) in the Great Asian-Pacific War **-** General **Douglas MacArthur** never had a "big land battle" by WW1 and WW2 standards. **MacArthur's** biggest land battle was Okinawa (April 1st-June 22nd 1944) where 12,521 US Troops were killed in 52 days of fighting (240 a day). For comparison - at the Battle of Kurst in Russia (July 5th-August 22nd 1943) the Russians lost 254,470 men in 48 days of fighting (5,302 per day).

General **MacArthur** won the war in the Pacific with a 100 knife wounds and very low US Troop deaths - a brilliant achievement in the world history of warfare.

One of the best things you can say about US Generals is they knew how to keep US Troop death rates low.

(452) Washington DC in the last 60 Years has Continually made the Mistake of going into Shooting War Conflicts that Washington DC should have Never Entered Into.

The USA today is on the downside of its Bell Curve. **One of the many classic characteristics of downside Nation's Leaders is they view any change anywhere in the world as something that by definition is bad** and must be militarily challenged.

General **Sun Tzu**'s advice was.

1. "All roads should not be taken. There are roads you should not follow because they are not worth following." **Sun Tzu** was saying the Leader has to do cost-benefits analysis with every war.

2. "Every opponent should not be challenged only to be challenged. There are forces that should not be confronted because they are not worth the risk of challenging."

3. "Every position of an opponent should not be attacked only because it is there. There are fortified areas that should not be attacked. Because something can be attacked is not a good reason to attack."

4. "Every change should not be challenged only because it is a change and different from the past. There are changes that should not be contested. There are changes that are good. There are changes that mean nothing. Knowing the difference is the Art of War."

5. "Every order should not be obeyed. Sometimes there are orders from your superior that are clearly wrong and should not be accepted."

"It is talent in the Art of War to recognize such situations and act correctly."

Sun Tzu's advice was to NOT automatically challenge every change in the world only because it is change. The USA's new Grand Strategy Team must analyze every change and see if it is strategically importance to the future of the USA's vital interests - or just another world equation power evolution that in the big picture means little or nothing for the USA.

On October 7th 2001 US Forces invaded Afghanistan on the false and phony charge that the Taliban had helped **Osama bin Laden** in the 9/11 attack. **There never was (or has been) an proof of that charge.**

At one point there were over 100,000 US Troops in Afghanistan. **Osama** left Afghanistan in 2005 and went to Pakistan but the US Troops stayed in Afghanistan. Why? Because **for the US Pentagon fighting a war anywhere becomes the justification for continuing to fight a war anywhere.**

For the first 160 years of USA history 98% of all Americans never knew there was an Afghanistan.

Many years after the 2001 US invasion of Afghanistan (and many years after the death of **Osama**) US Troops continued to be in Afghanistan. Why? Does the US Pentagon expect **Osama** to come to life? After decades of US Troops mucking around in Afghanistan (searching for a dead **Osama**) the Taliban continues to control 50% of Afghanistan. **Afghanistan was (is) clearly a stupid dead-end road that the US Military should have 100% ended when Osama left Afghanistan in 2005.**

Since 2005 Afghanistan has been a very expensive live-action playground for the Pentagon - and nothing more.

(453) **To Win Wars you need (1) A complete Understanding of your Opponent's Strategy, Desires and Objectives. (2) Doable ways to Defeat your Opponent's Strategy and Block his Desires. (3) A Highly-Defined Doable (and valuable-to-you) Written List of very Specific Objectives. (4) A Doable Way to Achieving YOUR Specific Objectives and to Defeat your Opponent's Strategy.**

The US Pentagon's organization is obsolete and its Grand Strategy analytical abilities are clearly NOT competent. The proof is the inability of the US Pentagon (since 1945) to clearly win wars. You can call yourself the "greatest golfer in the world" and have golf clubs made out of gold and diamonds (and a custom Rolls Royce for your Golf Cart) but if you never get your Golf score below 120 what are you?

(454) The Grand Strategist is Never a Slave to his Own Planning and/or to his Planning Timetable. If a Plan is NOT Working (and New Situations Develop) the Strategist Will Change His Plans As Needed.

The Grand Strategist always has one eye open for unexpected opportunities. If a valuable opportunity presents itself but was not part of "the Original Plan" **the Grand Strategist will nevertheless exploit the opportunity with great speed.**

One of **Napoleon Bonaparte's** (1769-1821) greatest talents was being able to quickly exploit opportunities. That talent combined with the superior speed of the French troops gave **Napoleon** the initiative both going to and on the battlefields. When the mistakes of his opponents presented themselves to **Napoleon** he quickly exploited those mistakes.

Napoleon (like almost all of the great military commanders of history) left this world with almost no written ideas about detailed battle plans and rules. Perhaps if **Napoleon** had written down a summary of his battle-strategy it might have said - *(1) Seize the initiative with speedy bold maneuvering that will (2) throw your opponent off balance and (3) cause him to make mistakes that (4) you will then immediately and quickly exploit to the maximum degree.*

CAVEAT. Always consider the percentage risk of the supposed mistake "opportunity" as being a possible trap. An unexpected opportunity should have the characteristics of a big advantage for little risk. **"Opportunities" that offer only a small gain for a large risk should be either avoided or approached only with extreme caution.**

For the Japanese - their 1941 attack plan against Pearl Harbor looked like an easy opportunity and a big win. However - it was a big trap. A trap that eventually destroyed Japan's extensive Empire.

The USA greatly expanding its attack against Iraq in 2003. It looked like a fast and easy win. However - it too was a big trap.

There are warmongers in Washington DC calling for a USA attack against Iran. **A Iran-USA war would (like Iraq) also be a trap.**

A USA-Iran War would greatly expand the influence and controls of both China and Russia in the Middle East - and all of Asia. The mistake of the USA attacking Iran (with a shooting war) would be **a giant step towards one of China's main Grand Strategy objectives. The objective of replacing the US dollar in Asia with China's currency.**

(455) Twenty-Five Centuries ago Sun Tzu Wrote that in Shooting Warfare Five Things will be attacked:

(1) Your opponent's troops.

(2) Your opponent's supply sources.

(3) Your opponent's transportation systems for moving supplies from their sources to the opponent's frontline troops.

(4) The finances of your opponent.

(5) The organization of your opponent's government.

Ideally you would be doing all 5 attacks simultaneously.

Twenty-five centuries after **Sun Tzu** we have to add:

(6) Your opponent's communications and military satellites. This is the Information Age of Warfare and without the information from all of the "spies in the sky" a military is shooting in the dark.

(7) Your opponent's electrical power plants. **Without electricity a modern society cannot function.** Most "money" in advanced Nations is electronic money and without electrical power there is no money. **There is no Internet. There are no phone services.** No TV no Wall Street Gambling Casino no government records no nothing.

(8) Your opponent's oil refineries, oil pipelines, oil shipping ports and main oil storage and distribution points. Without oil a modern society grinds to a stop.

(9) Your opponent's airports. Without airports (for people and vital products) a modern society cannot function.

(10) Your opponent's financial electronic centers like Wall Street and the locations that backup places like Wall Street.

(11) All of your opponent's Telephone and cable relay centers, internet centers and main broadcast centers.

(12 Railroads are of diminished importance in the USA but remain vital in other Nations like Russia and China. The railroad centers and main bridges would be attacked.

(13) Your opponent's ports and destroy the container loading and unloading cranes. Without the cranes the port is useless for freight.

(14) If it is a limited war without Satellite damage you must have a "Media Based Warfare Plan." **You must use every means of communications possible to sell the idea of making peace** to as many of your opponent's citizens and troop's as you can possible reach.

(15) Most of all you have to alter your opponent's desires and objectives for the war. Your must change your opponent's Strategy for achieving his desires. **If you block any chance of your opponent achieving his desires then you have defeated him in his mind.** Remember the core rule - **ultimate victory is always in exactly the same place - it is always in the mind of your opponent.**

Look at the above list of everything the USA's average person would lose in a full scale war with a major power like China or Russia. Those alive would be living back in the 1880's. There is NOTHING in the world worth the mass misery and hardship.

(456) The Shorter the War the Fewer Things There Are to Go Wrong with Your Original Plan.

Remember Murphy's Law - "If given enough time everything that can go wrong WILL go wrong." If you wanted to drive the same car 30,000 miles a year (for 30 years) everything that could go wrong on that car would go wrong.

The Never Ending wars that the US Military-Industrial-Complex loves have a very high probability of having everything that can go wrong eventually going wrong. Also - **in long wars your opponent will learn from you and will create both better soldiers and higher skilled opposing Generals and Leaders.**

The US Pentagon keeps mucking around in the Middle East with never ending wars decade after decade. This process has a high potential to eventually produce **a new Saladin that could decisively damage the USA in both the Middle East and the world.**

The Vietnam War lasted 15 years and was a disaster for America. The USA's scars from Vietnam have NOT to this day healed. **With shooting wars longer is NEVER better.**

Washington DC must have a new Grand Strategy Team with a new rulebook. With future proposed wars if Washington DC cannot clearly see (1) a doable road map to a (2) lasting sustainable advantageous peace (for the USA's people) in (3) under a year of shooting war fighting then (4) the USA should NOT go into that shooting war.

(457) The "War" that the USA Should be Fighting is the War Against the USA's Exploding Debt. The Victory the USA needs is the Victory of having the USA Creating Products and Services in an Equal Amount to what the USA Consumes.

The idea that it was WW1 and WW2 that made America great is dangerously FALSE. The facts are the opposite. The USA became the world's greatest power in the first half of the last Century because the USA was the VERY LAST great power to enter into WW1 and WW2. **"Making America Great Again" means returning the USA to being the very last Nation to enter into a shooting war.**

(458) The Big Winners in the Future's Wars are Going to Be those Nations NOT in Any Shooting Wars.

Modern wars for the USA are so massively expensive there is nothing that justifies the enormous war expenses for the average USA citizen.

From the end of WW2 to now Washington DC's shooting wars have been economic losses for the average American and for generations of US citizens to come.

USA Citizens can go down the list.

1. Suppose the USA never resisted in 1950 and today all of Korea was communist. Would that have changed life in the USA? No.

2. Vietnam eventually did become 100% communist. What changed for the USA? Nothing. A real "change" would have been the 58,220 Americans murdered in Vietnam to NOT have been murdered.

3. Suppose **Saddam Hussein** had never been killed. There never would have been a war in Syria or ISIS or the Civil War in Iraq. **Saddam** was anti-Iran and had a big Army that would (and did) fight Iran. The current Iraq Government is a de facto ally of Iran. The 4,424 US Troops murdered in Iraq would be alive.

4. Suppose US Troops never went into Afghanistan. What would change for average Americans? Nothing. The 2,372 US Troops murdered in Afghanistan would be alive.

(459) There Usually is a "First Attack Advantage." However - it Has a Limited Time Value.

If the major Grand Strategy factors are NOT on the side of the attacker (and the shooting War becomes a long war) then the original attacker has no advantage.

The power to pick the exact time to attack and start a war (with usually the advantage of surprise) gives the aggressor a substantial **short term advantage**.

Opening a war with an unexpected attack is like punching a person in the face when that person had no idea there was a fight coming. The person punched is stunned, mentally disorganized, emotionally confused and stumbling. If you immediately follow the surprise punch with many more punches you have the initiative and the other person is in serious trouble. However your first attack advantage has limited time value. **If the fundamentals are NOT on your side then the Nation attacked will recover and based on past world history of the last 120 years will in the long run win the war.**

Chapter 18

Religious Wars are the Hardest to Stop

Religious Wars (out of the 37 main types of wars) are the most difficult to stop. When two or more sides sincerely believe that their subjective views come directly from God then achieving peace becomes an especially challenging problem.

Unfortunately many of history's religious based wars had to "burn themselves out." They stopped only when both sides became tired of killing the other side in the name of God.

If at all possible Religious Wars must be avoid. There are zero absolutely vital interests for the USA in the Middle East. It is insane and criminal for the US Military - the Warmonger Lobby - and Washington DC to constantly be dragging the US population into the middle of the Muslim world's Religious Wars.

Many political movements in history have become semi-religions for their followers. The wars they started were some of the most difficult wars to bring to an end. Nazism was a semi-religion and when Red Army infantry in April of 1945 were inside of Germany's Capital Building (the Reichstag in the heart of Berlin) the Nazi soldiers kept fighting because for them it was a religious calling.

(460) Wars that are Motivated by Religion (or a Semi-Religion) Must be Avoided at All Costs.

Religion is a subjective belief system (see Chapter 10) and it can become impossible to reason with a belief system that is subjective. **How do you logically argue for or against a subjective belief that is not objectively provable one way or the other?** It is extremely difficult.

Part of the USA's persuasive offensive must be the use of the Internet (plus ALL forms of communicating ideas) to highlight the many similarities between Christianity and Islam. There must be a constant process of advocating the shared objectives of all sane and rational people worldwide.

(461) Ideologies that are Semi-Religions have Most of the Characteristics of Religious Wars.

Many ideologies are semi-religions to their followers. They include: Nazism, Stalinist Communism, Chinese Red Guard Communism, Zionism, Fascism, Fanatical Socialism, Totalitarianism and so forth. **All of the above are the same in that their followers completely believe that any type of crime is justified because they were creating a better world.**

The followers of the semi-religions (much like the extreme fanatical followers of a standard religions) believe that any type of crime today is justified to atone for the crimes that were committed against them (or against somebody) in past times.

The "Hall Mark" of all fanatical semi-religious groups (that were or are in semi-Religious Wars) is that they demand absolute power, absolute obedience and absolute control of everything.

Fanatics are NOT "reasonable people" with "common sense" or "common decencies." The **Fanatics will knowingly Lie about everything and knowingly distort the facts about everything because they sincerely believe it is all justified for "a higher loyalty to a greater good."**

It must be crystal clear in your mind that you are dealing with semi-religious fanatics who will tell you any Lie (and distort any fact) to achieve their ends while **sincerely believing that they are morally and ethically superior to you.**

To fight Fanatical Groups you have to keep correctly defining them and their true agendas. You have to correctly define their true anti-freedom intentions. You must keep demanding that the Fanatics prove that their anti-common sense and unworkable delusions are NOT the moral low ground and will not in the long-run result in National suicide.

Here is the ultimate test. If a group tells you that in order to create a better world you first have to murder many other people that is the time to become highly suspicious.

(462) When Faced with a Religion-Motivated Shooting Conflict the Number One Rule is to Do Everything Possible to STAY OUT of that War.

The best Strategy for the USA (with the Middle East's Religious Wars) is have small (and low profile) specialized military units (plus air power, intelligence and equipment help) for the Muslim side that the US Grand Strategy Team believes (will in the long run) be the most beneficial to the USA and the world.

Unfortunately most religious wars are only stopped with time. Sometimes it takes decades - and sometimes it takes centuries. **When you are dealing with religious wars do not be surprised if you die of old age before a peaceful equilibrium is reached.**

Christians and Muslims have been fighting on-again and off-again religious wars for over 1,000 years. **The Christians eventually win the wars that are in Europe and the Muslims eventually win the wars that are in the Middle East.** That is a historical reality that the USA's new Grand Strategy Planning Team must NOT ignore.

The Thirty Years War (1618-1648) was a religious war between Catholics and Protestants. As a percentage of lives lost the Thirty Years War was one of the worst wars in the last 500 years.

Large sized wars between Catholics and Protestants raged in Europe for 1524 to 1648. Religious conflicts continued into the 20th Century between Protestants and Catholics in Ireland, between Nazis and Jews, between Christians and Muslims in Bosnia and so forth.

The Religious War Crusades (1095-1291) were the worst military disasters in European history until the August 1914 to May 1945 wars (the 20th Century's 30 years war).

Historically **the worst place in the world for Christian Armies to be fighting is in the Muslim Middle East.**

In the year 1100 AD Paris was the Capital of France and a major European City with 50,000 people. The Crusade Wars in the Middle East killed over a Million (1,000,000) Christian soldiers - 20 times the then population of Paris. In today's numbers that would 160,000,000 dead Christian troops - and the Christians lost. Hopefully that would be a sobering number to the grand strategy morons in Washington DC Warmonger Lobby. However - it is very difficult to fanatics anything.

The US Military has been fighting in the Middle East since 1991 and at this point the US Pentagon has less-than-nothing of value to show for its (1) many Middle East wars (2) thousands of dead US Soldiers and (3) two Trillion dollars in real long term costs.

(463) The Muslim Middle East has Reset its Bell Curve. Despite the Current Religious Wars the Muslim Middle East is on the Upside of a New Bell Curve that will Once Again be World History Significant.

From the viewpoint of the USA (and Europe) the Middle East appears to be in chaos. However - the Muslim Middle East has reset its Bell Curve and will be on the upside of its Bell Curve for most of this Century.

The wars we see in the Middle East today are in large part growing pains. When Western Europe was in the process of colonizing and dominating the world (1492 to the late 19th Century) there were many wars in Europe.

The fall of the Roman Empire (in the West) happened in 476 AD. The Western Roman Empire disintegrated (never to be put back together again) because of too many stupid wars for stupid reasons. History has proven that when a Nation is on the downside of its Bell Curve wars only increase the speed of decent even if that Nation wins the wars.

The Roman Leaders (during the last 170 years of the Empire) made the same mistake that many Leaders in Washington DC are making today. The last Roman Leaders believed that it was Wars that made Roman great and it would be wars that would **make it great again.**

Today many Leaders in Washington DC believe that it was WW1 and WW2 that made the USA great. However - **it was the USA being the very last great power into WW1 and WW2 that made the USA the biggest winner in the world.**

Rome's fall was followed by the Dark Ages. Western Europe had crashed and burned. The low point was about the year 1,000 AD. **About the year 1200 Western Europe reset its Bell Curve.** First with the start of the Renascence in Florence Italy and then with Venice recreating Roman's pan-Mediterranean wide trade. By the 1890's European Nations (and Nations with European-cultures like the USA, Canada, Australia etc) controlled and/or dominated most of the world.

Early 1914 (before the start of WW1) was the high point in Europe's Bell Curve that had been reset upwards 600 years earlier.

Between the reset of Western Europe's Bell Curve (starting with the Renascence in the 13th Century) and European culture controlling most of the world by the 1890's - there were many wars in Europe (and a very large Civil War in the USA). Many of the wars in Europe were "growing pains."

Many of the current wars in the Middle East are growing pains for a Muslim World that is on the upside of a reset Bell Curve.

The absolutely vital and decisive Grand Strategy factor is that the USA is on the downside of its Bell Curve and the Muslim Middle East is on the upside of its newly reset Bell Curve. 4000 years of warfare history have proven that **Nations on the downside of their Bell Curves seldom win shooting wars against areas and Nations on the upside of their Bell Curves.**

The Holy Koran says, "Fight in the cause of Allah those who fight you." 2:190. And "Permission to take up arms is hereby given to those who are attacked because they have been oppressed - Those who have been unjustly driven from their homes only because they said, Our Lord is Allah." 22:39.

What the above scriptures from the Holy Koran mean for the USA Military in the Middle East is that **the more US Troops are sent into the Middle East the more Muslims will keep raise up in ever greater numbers to fight the US Military.**

Chapter 19

The Moral Bankruptcy of "Preventive Wars"

Preventive wars are by definition illegally and morally wrong. The Warmonger Lobby military idiots (criminals) insist that a shooting attack against another Nation is the one and only type of "preventive war" possible. That is a LIE. An ethical preventive war is using all of the many non-shooting tools available to remove or mitigate the areas of disagreement between Nations and peoples. A shooting war is 99% of the time the very last tool the ethical and competent Strategist would use.

(464) **A Shooting-war "Preventive Attack" should Not be Used unless one side is Absolutely Sure the Other Side is going to Attack and you are Absolutely Sure that You will Achieve a major Advantage and Save Lives by Attacking First.**

A preventive attack against an opponent that was never going to attack you destroys your moral high ground and will correctly be viewed by many in the world as a war crime.

Adolf Hitler murdered almost 6 million Jews - and starved to death 3 million Russian prisoners of war. **Adolf Hitler** (and the thousands of people who helped **Hitler** murder the Jews) believed that the murders were justified because they were killing their future enemies now and **preventing a war in the future.**

For **Adolf Hitler** (and his thousands of helpers) the Jewish Holocaust was preventive law enforcement. Killing people to supposedly prevent crimes is by definition both morally wrong (legally a crime) and completely insane.

The 2003 invasion of Iraq by US Forces was "sold" as a "preventive war" to stop the "weapons of mass destruction" that never existed. The USA must never again be con gamed into making the same moral (and criminal) mistake.

(465) **The National Leaders who insist on Dominating and Crushing Others are usually Suffering from Psychiatric Security Addiction Disorder (SAD).**

The National Leaders who insist on crushing other people (and destroying Nations) use the excuse of "preventive war." That group of "Leaders" have a mental sickness called Security Addiction Disorder (SAD). There have always been (and are now) such National leaders.

The National leaders with Security Addiction Disorder sincerely believe that the only way to achieve security is through the complete domination and mastery of others.

The Leaders and Nations who have Security Addiction Disorders are blind to their own aggressions and their own criminal acts.

Leaders and Nations (who have Security Addiction Disorder) when they appear before the TV cameras are friendly (and often keep telling people they love them) and appear to be sane and normal. However - **on the inside everyone is their enemy.** They are masters at pretending friendship but **they are constantly manipulating everyone because they are incapable of trusting anyone.**

(466) The Strategy of Using "Preventive Wars" is Highly Dangerous. All Shooting Wars are Highly Unpredictable and a War Once Started can Turn Against You.

During 2002 and 2003 the Warmonger Lobby kept telling the US Public that **Saddam Hussein** was planning to attack the USA and the USA had to attack first. The US Pentagon's invasion of Iraq in 2003 was supposed to be a fast, short, economical and easy for the US Military. **NONE of what the Post WW2 US Pentagon's usually incompetent war planners said would happen actually happened. It was the exact opposite - a long, muddled and highly expensive de facto US Military defeat that created more problems than it solved.**

In 2019 when this book was published US Troops were continuing their fighting in Iraq and the Iraq War's complete long term real costs for the USA's taxpayers had gone far over a Trillion Dollars.

Only a military morons start big shooting wars believing they know how the war will end. When WW2 in Europe started on September 1st 1939 no person in the world was smart enough to foresee the world's situation when WW2 in Europe ended in May of 1945. Not **Albert Einstein** not **Winston Churchill** or any British Generals or Admirals. Not **Franklin Roosevelt** or any US Generals or Admirals. Not **Joseph Stalin** or **Adolf Hitler** or any German Generals. Not General **Tojo** or any of the Japanese Generals and Admirals. **In 1939 nobody in the world came close to knowing what the world actually looked like at the end of the war.**

Napoleon never knew his wars would end with the Russian Army in Paris and him dying in jail. **Julius Caesar** never knew that his wars would end with him being stabbed to death in the Roman Senate. **Adolf Hitler** never knew that his war would end with him shooting himself in the head. This is why **you must at all costs avoid starting shooting wars because nobody can be smart enough to know how a big shooting war will end.**

The USA's Generals/Admirals and CIA/NSA etc who sincerely believe they know how a shooting war with Iran would eventually end have already been proven to be grand strategy incompetents who get the outcomes wrong.

In the last part of the 20th Century (and early part of the 21st Century) **the USA's biggest military disasters were "preventive wars."** The USA-Vietnam War (1959-1975). The expansion of the USA-Iraq War in 2003. The USA-Afghanistan War (2002- ?). They were all con games sold to the US voters as "preventive wars" and they have all been costly disasters for both the average American and the world.

Based on Washington DC's proven disastrous record of war crimes during the last 60 years the last thing the world needs is Washington DC starting a shooting "preventive war" with Iran and/or North Korea.

(467) The American People must Never Again let Washington DC Con-Game Them into Another "Preventive War."

A preventive war should only be started if you are 100% sure your opponent is going to attack you and the attack will be very soon.

The Nazis and **Hitler** murdered almost 6 million Jews using the excuse that they **might** have done something wrong in the future. Murdering large numbers of people because they **might do something wrong** in the future puts you shoulder to shoulder with **Adolf Hitler**.

(468) Preventive-Wars are Illegal by International Law. Illegal by the Laws of most Nations. Illegal by the United Nation's Charter and by both Christian and Islamic Ethics.

There are dozens of motivations for starting wars. The most insane reason for starting wars has been "preventive wars**." The idea that it is better to kill people now rather than run the risk of having to kill them later.**

(469) The Grand Strategist Never Believes that when One Undesirable Regime is Destroyed that Automatically Guarantees that the Destroyed Regime will be Followed by a More Desirable Regime than the Original.

One of the dozens of perpetual mistakes that Strategy incompetents make is the belief that if you destroy an "offending government" then the destroyed government will **automatically and magically be replaced by a new regime that will be to your 100% liking.**

During the 2011 war against **Muammar Gaddafi** (1942-2011) the US Military succeeded in killing **Gaddafi** and destroying his government. However- the undesirable regime of

Gaddafi was replaced by a collection of War Lords. Libya has became a haven for Radical Muslim Groups that did NOT a power base in Lybia when **Gaddafi** was alive.

Germany's **Kaiser Wilhelm** was replaced with a weak government that in 1933 caused the election of **Adolf Hitler** who was much worse than the Kaiser.

In Russia the weak and naive **Tsar Nicholas II** was replaced by the much worse **Joseph Stalin.**

On January 1st 1959 Cuba's (unpopular in Washington DC) Dictator **Batista** was overthrown and **Fidel Castro** became the new Dictator. **Castro** took Cuba from being South America's richest Nation to being its poorest Nation. 99% of the Cuban population was made "equal" - they are now equally poor.

If the American people are once again con-gamed into starting another "preventive war" (this time against Iran). What eventually replaces the current Iranian Government will in all probability be worse than the current Government.

Never assume that if you destroy one Government it will automatically be followed by something better. Many times in history an overthrown government has been followed by something far worse.

(470) The Best Grand Strategist is <u>Constantly</u> Fighting "Preventive Wars" - but NOT with the Shooting War Tools. The Correct Preventive Warfare Uses the Many NON-Shooting Warfare Tools Applied Wherever and However Possible.

The US Warmonger Lobby constantly says **"The USA has to be engaged with the world."** That is absolutely 100% correct. The USA must be engaged with the world to the maximum degree possible with ALL of the non-shooting tools of warfare.

The correct Grand Strategy is to (1) analyze the exact reasons that could motivate your opponent towards a shooting war and then (2) you apply every **non-shooting type of warfare** possible (and to the greatest extent possible) to alter and change your opponent's motivations.

The correct "preventive wars" use as many of the non-shooting war tools as possible to reduce crisis pressures from building to the explosion point and cause a shooting war.

(471) The Correct "Preventive War" the USA Should Have with Iran is a War Using the many Non-Shooting Warfare Tools.

One of the US Military's many big mistakes in the Vietnam War was giving only lip-service to the non-shooting tools of warfare. **The US Military was 97% fanatically devoted to only**

the shooting-war tools. When this book was published that great US Military flaw had NOT been corrected.

The US Military aggressively waged a vicious war in Vietnam for 1959 to 1973. **2,709,918 US Troops fought in the Vietnam War.** The US Military killed over 650,000 Vietnamese women, children, old people and men. The US Military's shooting war tools were (1) a complete failure for winning the war, and (2) in the opinion of most people in the world much of what the US Military did in Vietnam was a war crime.

Henry Kissinger's 1973 Vietnam Peace Accord was a disguised USA surrender - because the US Military had de facto been defeated. North Vietnam (and President **Nixon** and Secretary of State **Kissinger**) knew that when the US Army evacuated then after a few years time North Vietnam would occupy all of South Vietnam.

When this book was published there was a major lobbying effort in Washington DC for the USA to attack Iran as a "preventive war" and to "end terrorism." The Iraq War 2 in 2003 was sold as the war that was going to "end terrorism." The Afghanistan War was sold as the war to "end terrorism."

The decisive fact is that **here is no way for the US Military to envelop Iran exactly like there was no way for US Forces to envelop North Vietnam.** Once a USA-Iran shooting war started the US Military would have no way to end the war and it would be a Vietnam type defeat all over again.

The Basic Facts for a USA-Iran shooting "preventive war" are: Fact #1 - Iran will never surrender to the USA. Fact #2 - if the USA put 600,000 combat troops into Iran that would be **less than one US soldier per square mile of territory.** Fact #3 - if the USA put 600,000 troops into Iran that would be **less than one US soldier for every 138 Iranians.** Fact #4 - in a USA-Iran war both China and Russia would support and help Iran with massive amounts of arms and ammunition. Both China and Russia have railroad lines and car/truck roads going into Iran. Fact # 5 - the only way to cut off Iran's supply or arms would be for the US Military to start bombing Chinese and Russian supply trains. **A USA-Iran war would be another guaranteed major shooting war defeat for the American People.**

The correct Strategy with Iran is for the USA to try and become closer to Iran so the many non-shooting types of warfare could be used. The USA's objective must be the opening up of Iranian society to more Western exposure. **There are many "moderates" and "liberals" in Iran but they are not the majority at this time.** However - the Iranian liberals could be made into the majority if they were given helped from the inside.

The more "modern" Iran becomes (and the more Western the Iranian people become) the more they will elect moderate Leaders.

There are powerful pro-Western and pro-American forces in Iran. Give the Iranians more freedom and they will want additional freedoms. They would vote into power a pro-Western government and then the USA's objective of a more co-operative and safer Iran is achieved without a shooting war.

(472) Nations that Threaten "Preventive Wars" Because their Leaders are Security Addicted Paranoiacs Give other Nations Logical and Rational Reasons for Wanting to Build Their Own Atomic Bombs.

Washington DC always threatening to attack other Nations gives those Nations **rational and logical reasons for wanting to build atomic bombs as protection.** The US Military is always in somebody else's Nation bombing and murdering the local populations. Many of Washington DC's politicians and the Warmonger Lobby become more irrational with each passing year. **Washington DC's insanity gives all Nations everywhere a rational and logical reason for wanting to build atomic bombs as protection against Washington DC's growing insanity.**

Chapter 20

The Deadly Danger of Overreach

The most often made mistake in warfare from Platoon Tactics to Grand Strategy is Overreach. Grand Strategy overreach is often decisive for defeat in a war. Demanding "total victory" and the "total subjugation" of your opponent is overreach.

There are many types of warfare and Overreach can take many forms and be deadly in numerous ways. Nowhere in the Art of War and Peace is the talent of the Grand Strategist more valuable than in having the correct balance of avoiding Overreach while maneuvering to your maximum capacity.

You must capture and keep the initiative in warfare but you must simultaneously be protecting yourself from Overreach.

Most Generals and Admirals from Christian cultures have a mindset weakness for Overreach. The Christian Bible is about one great battle (Armageddon) deciding everything and that has had a great influence on Western military thinking for almost 2,000 years. It is one of the main reasons for General Carl von Clausewitz's war philosophy of total victory being popular in Christian Nations.

Judging what is Overreach takes exceptional military talent and often exceptional mental courage. In 1961 there were zero USA Generals and Admirals who had any idea whatsoever that the Vietnam War for the USA was going to be Deadly Overreach and a major military defeat.

(473) **The Grand Strategist Understands that Overreach (the Over Extension of Forces and Resources) is the single Most Often made Mistake from Grand Strategy to Platoon Tactics. Overreach is the Classic #1 Destroyer of Armies and Empires.**

Deciding what is overreach (and is not overreach) is one of the areas where Strategy becomes an Art. On one hand seizing the opportunities (Carpe Diem) is a core tactic for both short term and long term victories. **Seizing and keeping the initiative and forcing**

your opponent to always be reacting to your actions is vital in warfare, politics and business. On the other hand overreach has lost more battles and wars than any other single mistake. Overreach comes in many forms and it has been a major destroyer of great Nations.

In the 21st Century the USA's most dangerous overreach (and its greatest weakness) is its financial system. The USA is the most in-debt (in gross money amounts) of any government in the history of the world. USA total debt is increasing by over a Trillion dollars every year. **No Nation in history has been both a great debtor Nation and a great military power forever** and the USA is NOT in the long run going to be an exception.

(474) Permitting your Decisive Advantages to Atrophy but NOT Reducing Your Worldly Ambitions will Eventually Produce Overreach and Defeat.

During WW2 the great manufacturing capacity of the USA overwhelmed both Germany and Japan. Today most of the USA's WW2 heavy industry manufacturing (that decisively helped our Allies win WW2) is now somewhere else in the world (mainly China and Japan). However - **today's Washington DC desire to control the world continues to be too high.**

The above loss of US manufacturing is especially important because the next Age of Warfare (the Artificial Intelligence Age of Warfare) is going to be manufacturing intensive.

(475) There is a Thousand Year Old Japanese rule of Tactics and Strategy that says "The Long Shot Arrow has Little Force when it Lands."

Traveling a long distance to a war uses up much time and emotional energy just arriving at the battlefield. The above rule means that if your battle ground is too far away (and the area you want to control is too vast) then you will most likely be defeated.

Washington DC's Warmongers have dreams of controlling large parts of the Middle East. However the Middle East from Libya to Pakistan and from Muslim Kazakhstan to Muslim Somalia is almost **three times the size of the Continental USA in land mass** and has a much bigger population than the USA. It is also an average of 9,000 miles away from the center of the USA. The USA could never begin to envelop and/or control an area that gigantic.

The core of military battle Strategy is the concentration of decisive power onto a specific objective of value to you. The point of the above Muslim Middle East land area numbers is that if you put a cup of red dye into a bucket of water you get red water. If you put a truckload of red dye into the ocean you get nothing.

The probability of the US Military (using shooting war tools) to alter the Middle East so it fits Washington DC's delusions and subjective flawed fantasizes - that probability is less than ZERO.

(476) The USA Mainland is Invincible Against a Troop Invasion from An Outside Army. However that does NOT Make USA Military Troops Invincible in the Middle East.

General Sun Tzu wrote that sometimes a combination of both natural and military factors make some areas almost invincible against outside conquest. However **Sun Tzu** warns the Generals of those Nations to **not make the mistake of believing that because their Nation is invincible against outside conquest that automatically makes their army invincible wherever it might be fighting.**

(477) If your Nation (or your Positions) are Beyond Defeat and your Opponent is Far Away - then Wait for your Opponent to Come To You. When your Opponent is Overextended and Entangled in your Natural and Manmade Defenses then You Can Defeat Them.

If your opponent has to travel a significant distance to attack you (like the USA attacking Iran with a shooting war) then the defender has a enormous advantage. The defender is positioned next to his main supply areas but the opponent's supply areas are far way.

In the Russo-Japanese war of 1905 the Russian Fleet had to travel 18,000 miles around the world. There were no Russian Naval-base coaling stations along the way. The Russians had to buy their coal at regular ports and the Japanese knew exactly where the Russian Fleet was at all of the time.

The Japanese Fleet was waiting for the Russian Fleet only 100 miles distance from one of Japan's bigger Naval bases. The Russian Baltic Fleet in Japanese waters was overreach - an arrow shot too far. The Russian Fleet was destroyed and the strategy stupidity of losing the Russian Fleet produced a revolution in Russia.

Historically the biggest causes of Armies and Nations being destroyed are (1) **the overextension of grand strategy objectives** - both Germany and Japan made this mistake in WW2 by wanting objectives far beyond their capacities. (2) **The overextension of one or more of the 37 different types of warfare.** The USSR made this mistake when it tried to finance socialist Nations around the world. (3) **The overreach of putting your military forces into too far forward positions where your opponent can envelop your military forces and they are too far away to be rescued.** This is what happened to **Napoleon** in 1812 Russia and **Hitler** in 1943 Stalingrad Russia.

Washington DC's fantasy that it can control the Muslim Middle East (and achieve Washington DC's subjective and flawed delusions) by using US Military shooting war force is both deadly arrogance and dangerous overreach.

Chapter 21

Military Forces versus Police Forces

Those who do not understand either the Police or the Military think they are the same because they both have guns and they both work for the government. However the mindset of the Police (and the standard combat Military mindset) must be completely different. The objective of the Police is to protect lives and property and keep the peace (they are Peace Officers). The objective of the Military is to kill the enemy as quickly and efficiently as possible.

(478) Do Not Turn your Army into Police and Do Not Turn your Police into an Army.

If you turn your Army into Police you will lose them as an effective Army. If you turn your Police into an Army you will lose them as an effective Police.

The above rule is true with one giant and vital exception. In Guerrilla wars once a state of pacification has been achieved (and the active fighting has calmed down) then **the professional Army of hunter-killers who won the shooting war must be replaced with an army that is mainly friendly Police.** There are more details in Chapter 21 - Terrorism and Guerrilla Warfare.

Good military combat soldiers should be skilled hunter-killers. The military's job is to kill people as quickly and efficiently as possible.

A good Police Officer is the exact opposite of a combat soldier. A good Cop talks to people (and listens to them) and tries to help them. The objective of the Police is to save lives and protect property.

The mental set of good Police is to NOT kill anyone unless it is absolutely necessary.

If you start using your military as Police then **you distort your soldier's mindsets so they are confused about their mission.** If you start using your Police as military then **you confuse your Police as to what their exact jobs and missions are supposed to be.**

Currently in the USA there is a controversy over the US Military giving Police Departments military equipment. **It is NOT the equipment that is the core issue - it is the mindset of the people using the equipment that is the core issue.**

The old saying is 100% true - "Guns do not kill people - it is people who kill people." **If the Police have the correct mindset then the military equipment is only additional specialized tools to help law enforcement protect the public from the increasingly well armed and vicious bad guys.**

America's 14,000 Police Chiefs must NEVER deliberately create a "military mindset" in their Officers. **A military mindset at Police Departments should NEVER be encouraged.**

South America has a long history of using its military as Police. In almost all cases they ended up being both a bad military force and a bad Police force.

(479) Some Law Enforcement Departments should have Higher Wages and the Police must Keep Their Right to Confiscate Property Used by Criminals.

Some specialized Police like the Secret Service (that protects the President) must be good at everything. They have to be good soldiers, good spies, good detectives, good undercover police, good standard police, good CIA agents and most importantly good O.K. Corral gunfighters.

The Secret Service has a very demanding job of travel and they always having to be "on" and ready for action. They (plus the many other hard working Police Groups) should be paid accordingly.

The Police must keep the right to confiscate property used in criminal activities. **That is the lesser of two evils.** Back in the "good old days" (before criminal assets could be confiscated by law enforcement) when criminals were arrested they used the money and assets they had stolen to hire crooked lawyers who corrupted the Police, bought Judges and paid off Juries.

Confiscation of criminal assets is NOT 100% perfect and it can be and sometimes is misused and abused. However **it is the lesser evil**. The people who suffer the most from confiscations are the criminals and their corrupt lawyers. They are the ones who should suffer the most.

(480) The current US Military must be Exceptionally Cautious (and Completely Prepared) for Shooting War Situations that have a High Probability of Turning into Long Term Policing Jobs.

For the USA almost every shooting situation in the Middle East has a 99% probability of for turning into a Policing situation.

The US Military must have a major reorganization so it can successfully take on "Police Jobs." After the shooting battles are won the US Military must become Police with the Job of creating an improved environment (and better life) for as many people in the local population as possible.

The USA Military combat troops should (1) fight the battles and win the conventional war speedily. Then (2) replace the combat troops with semi-military forces that mainly have a Police mindset. Then (3) the US Police Force will help the local population in as many ways as possible to benefit the greatest number of people in creating a better, safer and more fun life. (4) When the local population wants the Guerrillas to go and the beneficial Americans to stay then the US Military has truly won the war. **This is the proven Roman Strategy to win the peace and most importantly to keep the peace for very long time periods - Centuries in the case of the Romans.**

There were dozens of dumb mistakes Washington DC made with Iraq War 2 (2003 to ?). One of the biggest mistakes was disbanding the Iraq Army and the Iraq Police and most of the Iraq administrative bureaucracy. **The US Military kicked out of office all of the Iraqis who knew how to run the country and had kept the peace between Sunnis and Shiites.**

The disbanded Iraq Army soldiers and officers had nothing to do (zero jobs) and many former Iraq soldiers drifted into joining radical Guerrilla and terrorist groups.

Suppose there was a USA-Iran War and by 1 change in 50 the US Forces actually won. What is Washington DC's plan for keeping the peace in a population of 83 Million Iranians in a landmass that is over 6 times bigger than Texas? When this book was published Washington DC (and the US Military) had FAILED at securing the USA's own border with Mexico. If the US Military cannot secure the USA-Mexican border then how would it successfully police all of Iran? The answer is it would NOT. There would be a continues Guerrilla War everywhere in Iran.

(481) Three of the Worst Disasters that can Befall a Nation are (1) a Corrupt Police Force and/or (2) a Corrupt Judiciary and/or (3) a Corrupt Military.

There is an old saying in the USA "Guns do NOT kill people it is people who kill people." A lesser known (but equal true) saying is "Laws do NOT protect people it is people who protect people."

It makes no difference what laws or Constitution you have - if the Police and Judiciary have been corrupted then the laws mean nothing. When the government is corrupt ordinary people have no protection and the powerful can do whatever they desire. Want an example? Mexico right next to the USA.

Everything the Nazis did was legal under German Law. The Holocaust was NOT a crime under **Hitler's** Nazi Government Law. Everything **Stalin's** KGB did (the murder of millions of people) was Legal under **Stalin's** USSR Government Laws. All of the crimes and brutality that the Red Guard did during China's "Cultural Revolution" was Legal by Chinese Law. The 5 Million Chinese civilians that the Japanese murdered during 1931-1945 was Legal under Japanese Law - and so forth.

If you have a corrupt Police and/or corrupt Judiciary "the Law" means NOTHING.

The Law is only a Tool and in the hands of evil people it becomes an evil Tool. "Laws" do NOT protect you - only people can protect you and if the Police and Courts are corrupted then YOU have No protection.

Corrupt or semi-corrupt Police and Judiciaries have been the normal situation during 80% of history. **Honest Police (and an honest Judiciary) are the historical Exceptions that must NEVER be taken for granted or as a "forever guarantee."**

Once your Police and Judiciary have been corrupted it becomes exceptionally difficult to change. The USA is on the downside of its Bell Curve and **a characteristic of downside Nations is they become more susceptible to both political corruption and political extremism.**

The USA is on the downside of its Bell Curve and must be on-guard against the systemic legal corruption of the fair play intent of the US Constitution. Recent USA events like the massive use of fraud-news are classic predictable happenings and NOT good signs for the USA's future.

Mexico (and a large part of Central and South America) have (1) corrupt Police and (2) a corrupt Judiciary and (3) a corrupt Military. The USA becoming more like South America means the USA becoming corrupt at every level. Many big cities in the USA are already far down that road.

A National ID card system is long overdue in the USA. Everyone should have a National ID card that tells the government what it needs to know including citizenship and the right to vote.

The Federal Government (and the world) should start a program were every birth certificate includes the baby's DNA code and that information should go into a master National and international file.

Every illegal (and legal) immigrant coming over the USA's borders should have their DNA Code on file.

The USA's left wing press keeps attacking Police Profiling as racist. However - **profiling is the core of good detective and police work. Profiling also has great value in both military intelligence and business.** Every good detective in the world is good because he has the skills to accurately profile people and events.

Google's and Facebook's profiling systems have market values of several Billion Dollars. Collectively "profiling" has a market value of Trillions of dollars because it is a proven powerful tool for achieving objectives.

If "profiling" is worth hundreds of Billions of dollars to businesses then WHY (according to the left-wing nuts) is it supposedly less than worthless for Police work?

The left-wing's upside down racist war against Police "profiling" must be both vigorously resisted and actively challenged at every level.

Chapter 22

How to Successfully End Terrorist/Guerrilla Wars

Terrorism is NOT a new type of warfare. Terrorism is the original type of warfare going back to the dawn of warfare. General Sun Tzu (544-496 BC) wrote about using Guerrilla Warfare and Terrorist Tactics.

The key to winning "Terrorist Wars" is to win the support of the local populations. You make life better for the local people to the point where the local population wants the "Anti-Terrorist forces" (perhaps the USA) to stay and to govern them. This is the proven "Roman Method."

HOW TO WIN AND END TERRORIST-GUERRILLA WARS

(482) **The Modern Rediscovery of How To Successfully Win Guerrilla Wars (and Successfully Win a long Lasting Peace) came from French Field Marshal Thomas Bugeaud and French Field Marshal Louis Hubert Lyautey.**

In the 19th Century French Field Marshal **Thomas Robert Bugeaud** (1784-1849) was the commander of the Algerian area of North Africa and Field Marshal **Louis Hubert Lyautey** (1854-1934) was (at a later time) commander of the Moroccan area of Northwest Africa. Both Field Marshals were faced with large Muslim forces using Guerrilla Warfare in the most difficult terrains of desert and mountains.

General **Bugeaud** followed by General **Lyautey** went back to studying the world history of wars and how past empires creating long lasting periods of peace.

At the beginning of General **Bugeaud's** Guerrilla War in North Africa he was making exactly the same mistakes that the US Generals have been making in Afghanistan since the USA invasion of Afghanistan started in 2001. The wrong strategy of **pushing the Guerrilla's around on the map and "winning" all of the battles for useless rocky mountains and deserts but NOT winning the war.**

In Algeria and the Western Sahara as soon as the French troops went away the Muslim Guerrillas were back in control. Exactly the same as today's Afghanistan when as soon as the US Troops (and US Mercenaries) are gone the Muslim Taliban is back in control.

Over two thousand years before Field Marshal **Bugeaud** the North African area was deadly "enemy territory" for the Roman Army. The Romans and Carthaginians fought the Punic Wars from 264 BC to 146 BC. Carthage (present day Tunis) controlled western North Africa.

The Carthaginian General **Hannibal** in 218 AD invaded Italy and defeated the Roman Army in the most studied Battle in military history - the Battle of Cannae, Italy in 216 BC.

Rome eventually defeated Carthage and occupied all of North Africa. Then **the Romans succeeded in keeping the peace in North Africa for the next 300 years without terrorist wars and without a large Roman Army in the area.**

French General **Bugeaud** wanted to know **how the Romans had kept the peace for 300 years** with no terrorism or rebellions against the Romans.

The Roman Method of Peace worked brilliantly in all of the Mediterranean with only one big exception. The insane Zionists in 66 AD declared war on Rome. The Zionists were insane 2,000 years ago and not much has changed.

The Romans created and maintained a prosperous peace in the entire Mediterranean for Centuries with only small local Roman garrisons everywhere except the Rhine River frontier. **The "Roman Peace" was one of the most brilliant political-military-strategy achievements in the history of the world.**

General **Bugeaud's** questions to himself was "How did the Romans do it? How did the Romans successfully govern all of the many different peoples around the Mediterranean (and turn them all into very valuable assets for Rome) and keep everything running smoothly for Centuries?" Did the Roman's have a massive police-state that kept the local populations in constant terror and de facto imprisonment? Was the "Roman World" a giant West Bank with Roman soldiers at every major intersection? The answer was "No."

Field Marshal **Bugeaud** studied the Roman's methods and **rediscovered their successful principles** for defeating terrorism and establishing a true very long lasting peace. The Roman's secret for success was simple and logical. **The Roman's made life better for everyone under their control. They established honest and effective governments. The Romans gave everyone logical laws and an honest justice system. The Romans gave the people access to a Mediterranean common market with standard business laws and a common currency. Everyone's National borders were safe from outside attack because the Romans controlled all of the borders. The Romans made life safer, richer, more productive and they made life more fun.** The Roman secret for long lasting successful peace is doing what is needed so the local people what you to stay and want you to be the Roman version of the USA's Federal Government.

The USA's Founding Fathers were very aware of the Roman method of peace when they formulated the US Constitution for "a more perfect union."

The Romans connected everything together with a Mediterranean Common Market. No Nation needed the costs of a big army because all their borders were safe from attack. Rome controlled all of the borders. The Mediterranean Sea was safe for travel and trade because Roman controlled everything. There was a common currency and common laws of commerce for all of the Mediterranean.

Local Leaders were given Roman Citizenship and some became Officers in the Roman Army. **The Roman system created greater prosperity, safety, peace and happiness for everyone everywhere in the Empire.**

The "Roman Peace" worked because all of the local populations around the Mediterranean wanted the Romans to stay and wanted the peace, prosperity (and safety from wars) to continue.

The British Empire (the greatest empire of modern times) was patterned after the Roman Empire. The British Empire (as with the Roman Empire) provided a common law for trade and business. The British provided a common reserve currency and a common worldwide British banking system. The British made life better, safer and more prosperous for everyone. **The people in the British Empire wanted the British to stay - until the British lost both their mystique and their value to their Empire's peoples in WW1 and WW2.**

It was stupid Civil Wars (and wars with the Germanic tribes) that destroyed the ancient Roman Empire. Roman rule was no longer valuable to the local peoples. The taxes to finance endless wars (at the end of the Roman Empire in the 5th Century AD) became a burden to all of the average people in the Empire.

What destroyed the Roman Empire also destroyed the British Empire. **It was WW1 and WW2 that destroyed the British and French Empires.** Wars greatly increased taxes in the colonies. Wars created forced conscription into the military for wars that many colonial peoples saw as NOT their wars and stupid wars that nobody should be fighting.

WW1 and WW2 destroyed the mystique of the British and French. Local people in their Empires started asking "If they are doing stupid things like this war then why are the ruling us?"

Likewise **it was the Vietnam War debacle that destroyed the mystique of the USA.** The USA's National Bell Curve was on its upside from 1775 to 1975. Year 1975 was the Vietnam War's official defeat for the US Military. The USA's National Bell Curve has (relative to the world) been on its downside since 1975 and will continue on its downside (relative to the world) for all of this 21st Century.

In the early 20th Century Field Marshal **Louis Hubert Lyautey** in the Morocco area built on the work of General **Bugeaud.** General **Lyautey** repeated General **Bugeaud's** methods used in Algeria (Rome's methods) in Western North Africa.

The vast area that General **Lyautey** had to pacify was almost exactly like Afghanistan - mountainous with few roads and peopled by Muslim tribes that hated all outsiders.

Like Afghanistan - the Muslim peoples of the greater Morocco area had been in rebellion against any and all outside forces for Centuries.

Marshal **Lyautey** created a long lasting peace without large numbers of French troops in never ending terrorist warfare. General **Lyautey** successfully defeated Terrorism in the ONLY way that Terrorism can long term be defeated - by creating a safer, better and more prosperous life for the local Muslims to the point where they want you to stay.

The British had basically did the same with their Empire.

The successful pacification of Muslim North Africa by **Bugeaud** and **Lyautey** was difficult. There were many bloody shooting battles in both the Algeria and the Morocco areas. It was Marshal **Lyautey** who said, **"Winning battles in a Guerrilla war means nothing. If all you are doing is winning a battle then as soon as you are gone everything goes back to before you arrived."**

Field Marshal **Lyautey** said that as soon as the shooting parts of the pacification combat are over - then your troops must change from being hunter-killer professional soldiers to being friendly Police. **The Police force must then help the local population in as many ways as possible. This is the core over 2,000 year old "secret" for decisively winning Guerrilla wars.**

In 1931 Field Marshal **Lyautey** was already a French National hero. In that year Field Marshal **Lyautey** made the front cover of ***Time Magazine*** as "The French Empire Builder" and "The Maker of Morocco."

In Afghanistan (and like situations) **the USA must work at winning the friendship of the local population in every way possible.** The US should start 3 TV stations in Afghanistan and run American movies and TV shows 24/7 (with Afghan sub-titles and dubbed). Have 3 American radio stations with American songs and bands. Give free radios and TVs to schools, hospitals, coffee houses and public gathering places. Have TV news programs - not BS but telling the truth about everything.

The US Military in Afghanistan should be providing as much free medical help as possible. The US Forces should constantly be advocating equal justice and equal rights for everyone. That will be "dicey" with women's rights but remember women are half of the population so if you get women on your side then you have 50% of the population on your side. US Forces must promote fair laws for everyone. Pay McDonalds to open up some stores in Kabul. Pay some US carnivals to spend the summers in Kabul. Open a Shopping Mall and a USA-like Super Market. Open a USA High School in Kabul. Open a USA College in Kabul with credits applicable in the USA. The US Military in Afghanistan should constantly be asking the local people (all of them) anywhere and everywhere what the USA can do to make their lives better. **Making yourself needed, wanted and loved is the proven road to long term success in both Guerrilla Warfare and in maintaining long term peace everywhere.**

The same proven "Roman strategy" applies to Iran. The USA has to get inside of Iran and make itself valuable to Iran. That is the ONLY road to a long term successful Iran-USA working relationship. If the USA is protecting Iran and the Iranians believe in the protection then they will NOT build atomic bombs exactly like Japan and Germany have not built atomic bombs.

(483) The American Press going Hysterical over Every "Terrorist Act" is the Wrong Response.

The main defense against "Terrorist Warfare" has always been the same - do NOT panic. "Terrorism" depends on psychological effects and emotional overreactions. When the USA has a terrorist event we must stay calm and say - "We are not soft and we are not weak. We do not panic at the idea of a little pain. The terrorists are NOT big enough to ever significantly hurt the USA. On a relative basis the actual damage was insignificant."

In the USA over 100 people are killed PER DAY in vehicle accidents 365 days a year (37,000+ a year). In the USA 7,600 people die PER DAY from all causes - the USA survives.

Many times what is primarily motivating the Terrorists are things that have no value to a Great Power like the USA. In that case ending terrorism against the USA could be accomplished by giving the Terrorists what they want.

(484) All Terrorist Attacks Happen for a Reason. If you can Change the Reason for the Attack you can Stop the Attacks.

One of the main keys to stopping terrorist attacks is to completely and honestly find out what is motivating the terrorist's aggression and then change or alter the source of the motivation.

The first Muslim "terrorist attack" in the USA was January 25th 1992 when the CIA Headquarters in Langley Virginia had a Muslim shooting incident. **From 1776 to 1992 the USA had 216 years of no Muslim terrorist attacks. So what changed? What USA alliances**

are there now that were not there for almost 200 years? Remember the axiom - when you take on a new ally you also take on the enemies of the new ally.

If Muslim terrorist attacks happen because "The Muslims are jealous of America's freedoms" like military moron President **W. Bush** said then WHY did it take the Muslims 216 years to become jealous?

To solve problems you have to first open your mind (and be intellectual honest) and see the true sources and real causes of the problems.

If there is a problem and all of your explanations for why the problem is happening are lies and falsehoods - then you are NOT going to find any solutions.

Washington DC is NOT finding any solutions to terrorism in the USA because it is too weak and too cowardly and too controlled by the Warmonger Lobby and Military Industrial Complex to acknowledge the core motivations of the Muslim terrorists.

(485) Terrorist Type Attacks Usually Cause Those Being Attacked to Resist and Fight Harder.

During the Blitz of London (September 1940 to May of 1941) every night was a massive terrorist attack. The British stayed calm and carried on. They were NOT weak and they were never defeated in their minds. The British people had unshakable courage despite the massive nightly terrorist attacks that killed 32,000 civilians and seriously wounded 87,000 other British citizens.

(486) It is Schizophrenic and Insane to Believe that Killing People with standard Military Weapons (Like Jet Airplanes) is Ethical Murder but Killing people with Guerrilla Type Weapons (such as Roadside Bombs) is Unethical Murder.

Calling Guerrilla warfare a crime is propaganda for the masses and NOT for the thinking and planning of the Grand Strategist.

In Strategy the big difference between standard warfare and Guerrilla warfare are the weapons used. Guerrillas use mainly small arms and explosives because that is usually all they have. The US Military uses bigger and far more expensive tools because they have $700+ Billion dollars per year.

If you are killed with a $10 car bomb - or with a $50,000 "smart bomb" dropped from a $40 million dollar Jet Airplane you are exactly the same - dead. The rich Nations insisting that you can only make "ethical war" by using expensive tools - all of that is self-serving Schizophrenic nonsense.

Everything that was done during WW2's Holocaust of murdering almost 6 million Jews, and 3 million Russian soldiers and 8 million non-Jewish USSR civilians was all 100% legal by German Law. The murder of 4 million Chinese civilians by the Japanese during WW2 was all legal by Japanese Law. However it was all a war crime despite being legal by German and Japanese Law. The US Military in the Middle East murdering Muslims is all legal by USA Law - but **it is NOT seen as legal or ethical by about 4 Billion people in the world.**

The fact is that Washington DC has NEVER supported the International Court of Justice at the Hague - and has actively fought against the Court by threatening and sanctioning many Nations who support the Court. Washington DC does NOT support the Court (or recognize the Court's decisions) because Washington DC knows that many of its actions would meet the International Standards and Legal Definition of War Crimes.

The Guerrilla groups are never going to play by the USA's rich kid's games with the rich kid's rulebook and the rich kid's expensive military toys. **The Guerrillas are never going to accept the rich kid's idea that killing people with expensive weapons (like Drones) makes everything ethical and legal but killing people with cheap weapons makes everything a crime.**

(487) One of the Classic Ways to End a Terrorist Guerrilla War is to Declare Victory and then Go Home.

The British fought Guerrilla wars in Ireland (1569-1921, 1968-1998) in Afghanistan (1839-1842, 1878-1880, 1919-1921, 2001-2014) in Iraq (1921-1932, 2003-2007) in South Africa, Rhodesia, Sudan, Aden, Somaliland and North America. The British fought rebellions in India (1919-1947) and vicious Guerrilla wars in Kenya (1950-1963) and in Burma (1945-1948). **The British "won" many of the above wars by declaring victory and then withdrawing their troops.**

A classic Guerrilla war of the mid 20th Century was the Algerian War. After WW2 White man colonialism was the walking dead everywhere in the world. The very bitter Algerian terrorist war for freedom was from 1954 to 1962. **The French won that war by declaring victory and giving Algeria its independence.**

The American Revolution was in large part a Guerrilla war against the British. **Like the Viet Cong in Vietnam the Americans had an unending wilderness at their backs to retreat into and they could never be enveloped.** The British could never enveloped the Americans and the Americans made it clear that they would never stop fighting.

In a USA-Iran shooting war the Iranians would never surrender and would have enormous amounts of land and mountains to retreat back into where they would be supplied by China and Russia - and by Millions of Muslims in the "stan" Nations (Pakistan, Turkmenistan, Uzbekistan, Kazakhstan, Kyrgyzstan, Tajikistan etc).

(488) The "Water Strategy" is Applicable to Guerrilla and Terrorist Warfare.

The civilian populations in every area are mainly not Guerrilla fighters. **The civilian populations are the Sea that the Guerrillas swim in.** The Guerrilla is usually just another innocent looking fish in the Sea until it is time for him to become a Guerrilla fighter. After the attack the Guerrilla goes back to being just another fish swimming in the Sea.

The US Military's key to success in Guerrilla wars is to use the proven Roman Method of make the US Military more desirable and beneficial to the civilian population (the Sea) than the Guerrillas.

A military example of making yourself more valuable is Lawrence of Arabia. Lt. **Thomas Lawrence** (1888-1935) was a WW1 British Army Officer (and Christian) that **the local Arab Muslim population chose to follow Lawrence over the Muslim Turks.** Why?

Why would the Muslims follow a Christian (and British Army Officer) over Muslim leaders - the Turks? Washington DC understanding the reasons for "why" is absolutely vital for the US Military's current "Insurgent" wars in the same areas where **Lawrence** was a great Leader of the Arabs.

The Muslims followed a Christian Leader because **Lawrence was promising the Arabs what they wanted.** The Arabs wanted independence and **Lawrence** promised them independence. Short term **Lawrence** backed up his promises with guns, money and military support. The **Lawrence** of Arabia Rule (the universal Rule) is - **Give (or promise) people what they dream of having and they will follow you.**

After WW1 the British and French Leaders in London and Paris double crossed the Arabs (betrayed **Lawrence's** promises) and imposed White man colonial rule. In a 100 years not that much has changed. The betrayal of London and Paris has been replaced by the betrayal of Washington DC protecting and financing White man colonial rule of Muslims.

Lawrence never returned to the Middle East. He joined the Royal Air Force as an "aircraftman" under a phony name. In 1935 Lawrence crashed his motorcycle and died - it was probably a suicide.

In his Book "The Seven Pillars of Wisdom" Lawrence describes the Arabs as - "excellent defensive fighters but lacking the organizational discipline to be a war winning offensive force." The Arabs lost all of their shooting wars against Israel. However in 2018 the Dubai Airport with 89 million passengers per year was the 3rd biggest Airport in the world. Arabs clearly can manage big organizations. (Note. Atlanta was #1 Airport in the world and Beijing #2).

(489) One of the Best Defenses against Terrorist Attacks is to Not Engage in Unprincipled Activities. Stay on the Moral High Ground and Try to Help Everyone.

The reverse of the above rule also applies - do NOT supporting those who are on the Moral Low Ground and committing crimes.

In every legal system **the accomplice to a crime** is almost as guilty as the actual perpetrator of the crime. **When the close Allies of the USA commit crimes against justice and equality (using the USA's military equipment and money) that makes the USA equally guilty in the eyes of the world.**

The World's historical record clearly shows that if you are NOT in other people's Nations attacking and killing the local people then there is a very high probability that those people will not be attacking and killing you. **What goes around comes around. What does NOT go around does NOT come around.** Places like Norway and Japan have very few terrorist attacks.

(490) The Grand Strategist always Treats Captives as Humanly as Possible.

Good treatment of prisoners will hopefully cut down the degree of collected-revenge that could fuel the next war. Good treatment will ideally turn former captives into possible future allies.

The Strategist will have programs to hopefully turn his captive's ideas around so when the captives are repatriated home the captives will become salesmen for the Strategist's side.

An example of exactly what NOT to do was the US Army's prison at Abu Ghrib during Iraq War2. The USA has regressed and become worse since WW2 when the USA treated war prisoners with common sense and moral correctness.

Another major Washington DC mistake and moral devolution is the prison at Guantanamo Cuba. Guantanamo represents a Fascist military mindset. It is what Stalin and the Chinese Red Guard did.

America's sense of "fair play" has deteriorated greatly since the end of WW2. **Washington DC's never ending murder of other people inside of their National borders has gone on for so many decades it has become the US Military's new normal.**

The USA is a Nation with collapsing moral standards. If the USA is to survive it must reverse the trends. A big step in the correct direction would be for Washington DC to STOP killing people in other people's Nations. Especially killing Muslims in the Middle East to make the Middle East safe for Apartheid and colonial oppression.

(491) If you want More of your Opponent's Troops to Surrender then Acquire a Reputation for Treating your Captives Humanly.

It is much easier (and less costly) to take 1,000 troops captive than to kill 1,000 troops on the battlefield.

During WW2 (by the Spring of 1944) the Americans (and in particular the British) had acquired a reputation for treating their German prisoners well. Consequently after the D-Day invasion it was possible to find thousands and thousands of German troops willing to surrender to the Americans and British. That saved the lives of many American and British soldiers.

On the Eastern Front the German troops usually did not easily surrender to the Russians. The Russians often killed their prisoners and that reputation cost many additional Russian lives on both the battlefields and in the German prisoner camps because the Germans were starving their Russian prisoners.

Places like Abu Ghrib and Guantanamo are giant steps downwards and backwards for the US Military and thus for all Americans. The USA's Founding Fathers would NOT approve. Washington DC's ethical corruption must end.

(492) One of Military History's On-Going Strategy Decisions is about a "Far Forward Defense" Versus a "Home Fortress Defense."

In the 21st Century both Washington DC and the US Pentagon are using the "far forward defense" philosophy as an unjustified excuse to be in other people's lands killing local people.

Against an opponent that has the ability to decisively attack your homeland there are very strong arguments for a far forward defense (like nuclear firing submarines off the cost of Russia and China). However against an opponent that has no ability to decisively attack your homeland (as is true in the current war between the USA and the Muslim World) there is no need (or ethical justification) for a far forward war with USA ground troops on the other side of the world. The Taliban's Navy is NOT going to invade and conquer California.

(493) To Win a Guerrilla/Terrorist Wars the USA must Use All of the Information Warfare Tools. Every Category of Media Communications Possible must be Used to SELL the IDEAS that can Counteract the Ideas of the Radical Muslim Groups.

To envelop a terrorist group that is spread over a large area (perhaps several Nations) **you must concentrate on enveloping the IDEAS that are motivating the group's actions.**

The people in Washington DC and US Pentagon who sincerely believe that there are US shooting war solutions in the Middle Ease are both history and strategy idiots.

Standard military tools (Rifles, Tanks, Artillery etc) are useless for successfully attacking guerrillas-terrorists hiding in large populations over vast areas where a significant part of the civilian population sympathizes with the terrorists.

To defeat the terrorists you must very aggressively use all of the media tools possible to "sell" the advantages of your ideas to the majority of the local populations.

Can the US Military free its mind from the past? That is the decisive question.
"Salesmanship" was not needed or used by the US Military in WW2 - so it is NOT a tool in the US Pentagon's cultural mindset tool box.

(494) The General who has a Professional Army (i.e. the USA) should Never Make the Arrogant Mistake of Believing that He will Automatically Win because his Opponent Has an Irregular Army.

In the American Revolution (1775-1783) the British had a world class professional army and **George Washington** had a rag-tag citizens army. **George Washington** (and most of the American Generals) had very little (if any) formal training as higher ranking Officers.

The British opinion was that much of the "American Army" (like **Ethan Allen's** Green Mountain Boys) were Terrorists. However the motivation will-to-win of the American rag-tag army was much higher than the motivation will-to-win of the British "professionals." The USA's superior will-to-win won the war against England's superior financed army.

In the Vietnam War (1959-1975) the US had a standard WW2 Army with basically **Clausewitz** Tactics (i.e. defeat the enemy in battle). The Viet Cong had a citizens irregular army.

The US Pentagon had a 1,000 to 1 advantage in gross firepower and a 1,000 to 1 advantage in money spent on the War. However the motivation of the Viet Cong was much higher than the Pentagon's extravagantly financed **Clausewitz** Army and the Pentagon was defeated.

The Vietnam War <u>disproved</u> the US Pentagon's core assumption that grossly bloated military budgets guarantee that the US Pentagon will win all of the wars.

(495) So called "Terrorist Attacks" Must Never be Used as Justification for the Oppression of Basic Human Rights and Freedoms.

After the 9/11 attack (2001) then President **George W. Bush** put great effort into selling "The War on Terror." The word "Terror" was twisted and spun to give it a sinister meaning like some form of brand new deadly virus had just been spawned. A brand new deadly creation that had to be aggressively attacked and stopped at any cost in USA blood, treasure and **a high cost to the USA's moral/ethical reputation** in the world.

"Terrorist Warfare" is as old as the history of warfare. Warfare started as small groups of fighters traveling fast and using surprise-raid tactics against outer groups. Then fading away back into the wilderness.

Calling something "Terrorist Warfare" is almost always a subjective opinion (read Chapter 19). The French Resistance of 1940-1945 was "Terrorist Warfare" for the Germans. However for the French it was fighting for freedom and independence against a brutal invader from another Nation.

The Chinese Guerrilla warfare of 1931-1945 was "Terrorist Warfare" for the Japanese. For the Chinese it was fighting for freedom against a brutal invader that was trying to colonize their Nation.

For the British much of what happened during the American Revolution (1775-1783) was "Terrorist Warfare."

For the British Army (in 1947 Palestine) terrorist **Menachem Begin** was blowing up civilian buildings and murdering British soldiers. When **Menachem Begin** (1913-1992) became Israel's Prime Minister (1977-1983) the Palestinian Muslims fighting for freedom were defined (in the USA's Press) as "Terrorists."

Today's Palestinian "Terrorists" are fighting for what **Menachem Begin** was fighting for before 1947 - freedom from a colonial overlord.

Looking at events in morally schizophrenic ways is a major detriment to effective Strategy thinking and planning. Moral and ethical schizophrenia is a major detriment to Washington DC's ability to resolve anything in the Middle East. **Washington DC calls the groups that are fighting for Palestinian human rights and equal freedoms "Terrorists" when 80% of everyone else in the world is calling them freedom fighters.**

Was the US Air Force's firebombing attack against Tokyo on the night of March 9-10 1945 a "terrorist attack?" A firebombing attack that burned to death over 100,000 Japanese: women, old people, babies, sick people in hospitals, children plus burned alive their cats, dogs, horses - everything. Was that a "Terrorist Attack"? All of the dead children and old people would say - "Yes."

In March of 1945 Japan was already defeated. 90% of its oil was cut off. Half of the population was on the edge of starvation. Japan's industry was grinding to a halt. US Forces did NOT need to attack the mainland - all they needed to do was wait. The Ethical Rule is you must know when you have either won or lost and then stop killing people (including your own troops) to do little more than be killing people.

Chapter 23

The Strategy Democracies Must Follow

For thousands of years wars were between Kings. If the King wanted a war then there was a war. For thousands of years "public opinion" meant little or nothing towards starting wars or the outcome of wars. However in the 21st Century "public opinion" has become decisive in warfare and peace.

The "people" (especially the American people) now have the power to end wars if they learn how to channel their votes.

The warmongers have gained the upper hand in recent years and have started wars. However - in a Democracy (like the USA) the voters have the power (if they use it) to end the stupid wars that are being fought in stupid places for stupid reasons.

The objective of this Book is to educate average Americans in military strategy to the point where they are NOT "cowed" by "military authorities" (and the "Warmonger Lobby") that makes hundreds of Billions of dollars from shooting wars despite the US Military losing those wars. Every shooting war since WW2 has been a big "win $$$" for the Warmonger Lobby.

(496) **In the current Information Age of Warfare if the Generals and/or Leaders Lose their Ability to keep Selling a War to the Voters - then the Military will Lose that War.**

From the dawn of organized warfare 4,700 years ago up to the invention of mass newspapers the "populace" usually knew only what their Kings told them - and most of the time the Kings told them nothing. (Note. the 1st recorded war was in 2700 BC between the King of Sumer and King of Elan in Mesopotamia).

For thousands of years there might be a major battle and people 30 miles away did not know it was happening.

In the 21st Century there are a few Nations like North Korea and Cuba where "the TV News" is 100% controlled by the government. However - even in Cuba most people know what is happening in the world.

Starting in 1803 (with the **Napoleonic Wars**) public opinion became decisively important for France's military success. **Revolutionary France turned warfare into being total National Warfare.** The Napoleonic Wars were no longer a King's wars. They were all of the French people's wars.

One of Napoleon's biggest advantages (in his early wars) was the complete enthusiastic support of the French population and the support of all of the many assets of the French Nation.

In Napoleon's early years he had the great advantage of fighting a total war on his side while his opponents were continuing to fight "the King's war."

In the early 19th Century modern newspapers were born and by the 1840's in Europe (and in the USA) big city newspapers had acquired the ability to change "public opinion." However up to and including WW2 the newspapers everywhere reported only what their governments wanted reported. Almost all of the newspapers were either government censored or self-censored.

As late as WW2 the US and British newspapers only reported the "war news" that their governments wanted them to report. During WW2 the New York Times and other US media did NOT report on the mass murder of Jews and Russian prisoners because President **Roosevelt** did not want to be rushed into the D-Day invasion. President **Roosevelt** agreed with **Churchill** who believed that the more D-Day was delayed the more the Russian Army would destroy the German Army.

The power of post WW2 public opinion can be seen in the fact that **it was public opinion that defeated the US Military in Vietnam.** The US Military and the White House (despite **McNamara's** and President **Johnson's** lies and extensive falsifications of the facts) could no longer sell the Vietnam War to the US voters.

In the current Information Age of Warfare (1960 to ?) with today's Internet and Social Media almost anyone who wants to know the true facts can learn the truth about almost everything. What this means for the Grand Strategist is that "the Leader" must use the correct honest facts and honest arguments to sell his war to his voters and the world.

Twenty-five centuries ago **Sun Tzu** put great emphasis on always being on the moral high ground. Today being on the moral high ground is decisive. **With today's Internet and social media everyone knows what is happening. Your wars MUST be on the moral high ground or your opponent will receive help from all over the world.**

You can publically say you are for freedom (and you can sincerely believe your own lies as much of Washington DC does). However - if you are supporting colonial slavery and anti-freedom racist Apartheid oppression then 80% of the thinking people in the world will know that you are (1) a double-talking phony or (2) delusional-crazy or (3) both #1 and #2.

Remember the three rules - (1) Do NOT become a victim of your own lies. (2) Do NOT become a victim of your own con game frauds. (3) If you cannot "sell" your war with the truth and honest facts then you should NOT be in that war.

(497) The Grand Strategist must NEVER let Any of His Allies' Moral Low Ground Actions Undermine the Strategist's Moral High Ground Positions.

The old saying is "You are known by the company you keep."When the USA's Allies are on the moral low ground then they are also dragging down the USA's reputation in the eyes of the world.

The US Military had stunning successes in WW2 because the USA was clearly and honestly on the moral high ground fighting against slavery and oppression. However - today there are times when the US Military is fighting to protect slavery and oppression. A USA-Iran War would be a Washington DC war to make the Middle East safe for Apartheid.

(498) Starting Useless Wars with Frauds and Deceits to Trick and Con Game the Voters into the wars - and then Continuing the Wars With More Lies and Falsehoods - that Should be an Act Of Treason and Punished Accordingly.

The Pentagon Papers (of 1971) proved that President **Lyndon Johnson** (President from November 22nd 1963 to 1969) and Secretary of Defense **Robert McNamara** (Secretary of Defense from 1961 to 1968) had lied significantly about both the need for the Vietnam War and then lied about the lack of progress in the Vietnam War.

The lies that President **Johnson** and Secretary of Defense **McNamara** told to both the American public and Congress did NOT prevent the Vietnam War from becoming a defeat. The lies, deceits and frauds of President **Johnson** and **McNamara** continued a de facto lost war for many additional years and in the process murdered tens of thousands of additional young Americans and hundreds of thousands of additional Vietnamese. **Washington DC should have closed down the shooting part of the Vietnam War in 1963 rather than 1973.**

The Post WW2 standard operating 5 Rules in the Washington DC Swamp are (1) Admit to Nothing (2) Deny Everything (3) Accuse Your Accuser of Doing What You Were Doing. (4) Always Promote "Insiders" those who where Disastrously Wrong to "Prove" that They Were Never Wrong. (5) Never Charge an "Insider" with a Crime because Insiders are Always Correct.

The Washington DC "swamp" will never improve until the lying con artists who falsify the facts about the wars being fought start going to jail.

(499) There are valid Military Arguments for Giving your Troops some Bloody Combat Experience Now and Then but NOT Wars that Go On Forever and Cost Trillions of Dollars.

The USA is already the biggest gross debtor Nation in the history of the world (by gross amount of money). When this book was published the Federal National Debt had been going up by a Trillion dollars a year for over 10 years straight. The average American does NOT need more stupid wars in stupid places for stupid reasons that cost Trillions of Dollars.

(500) Historically it is usually Bad for a Nation when the Military Becomes a Government Inside of the Government.

The US Military (and their first cousins like the CIA, NSA, FBI and so forth) should NOT be a Government inside of the Government. **Washington DC is full of proven failures and incompetents who nevertheless have bloated opinions about their own intelligence.**

You would think that the Washington DC rancid bureaucracy that has been so horribly wrong (about so many important actions for so many decades) would have a little more **humility about their own disastrous lack of strategy competence?**

The USA's Founding Fathers were the greatest collection of political geniuses in the history of the world. In the US Constitution the Founding Fathers made sure that the military was always controlled by the elected political branch of government. The USA's Founding Fathers also gave each State and City its own independent Police Force so one National Police Force (or State Police Force) could never seize National power.

The problem is that the current USA military-industrial complex (and the USA's Warmonger Lobby) have so much money (and are so big and complicated) **they have become governments inside of the US government.**

(501) In Desperate Situations the Grand Strategist Will Do Whatever is Needed to Win and/or Avoid a Crushing Defeat.

If you are hopelessly losing a war then have the courage to be honest with yourself and save what can be saved. If you are losing but some very difficult actions could reverse the situation then have the courage to do what is needed.

Chapter 24

Improving the US Pentagon for Fewer Shooting Wars and More Peace

It is exceptionally challenging to change the mindsets of old, bloated, organizations like the US Pentagon.

The fact that your military happens to be big means nothing. In 1989 the USSR's military was enormous in the size and numbers of its troops and their equipment. However the USSR fell apart because its fundament political and economic systems were decisive contradictions to reality.

In the 21st Century the USA's economic system is built on ever increasing massive amounts of debt (over a Trillion dollars of new Federal Debt per year). The USA's debt driven economy is a long term contradiction to proven historical standards for long term stability and the long term ability to faction effectively.

The classic conundrum problem is that those who are on the top are NOT motivated to change despite the totality of the system becoming more and more long term obsolete and unworkable. Like Tsarist Russia's nobility (or the USSR's top Leaders in the 1980's) those on top were getting what they wanted and they fought all of the needed changes.

Five thousand years of world history has proven that 95% of the time those on top are motivated to keep everything the same. The USA's power elite have wealth, control, comfort and status and they do not want to change anything.

(502) Many Times in World History the Organizations that Most Needed to Changes were Also the Ones that Fought Change the Hardest.

Large old organizations with many levels (that include the US Military, US State Department, US Congress and so forth) are highly resistant to attitude and mindset changes.

The above Rule is one of the reasons for WHY Nations (and business companies) have Bell Curves. Eventually defending and maintaining the status of the Leaders becomes the #1 priority. **Historically real change comes only when the Bell Curves crash and burn.**

(503) The Last Thing the Grand Strategist Should Do is Stand Up Before a Military Organization of Senior Officers and say, "We are Going to Change." The Strategist must Change the Organization through Skillful Stealth and Covertness.

If you stand up before a group of people who already have power and status and say " We need to change" you will immediately run into a high brick wall of resistance.

The change Strategist has to "sell" a new idea as something that is already happening and only a different way of doing what is already accepted practice.

The inventor of modern Tank warfare (Lieutenant-Colonel **John F. C. Fuller)** in 1915 saw and understood that the new Tank concept (and Tank weapons system) was the key to breaking the nightmarish Trench War deadlock created by machine guns and barbed wire.

During the 1914-1917 period of WW1 (on the Western Front) all of the Generals used mainly mass infantry attacks against their opponent's frontlines. Infantry attack tactics that went back to the dawn of organized warfare.

Using obsolete infantry tactics against Industrial Age of Warfare machine guns and barbed wire was a bloody disaster. By 1917 millions of soldiers had been slaughtered.

Colonel **Fuller's** problem was that the Generals in 1916 did NOT understand steel Tanks driven by gasoline engines. That was a "new" concept and **truly new concepts are usually resisted by the senior military leadership of all large established old military organizations.**

However - all of the Generals in WW1 loved horse cavalry. When WW1 started (in 1914) England had 8 cavalry divisions, Germany had 11 cavalry divisions, Austro-Hungary had 9, France had 14 and the Russians had 28 horse cavalry divisions.

What is amazing and revealing about the mental set of all of the Generals in WW1 was that in 1918 the English, French and Germans had almost as many horse cavalry divisions as at the start of the war in 1914 despite horse cavalry having been proven (from the start of the war on the Western Front) to be less than useless. **Once a military organization falls in love with a weapons system its evolution into war winning uselessness is irrelevant.**

Lt. Colonel **J. F. C. Fuller** (1876-1966) went to the British General Staff but did NOT present the idea of Tanks as a radical new idea. Colonel **Fuller** (later General **Fuller**) told the Generals that Tanks were a return to the days of heavily armed knights. Lt. Col. **Fuller** told

British Army Commander Field Marshal **Sir Douglas Haig** that the only difference between a Tank cavalry and Horse cavalry was the mode of movement. That was NOT a completely new concept in the eyes of the Allied Generals and the British and French commanders bought the Tank idea.

On the other hand - **the German General Staff in WW1 belittled the concept of Tanks and were highly resistant to the idea.** During WW1 the German General Staff never gave up their love for horse cavalry. During WW1 the Germans built only 30 Tanks of a highly flawed and unworkable on a battlefield design and the Germans used only a dozen in actual combat.

On the other hand - the British during WW1 built 2,818 Tanks and the French 4,800 Tanks. The USA built 84 Tanks. The over 7,000 British and French Tanks proved to be the 1918 decisive weapons system for defeating the WW1 weapons system of machine guns dug-in and behind barbed wire.

One of Strategy's Rules is that - **After a war the losers are motivated to learn much more than the winners.**

After WW1 the German Generals 100% embraced the idea of Tanks and by 1939 were the world's leaders in Tank tactics warfare. During WW2 the Germans built the best Tanks. The USA's Tanks never approached the quality of German Tanks until late 1944. However Russian, USA and UK Tank gross production during WW2 compared to Germany was 5 to 1 and the Allie's Tank manufacturing completely overwhelmed the Germans.

(504) Updating the US Military's Mindset is Going to be Difficult but it Must be Accomplished.

The vital First Step in Solving a Problem is to Admit that There Is a Problem. The US Pentagon's Officers keep insisting that there are no fundamental mindset problems within the US Military despite the fact that with the biggest military budget in the world the US Pentagon has NOT clearly won a significant shooting war since 1945.

The Chinese Military has already openly said that one of their highest objectives is - "To own the soon to arrive Artificial Intelligence Age of Warfare." By 2015 China was the number one manufacturing Nation in the world and already significantly bigger than the USA in advanced product manufacturing. China already manufacturers most of the world's smart phones/computers and China already manufacturers 120% more vehicles per year than the USA.

(505) The USA's Military/Washington DC Leadership Must Understand that 21st Century Warfare is One Nation's Complete-Culture (in All of its Parts) against Another Nation's Complete-Culture (in All of its Parts).

21st Century warfare has expanded to include everything: economics, banking, trade, transportation, population growth, National IQ quality, communications - EVERYTHING. **Warfare now encompasses EVERYTHING and the Grand Strategy planning equation must include EVERYTHING.**

The big picture is that China is racing forward with advanced manufacturing on a massive scale while the USA becomes more and more of a South American Nation (on the downside of its Bell Curve) and concerned mainly about: sports, entertainment, food, the Wall Street gambling casino, sex, drugs and how much pleasure a person can stuff into their bodies.

(506) **Remember the basic Rule of Management - Do More of What is Working and Less of What is NOT Working.**

Washington DC (since the 1970's) has mainly done the opposite of the above axiom # 506. **Washington DC takes proven failures and then makes the failures bigger in attempts to prove the original failure was not a failure.** That process 95% of the time produces a bigger failure. Only being able to do what is already a proven failure is one of the many signs of the USA being on the downside of its Bell Curve.

Everything in the Universe has a Bell Curve including the Universe.

Sears Roebuck & Company (founded 1892) was the biggest retailer in the world after WW2. **Sears** was the largest retailer in the world. However because of its size it became stuck in the status quo and its Bell Curve turned downwards relative to other brand new retailers like Walmart. Today Sears is far into the downside of its Bell Curve and **drowning in debt**.

Nations on the upside of their Bell Curves create profits and those on the downside create debt.

Howard Johnson's (founded 1925) was in the 1960's the biggest restaurant chain in the world. Today there is one (1) **Howard Johnson** restaurant in the world.

Everything in the world has a Bell Curve. Based on 4,000+ years of history the US Military has a very high probability of being **Sears** by year 2045.

When Sears was already far into the downside of its Bell Curve it tried to "prove" that everything was wonderful by building the Sears Tower (in Chicago 1970). When built the Sears Tower was the tallest building in the world. Competent historians in world history know this is the "classic pattern" for the downside of Bell Curves. **The biggest spectacular "showboat" achievements always happen at the topping out process of Bell Curves. The big showboat spectacular is historically a sign that the downward decent has started.**

The USA landed a man on the Moon in 1969 - almost at the 1975 top of the USA's National Bell Curve.

Egypt built the great Pyramid at the top of its civilization Bell Curve.

Roman Catholics started building St. Peter's Basilica in 1506 and 11 years later in 1517 **Martin Luther** started the Protestant reformation.

The Taj Mahal was built in 1632 at the top of the Mughel Empire's Bell Curve. The British East India Company was started in 1612 and the English were already taking over India.

The Sears Tower was not the symbol of Sear's success but the symbol of Sear's obsolete concept thinking. It only increased the speed of Sears' decline.

Today the US Navy's 10 super Aircraft Carriers are symbolic floating Sears Towers. They are classic textbook examples of the US Military's decline. Obsolete weapons systems for a Pacific Aircraft Carrier War with China that will happen only in the Pentagon's dreams of refighting WW2.

Through 4,000 years of recorded history when Nations or groups reached a very high point they all believed that God had put a stone floor under them and they would never ever go down.

Every Nation and Army that believed they were on top forever has been proven wrong. The USA and US Pentagon are NOT going to be exceptions. Both the USA and US Military are into the downside of their Bell Curves.

The USA's decline (relative to the world) is already happening and the more stupid wars that the USA's Warmonger Lobby gets the USA into the faster the USA will decline.

The USA should follow England's example. Cut the military to defensive only. No shooting wars anywhere. Put maximum effort into increasing exports and cutting down imports. The USA does NOT need to be murdering people in somebody else's Nation to be happy.

(507) The Grand Strategist needs Excellent Persuasive and Management Talents to Keep the Needed Changes Moving Forward.

For thousands of years the world was in the Sword and Spear age of Warfare. A talented General in 200 BC could have successfully commanded an army and won wars 1,400 years later in 1,200 AD.

Today's tactics, **grand tactics and strategy are all changing at an increasing rate of speed.** Can America's Military Officers make the many mindset changes needed? Or will they all be WW1's 1915 Generals - one age of warfare behind with obsolete thinking, obsolete planning and obsolete actions that produce military disasters?

(508) Every Organization has "Rough Edges." The Grand Strategy Team must be Constantly Working hard to Continually be Rounding Out the Rough Edges.

The rough edges are often problems of subordinate commanders and/or different branches (Navy and Army) not understanding how to effectively work together as one unit.

General **Sun Tzu** said - "A square stone cannot roll. But a rounded stone can crash down a thousand-meter mountainside with great force."

What **Sun Tzu** was saying is that military leaders have to "round out" their ideas and actions and attitudes. They cannot be ridged with square sharp corners. 24 Centuries ago **Sun Tzu** was saying "Don't be a square."

The ideal supreme commander knows how to organize his total military force so there are no rough edges (no part acting alone and/or interfering with other parts) and thus little self-caused resistance to **the very important factor of acquiring and sustaining operational momentum.**

Your ideal organizational objective (as a Strategist) is to have your total military organization grow to the point where all branches are fighting together as if one unit. **Then the organization has more power than the sum of its parts.**

On the other hand - **a military organization that does NOT know how to work together has LESS power than the sum of its parts.**

(509) If You have a Really Big Problem 90% of the Time You Must Bring in New People to Solve that Problem.

Dr. Peter F. Drucker (the 20th Century's greatest management teacher) and **Albert Einstein** both said - **"The group that created the problem is seldom the group that can solve the problem."**

The US Military (and Washington DC) are incapable of collectively saying "We were wrong." However - **the first step in fixing a problem must be the admission that something is wrong and needs to be fixed.**

One of the US Military's problems is that the US Military-Culture demands a General George Armstrong Custer "Can Do" attitude. An attitude of "Where the US Military goes the US Military wins." If you have valid and accurate concerns about the numbers and doable logistics of a proposed action then you are labeled "a defeatist" and NOT promoted in the US Military. In the US Military "pride" wins out over realistic competent planning.

If you were a US Major in the US Army and said " There is no way in Hell the US Army can win a land war in Iran." Would you ever make it past being a Major?

(510) The Successful Grand Strategist Avoids the "Not Invented Here" Common Mistake of Organizational Arrogance.

In many organizations (especially large and blotted organizations like the US Pentagon and Washington DC) if an idea does NOT come from inside of the group then it is de facto dead on arrival.

The Washington DC "insiders" and US Warmonger Lobby and Military Industrial Complex are NOT the solution - they are the problem. If the Washington DC "insiders" and warmongers had the solutions the USA for the end of WW2 to NOW would have been winning wars everywhere rather than losing significant shooting wars everywhere.

(511) Change is Not Necessarily Bad only Because it is Change.

One of the many characteristics of a Nation that is on the downside of its Bell Curve is that **downside Nations tend to view every change as a threat** and something that must be resisted and fought.

Washington DC viewing every change in the world as a threat has produced highly destructive bad decisions and never ending muddled wars since 1959 when the Vietnam War started.

(512) When a Nation does NOT have the Courage to Change when History Demands Changes then the Refusal to Change can Often be Disastrous.

America's Military did not destroy the USSR. The USSR's Communists **who refused to change** destroyed the USSR from the inside out.

There has never been (and never will be) a <u>forever</u> of anything. When the environment (i.e. God) demands change then your options are: (1) Change as needed to successfully meet the new challenges. Or (2) join the very dead Dinosaurs.

(513) A Tactic that MUST be Used Periodically for Measuring the Need for Change is Stepping Back and Looking at the Big Picture Record of Success or Failure.

Suppose there was a professional baseball team that was winning about 50 games out of its usual 160 games a year. Suppose a fan of the team asked the Coach if his team needed any changes - and the Coach answered "Not really."

If you are losing a third of your games and everyone in the organization is satisfied with the status quo what is wrong with that picture?

Since the end of WW2 the US Pentagon has not clearly and unquestionably won any significant shooting wars anywhere. However there have been clear and significant defeats during that time with the Vietnam War (1959-1975) being the biggest.

The correct answer to the above question is - "Yes, if it's broken then do fix it."

The only thing that is constant in the world is change and that definitely includes new military ideas and new Strategy concepts. The difficult part is convincing both Washington DC and the US Pentagon to **embrace the cycle of destruction and creation** and to move past their current obsolete and clearly flawed mindsets.

Chapter 25

The Vital Importance of Air Power - & Space Power - & Communications Power

Air Power (and Space Power) are on the upside of their Bell Curves. Air Power in all of its forms is now the most important part of the USA's military matrix. Air Power and Space Power (in its many forms) will continue to increase in importance for all of this Century.

The 21st Century axiom is "He who rules the minds of the world rules the world." The USA has a decisive need for 2 new branches of the US Military Service "The U.S. Communications Force" and "The U.S. Space Force."

In 1900 "Air Power" did not exist. There were only a few people in the world who thought it would ever exist. At this point in history (and beyond) Air Power (plus Space Power) will continue to be the #1 defensive weapons systems and #1 offensive weapons systems for as far into the future as can be seen.

How many Generals and Admirals in 1900 would have said that Naval Power in the 21st Century would be 75% obsolete and Naval Power would have been mainly replaced with Air Power and Space Power? The answer is - "none." Today's Generals in today's Pentagon have the same highly dangerous lack of imagination when it comes to the importance of Space Power and Communications Power.

In the 21st Century the 2 correct axioms are - (1) "He who Rules the Air and Space Rules the World," and (2) "He who Rules the Minds of the World Rules the World."

The reactionary US Pentagon is digging-in its feet on what are the 2 most important branches of the military for future wars. The US voters must keep pressuring Washington DC to get the USA's National defense out of the obsolete past and into the future.

The Chinese have no fear of the US Navy but they would have deep respect for a first-rate US Space Force. The Chinese have a big economic edge over the USA but the USA has a

potentially far more powerful communications mind-changing edge over China IF IT USES THAT ADVANTAGE.

Yes the new USSF and new USCF will NOT look like the old military organizations. They need a new type of organization - a new ranking system - and a new mindset. President Trump should jump start the process and change NASA into the United States Space Force. YES there would be great confusion but there is time for the resulting organizational confusion to "work itself out" and create the new type of organizational structure needed.

The two greatest Strategy advocates for "Air Power" (in the first third of the 20th Century) were Italian General **Giulio Douhet** (1869-1930) and US Colonel **Bill Mitchell** (1879-1936). Both were amazingly correct in some of their predictions. However **both were very wrong** in believing that (1) air bombardment alone could destroy a Nation's morale and (2) air power by itself could achieve victory.

Colonel **Billy Mitchell** (a combat pilot, brilliant organizer, strategy visionary and WW1 hero) was the US Military's best Grand Strategy analyst of the 20th Century. During WW1 (in France) **Billy Mitchell** organized the first large Air-Ground coordinated attack in history. In the 1920's Colonel **Mitchell** put great emphasis on the ability of ground attack air power being a decisive factor for most future ground attacks. **Today air-ground attacks in cooperation with the infantry are the core of infantry tactics everywhere in the advanced world.**

Billy Mitchell also correctly analyzed (in 1921) that Air Power had already made Navy Battleships obsolete. **Billy Mitchell's** line of analysis and reasoning was correct up into the 21st Century. Air Power (and now Space Power) have made **ALL** large Navy surface ships obsolete.

The US Military's High Command did NOT like what **Billy Mitchell** was advocating and saying. The US High Command especially disliked **Mitchell** saying that their beloved Battleships were obsolete.

Military organizations do NOT like their comfortable old ideas challenged. Colonel **Mitchell** was charged with non-conformity-thinking. There was a Court Martial in 1925. **Mitchell** was found guilty and thrown out of the US Military. At age 56 (in 1936) he died of heart failure (i.e. a broken heart). If he would have lived he would have been a brilliant WW2 General.

After **Mitchell's** Court Martial in 1925 the US Military never let **Billy Mitchell** return to military service. However - by 1942 all of **Billy Mitchell's** ideas had in battle been proven to be correct. In particular **Mitchell's** insistence that airplanes can easily sink Battleships (i.e. Pearl Harbor). The US Military in 1942 (to make its 1925 small minded stupidity look better) promoted **Bill Mitchell** to Major General 6 years after **Mitchell** had died of a broken heart.

In 1925 the US Military disliked anyone saying their beloved Battleships were obsolete. Today in the 21st Century the US Military dislikes anyone saying that their beloved Super Aircraft Carriers are now far more obsolete then WW2's Battleships were during WW2.

President **Trump** has proposed the creation of the United States Space Force (USSF). The Royal Air Force (RAF) was established as a separate military organization in 1918. The US Air Force was not made a separate branch of the US Military until 1947 - so this time around the US Military will hopefully not be super late with its organizational updating. Unfortunately (but absolutely predictably) the vitally needed creation of a US Space Force is running into considerable resistance at the Pentagon's obsolete and reactionary and military grand strategy incompetent Joint Chiefs of Staff.

President **Trump** should side-step the Pentagon's war losing reactionary dead-heads and boldly re-name NASA the United States Space Force. All of the detailers can be worked out over time (and at this point we have time). Do NOT worry about it being messy at the start - jump in and get it started.

What is certain is that the new US Space Force is going to need a new and highly innovative type of organizational structure. The old "ranks" and organization from the ancient ages of warfare must be updated into a 21st Century organization.

(514) General Giulio Douhet's Idea that Air Power by Itself could Win Wars was Proven Wrong both in WW2 and Especially Proven Wrong in the Vietnam War.

The influence of Italian General **Giulio Douhet** (1869-1930) was far-reaching between WW1 and WW2. General **Douhet** wrote a book in 1921 (revised in 1927) titled ***The Domination of the Air***. The two major assertions of the 1921 "Douhet Theory" were:

1. "Aircraft are instruments of offense attack that have incomparable potentialities against which no effective defense can be foreseen."

2. "Civilian morale will be shattered by the bombardment of centers of population."

The extensive mass bombing of civilian areas in WW2 (by all sides) can be traced back to **Douhet's** Theory. However the massive bombings of civilians in WW2 (and in the Vietnam War) proved that **the bombing of civilian targets is almost always counterproductive** - in addition to being immoral.

The record of WW2's mass bombings of German civilian targets proves that it had little effect on the overall military production ability of Germany. Many German factories kept increasing their production until they were overrun by allied ground troops.

The German production numbers for Tanks and Guns and Ammunition in 1944 was (after 4 years of Allied bombing) many times greater than the German production numbers in 1940 (the first full year of war). In 1940 the Germans produced 1,888 Tanks and in 1944 they produced 18,956 Tanks. In 1945 when much of Germany was overrun by advancing Allied forces the Germans produced 1,946 Tanks.

In 1945 - when Allied forces already occupied large sections of German (and after 5 years of Allied intensive bombing of Germany) the Germans produced 2,798 Messerschmitt Bf 109 fighters and 2,734 Fock-Wolf Fw 190 fighters.

During WW2 **Adolf Hitler** told the German people that the Russians and Americans wanted to kill them all and that was why they had to keep fighting to the death. The mass bombing of civilian targets only "proved" to the German people that **Hitler** was telling the truth.

The Japanese leadership in 1944 and 1945 told their people exactly the same - that the Americans would kill everyone. With the mass firebombing of Japanese cities by the US Air Force (and the murder of hundreds of thousands of Japanese civilians) the Japanese people believed their leaders and were determined to keep fighting.

During the Vietnam War the US Air Force reproved that bombing alone cannot win wars. The US Air Force dropped 7,662,000 Tons of bombs onto Vietnam (five times the Tons of bombs dropped on Nazi Germany) and in Vietnam murdered 650,000 Vietnamese including women, children and old people.

In addition - the US Military in Vietnam had the biggest Chemical Warfare attack since WW1. The US Military used 20,000,000 gallons (75,600,000 liters) of deadly poisons including 13,000,000 gallons of Agent Orange. Agent Orange's main poison is Dioxin (a substance that breaks downs very slowly in Nature) and once in a human body can last for decades. The US Military's Chemical Warfare Attacks in Vietnam continues to this day to cause cancer, birth defects and death in Vietnamese humans and animals.

The US Veterans Affairs (VA) has received 280,000 claims for Agent Orange disabilities from Vietnam US Military Veterans (check the facts on Google).

The "Pentagon Papers" that were made public in 1971 proved that the massive bombing by the US Military (and the bombing's failure) had been kept secret from both the American voters and the US Congress.

When this book was published there was much lobbying in Washington DC towards starting a shooting war between the USA and Iran. The con game argument of the

Warmonger Lobby is that - "All the US Military has to do is drop a few bombs onto Iran and the Iranians will be screaming for mercy and surrender." This is General **Giulion Douhet's** wrong idea that was proven and reproven to be wrong in both WW2 and in Vietnam.

(515) General Sun Tzu wrote that One of the Worst Tactics You Can Use is to Attack the Civilian Population of Your Opponent's Cities.

General **Sun Tzu** 25 Centuries ago said that attacking your opponent's cities for no military purpose other than killing people will (1) harden your opponent's population against you and (2) put you on the moral low ground.

In WW2 (and the Vietnam War) the advice of Sun Tzu was proven to be absolutely correct. The more the US Air Force bombed Vietnam (and the more they murdered Vietnamese) the more determined the Vietnamese became to never stop fighting.

A USA-Iran war would be exactly the same. Iran would never stop fighting - the US voters would turn against the war - and the US Military would have one more defeat to add to their Post WW2 very long list of shooting war defeats.

General **Sun Tzu's** advice was the exact opposite of General **Giulio Douhet's** advice of destroying your opponent's cites. WW2 proved **Sun Tzu** to be correct and proved **Giulio Douhet** to be wrong. The mass attacking of cities (and the murder of hundreds of thousands of civilians) had a **non-decisive effect on the outcome of the WW2, the Korean War and the Vietnam War** for what was an enormous cost in Air Force lives (in WW2) - plus a horrible moral-ethical degrading of the USA in during the Vietnam War.

General **Douhet** wrote that there was no effective defense against "Air Power." **Douhet** was very wrong - **during WW2 the single most dangerous US Military job was turret gunner on a Bomber**. The other most dangerous jobs in order of percentage deaths were all of the other crew positions on a Bomber. An American Bomber crew in early 1943 Europe had an 85% probability of NOT surviving 25 combat missions into Germany.

During WW2 Europe the desired attack by the German fighters against US and British bombers was a head-on attack. German Radar would vector the German fighters for a head-on attack against the Bombers. About 3 miles from the US Bomber the German fighter would start shooting its machineguns and keep shooting. The German fighter was going towards the Bomber at about 400 mph. The Bomber was going towards the fighter at about 280 mph. **The Bomber would fly into a hail of hundreds of bullets.** Usually the Bomber's front turret-gunner was killed before he could fire a shot and often the pilot and co-pilot were killed. The German fighter would pass over the Bomber at a relative speed to the Bomber of 680 mph (or 1,100 km/h). At that speed the back and side defensive guns of the Bombers were useless.

The German fighters being 100+mph faster than the Bombers would circle around and do it over again to the surviving Bombers.

(Technical Combat Note) The 7.92 mm Machine Guns on a German WW2 Focke-Wulf 190 fighter had a muzzle velocity of 1,700 mph. Adding 680 mph to the relative speed of the bullets increased the already powerful penetration power of the German bullets by 40%.

In warfare it is seldom good to be "first." It is almost always best to be last. If you were a British fighter pilot when WW2 started in September of 1939 there was an 85% probability that you would be dead (or seriously wounded or captured) by the summer of 1941. If you became a British or USA fighter pilot in the summer of 1944 you had a 96% chance of surviving the war.

(516) One of the Lessons from WW2 is that an Air Force of Many Smaller (but Very Fast) Bombers Going After Targets of Proven Value would have Been Better than Many Big Slow Bombers Destroying Cities and Murdering Civilians.

During WW2 the British, American and German Air Force Generals were all following the wrong theories of Italian General **Giulio Douhet**. They all believed that the mass bombing of cities would force a Nation to surrender.

The big and fat (and slow) typical 4 engine WW2 bomber was both an advantageous target for fighters and for ground anti-aircraft guns. The typical WW2 four-engine big bombers used by the British and USA in Europe had a speed of about 280 mph but the German fighters had speeds of 400 mph.

If England (and the USA) had concentrated on building very fast 2 engine bomber-fighters that could carry a 2,000 pound bomb (and concentrated on targets of tactical and strategic value) the results would have been (1) more affective against the opponent's war effort. (2) Faster meaningful results. (3) Had a much smaller cost in Allied Airman lives and (4) fewer deaths of the opponent's women, children and old people.

The British 2 engine de Havilland Mosquito fighter-bomber could fly a little faster than the Messerschmitt Bf 109 fighters - and as fast as the German Focke-Wulf fighters. The Mosquito could carry a 2,000 bomb 1,400 miles and the Mosquito was well armed and fast enough to not be an easy target for German fighters.

Two hundred Mosquito fighter-bombers coming in low and fast for pinpoint bombings of high value targets would have been far more of a problem for the Germans than 600 big, fat (and slow) bombers attacking a city with little industry and few war-age men.

The same idea applies to the US Navy in the 21st Century. **A US Navy with many smaller ships (but ships that have nuclear missile firing capability) would be more dangerous (and much harder targets to destroy) than the current 10 USA Super Aircraft Carriers.**

All of the USA's Super Aircraft Carriers are Super-Targets in a shooting war with China or Russia. They would probably all be destroyed by the end of the 2nd week of the war.

All "Super Ships" have the same tactical usage problems. **The bigger the ships the more cautious and restricted the commanders become with their use.** In a shooting war with Iran the US Navy would probably never send a Super Carrier into the Persian Gulf - so what good are they?

(517) One of the Major Developments in WW2 was (as Billy Mitchell had Predicted) Air Power Became a Decisive Part of Both Ground Defensive and Ground Offensive.

On D-Day (June 6th 1944) German troops on the ground in France greatly outnumber the allies who landed 156,000 troops on the first day. The Germans had 50 Divisions in France and the Low Countries. The ground superiority of the German forces in France continued for the first 3 month after the D-Day invasion. However **the allies had almost complete control of the air**. **The Allied Air Force's support for the ground troops was enormous and decisive.**

Without Allied domination of the air over the battlefields the German forces would have pushed the Allies back into the Sea during the first few weeks of fighting.

One of the many grand-tactics that was decisively proven during the 1944 invasion of France was that **having an effective ground defense must include an effective defense of the air over the battlefield** exactly as **Billy Mitchell** had foreseen in WW1.

(518) Bombing Civilians almost always Makes the Population Being Bombed more Determined to Fight.

What happened on December 7th 1941 when Japan bombed Pearl Harbor? Was everyone in the USA screaming "We surrender we surrender - don't bomb us again!" No. On December 6th 1941 the US voters were 75% strongly anti-war against anyone. On December 8th 1941 95% of the voters were demanding a war against Japan.

What happened in 1940-1941 England when the Nazis blitzed England and over 43,000 English civilians were killed and over 139,000 were wounded? Were the British people screaming, "We surrender we surrender - don't bomb us again!" No. The English people and **Winston Churchill** said - "We will NEVER surrender."

At the WW1 Battle of Verdun (February to November 1916) the Germans pounded the French troops with massive artillery barrages for week after week and month after month (killing 163,000 French Troops) and the French front line Troops said - "We will die rather than surrender Verdun."

During the Vietnam War (1959-1975) when the US Military dropped 7,662,000 tons of bombs onto Vietnam (5 times the tons of bombs dropped on Nazi Germany) plus 20,000,000 gallons of chemical warfare poisons (the biggest chemical weapons attack since WW1) - were the Viet Cong screaming "We surrender we surrender - don't bomb us again!" The answer was - NO. North Vietnam's commanding General **Vo Nguyen Giap** (1911-2013) said - "Surrender was never in our vocabulary."

If the US Warmonger Lobby con games Washington DC into bombing Iran the US Military's bombing will unit all of the Iranian people behind the most reactionary forces in Iran. The Iranians will never surrender and the US Military will once again be defeated.

SUMMARY

The **US Air Force** (with the addition of the new **US Space Force**) continue to be on the upside of their Bell Curves that extend far into the next Century. On the other hand - large surface Navy ships have been on the downside of their Bell Curves since the Vietnam War (where they were already almost useless).

In the 21st Century the bigger the US Navy's surface ships become the more they become death traps. The "Sea of Water" has been replaced in importance with the **Sea of the Air** and the **Ocean of Space.**

The US must have the best US Space Force in the world. The US Air Force (and US Space Force) are both the current (and future's) decisive shooting war military branches and shooting war tools.

The World in the 21st Century has a new #1 pan-military axiom. The 21st Century Axiom is - **"He who Controls the Air and Space will Rule the World.**

Chapter 26

The Obsolescence of Big Ship Military Sea Power in the 21st Century

For almost 500 years (1492-1960) the motto "He who rules the seas rules the world" was absolutely correct. However in the 21st Century large surface navies (like the US Navy) are already far into the downside of their effective Bell Curves. Big ship surface Navies today are what horse cavalry was in 1913.

The US Navy has as many large Aircraft Carriers (10) as all of the remainder of the world combined. However against a first-rate military power like Russia or China the USA's Super Aircraft Carriers are obsolete and floating coffins.

Against a 2nd or 3rd rate Nation like Libya or Somalia possibly one or two US Aircraft Carriers might at some time in the future be useful. All of the other Aircraft Carrier owning Nations in the world are strategically correct to only have 1 or 2 Aircraft Carriers.

The 19th Century's greatest American born writer about Grand Strategy was US Navy Admiral **Alfred Thayer Mahan** (1840-1914). Admiral **Mahan's** classic 1890 book ***The Influence of Sea Power Upon History*** had a gigantic influence around the world when it was first published.

In 1890 Admiral Mahan's conclusions were absolutely correct - the Nations that had ruled the Seas for the past 400 years had been the Nations that ruled the world: Spain, England, France the Dutch, Portugal and eventually the USA, Germany and Japan.

In 1890 (and for thousands of years before 1890) if you wanted to travel any distance in the world 95% of the time the only way to do it was by Sea. The only way to communicate world information over great distances was 95% by Sea or by River. However **everything has a Bell Curve including military "Sea Power."** Military Sea Power has been eclipsed by Air Power and Space Power.

Yes -Aircraft Carries in the great Asian-Pacific War of 1941-1945 were **the USA's most valuable weapons system in the Pacific**. During WW2-Pacific the biggest advantage the American Carriers had was secrecy of location. In the vast Pacific the Japanese 97% of the time did NOT know where the US Aircraft Carriers (i.e. floating airports) were located - and where they would attack next. **The US Carriers had captured the initiative at the Battle of Midway and for the remainder of the war kept the initiative and kept putting the Japanese on the "Horns of Dilemma" everywhere in the battle zones.**

Everything is different today in the 21st Century. The Russians and Chinese know where the USA's Super Aircraft Carriers are located all of the time. The Aircraft Carrier's decisive advantage of WW2 is GONE.

In the 21st Century the Russians and Chinese have targeted the USA's super Aircraft Carriers with ICBM's that can hit the Carriers no matter where they are located in the world.

Both the Russians and Chinese now have power reentry ICBMs. The USA's Super Aircraft Carriers (in the open ocean) have zero defense against the new power-reentry nuclear ICBM's coming down on top of the US Aircraft Carriers at 24,000 mph. (A bullet from a high powered rifle travels at 2,000 mph).

(519) Sea Power from 1492 to 1960 was Decisively Important in World History. However in the 21st Century (when the Great Worldwide Colonial Empires are Gone) the Need for Military Surface Sea Power is Obsolete.

In the 21st Century the USA maintains (at great cost) a large and very expensive surface Navy for **ZERO strategic reasons**. What are the facts? (1) The USA has zero colonies to protect. (2) Most of the "sea trade routes" are of value to somebody else's merchant ships. (3) The USA is a gross importer of goods and consistently runs enormous Trade Deficits. The US Navy is protecting somebody else's trade profits. (4) The USA Merchant Marine has about 350 ships (most are small) but China/Hong Kong has almost 4,000 large ships - many are the largest in the world. (5) The USA has several Islands in the Pacific but they are all better protected with Air Power rather than ship Sea Power. (6) China has 175,000 ocean going fishing boats that are keeping track of the US Navy's ships the old fashioned way with eyes and radios. In a US-China shooting war if China lost all of its spy satellites China would nevertheless know where all of the US Navy's big ships were located.

If the objective of the US Navy is to "keep the Sea Lanes open" then the US Navy is keeping them open mainly for Chinese and Japanese ships.

In the 19th Century the British Navy keeping the Sea Lanes open was 100% logical because England had the biggest colonial Empire in the world (almost 25% of the total world's population). England had close to 8,000 merchant ships flying the British flag. However -

for the USA a big surface Navy makes no economic sense. The US Navy having more Super Aircraft Carriers than all of the world combined makes no military strategic sense whatsoever.

(520) The 21st Century Axiom is "He who Rules the Air and Space Rules the World."

Sea Power has been eclipsed by Air Power. Sea Power is also being replaced by a resurgence in Eurasian railroad land power promoted, financed and controlled by the Chinese. The Eurasian Continent is the world's biggest island with 65% of the world's population and a collected Gross National Product much larger than the USA's.

The Chinese are investing Trillions (in US Dollar value) in **the "One Belt One Road" system that will be able to connect markets all over "the world's biggest Island"** (and soon to all of Africa) without ever going to Sea. The Chinese will then have land routes to 3 Billion customers for Chinese products.

Despite the US Navy's showboat Aircraft Carriers from yesterday's wars the US Navy is already outclassed. When this book was published the Military Navies were: China's Navy 714 ships, North Korea's Navy 907 ships, Iran's Navy 398 ships, Russia's Navy 352 ships. The US Navy has 415 ships. The immediate collected "opposition" to the US Navy are 2,371 "enemy" ships versus 415 US Navy ships - a 5.7 to 1 ratio.

For comparison - at the start (for the USA and UK) of the Great Asian Pacific War on December 7th 1941. The "Allies" (USA, England, UK) had a 2.4 to 1 military-ship favorable ratio over the Japanese Navy. In reality the Japanese Navy never had a chance at victory - as **Winston Churchill** and President **Roosevelt** know.

(521) The Strategic Value of the US Navy is Declining and will continue to Decline during All of this Century.

During the Vietnam War (1959-1975) the US Navy was the largest and most expensive Navy in the world. North Vietnam had NO Navy - but North Vietnam won the war.

The US Navy was useless in influencing the eventual outcome of the 1960's Vietnam War. So why would the US Navy not be more useless in a 21st Century War?

(522) Military Grand Strategy Success is More than Throwing Money at the Military-Industrial-Complex.

The cost of the USA's Military is $700+ Billion dollars every year - a cost that is more than the combined yearly military budgets of Russia, China, Saudi Arabia, France, United Kingdom, India, and Germany combined.

The dollar costs of the US Military are so massive that people in Washington DC constantly keep asking and re-asking "Why can't we use this massively expensive machine for something of value?"

The answer is "As a machine to achieve many of Washington DC's usually unrealistic subjective desires for world control the US Military will continue to be impotent."

In the 21st Century the US Navy's nuclear missile firing Submarines and missile firing Cruisers **DO HAVE great deterrent value** against Russia, China and many other Nations. The US Navy's missile firing Submarines and Cruisers have great strategic value and are too small to be significant ICBM targets.

On the other hand the US Navy is wasting its money, time, effort and work on obsolete worthless status symbol super Aircraft Carriers.

The US Navy should be building dozens of nuclear firing Cruisers and Submarines and smaller ships that can launch highly intelligent weapons of many types for many specific objectives.

(523) Wasting large Amounts of Money on Yesterday's now Obsolete Sea Power Strategy is Highly Dangerous Overreach.

If spending money on wars won wars then the US Military would have won the Vietnam War by 1962. It would have won the Korean War 6 months after it started. It would have won Iraq War 2 and Afghanistan in under 5 months. **The proven and reproven FACT is that throwing Trillions of dollars at the US Military Industrial Complex does NOT win shooting wars for the average American.** Since the end of WW2 the ONLY winners in US Military shooting wars have been the US Military Industrial Complex and the US Warmonger Lobby.

(524) The Number One Greatest Strategy Benefit of a Large Navy was the Ability to Blockade your Opponent's Ports.

For centuries of warfare one of the decisive values of a large Navy was the Navy's ability to blockade an opponent's ports. Well into the 20th Century 85% of international trade and travel was by ship. If you could blockade a Nation's ports then you would cause great hardship to that Nation.

During WW1 the British Navy blockaded Germany and by 1918 tens of thousands of Germans were starving. Germany also suffered a critical lack of many vital war materials. **The British Navy's ability to blockade Germany from 1914 to 1918 contributed as much (or more) to Germany's 1918 surrender as the British Army's very costly in lives battles in France.**

It was France's Naval blockade of the British at Yorktown (1781) that won the American Revolution for **George Washington** and the USA.

During the US Civil War (1861-1865) the Union's Naval blockade of the South made it impossible for the South to sell its cotton to Europe. **Having slaves to grow cotton that could NOT be sold was a Union Navy highly successful <u>indirect attack</u> on the South.**

During WW2 Pacific the USA's Submarine Warfare was highly successful against the Japanese. The US Navy progressively blockaded Japan and by July of 1945 Japanese merchant sea trade was 93% dead and much of Japan was on the edge of starvation.

(525) Sea Power in the 21st Century has been Displaced by Air Power. In Eurasia - Sea Power is also being Displaced by Chinese Railroad Power.

Aside from the already great importance of Chinese world air transport power **(China has the world's 2nd largest fleet of commercial airplanes).** The Chinese are also working hard to expand Railroad Power as a further block to the USA's Sea Power. **You can already go from Shanghai China to London England all 100% by railroad.**

Air Power can go over both Sea and Land. Air Power can go anywhere in the world - and do it many times faster than ships. **Every new airport that is constructed in the world (or old airport that is expanded) makes Sea Power more obsolete.**

In the 21st Century most ships (and shipping) have value mainly for the movement of bulk cargos: oil, coal, iron ore, lower cost heavy items and so forth. In addition - 97% of the world's cargo ships are NOT USA ships - so **what is the US Navy protecting?**

WW1's and WW2's large conscript armies (that numbered in the millions) had to be moved by ships but mass armies in the Millions of men are now obsolete and **will never be seen in battle again**.

Today's cargos of US Troops (to the Middle East) are moved entirely by airplanes. The US Navy's last "Troop Ship" was taken out of "mothballs" and scraped in 1985.

The Russians and Chinese are constantly advancing and improving their ground to Sea nuclear ICBM's and improving their "spies in the sky" ship location systems - **especially the location of prime targets like US Aircraft Carriers.**

Suppose in a war with China the US Space Force succeeds in taking out all of the Chinese satellites. Then what? The Chinese have 175,000 fishing boats all over the world's Oceans. The Chinese boats can quickly and easily be disguised as boats from other Nations. The Russians have 2,500 large Ocean going fishing boats that are also part of the Russian Navy's intelligence. The US Navy has no such system and the USA has few fishing boats anywhere.

If the US Space Force takes out all of the Chinese and Russian "spies in the sky" the Chinese and Russian fishing boats can (and will) track the US Navy's ships and using old fashioned radio inform the Chinese and/or Russian Space Force.

If the US Space Force takes out all of China's satellites **we have to assume that the Chinese Space Force (and perhaps future Russian Space Force) will have taken out all of the USA's satellites.** That means NO internet, no cable TV and NO smart phone service probably for years.

The deadly mistake that the US Military is making is that money wasted on yesterday's weapons systems for yesterday's wars is <u>money NOT being spent on tomorrow's weapons that will be decisive for winning tomorrow's wars</u>.

The US Pentagon today is much the same as the European Armies immediately before WW1 that were putting large amounts of money, time, resources and effort into Horse Cavalry Divisions while 96% ignoring machine guns, airplanes and motorized transport.

(526) The USA's Strategists MUST be Aware of the fact that the Chinese have Resurrected and are Using the Geopolitical Strategies of: (1) Halford Mackinder, (2) Friedrich List, (3) Nicholas Spykman and (4) Friedrich Ratzel.

Sir Halford Mackinder (1861-1947) was the director of the London School of Economics and was famous for his book ***The Geographical Pivot of History*** (1904).

Mackinder said that the great Eurasian land mass (Russia, China, Europe etc) is the "Heartland of the World." Mackinder's summary was **"He who controls the Heartland has command of the World's Biggest Island. He who controls the World's Biggest Island will command the world."**

The Chinese are working hard to greatly expand their already large Eurasian railroad system and dominate "The World's Biggest Island."

The Chinese military objective is to make the US Navy in Asia and Europe irrelevant and useless and the Chinese are succeeding.

The Chinese are NOT going to play the US Navy's game with the US Navy's WW2 rulebook. The Chinese are NOT going to try and match the US Navy's strategically obsolete Aircraft Carriers for an Aircraft Carrier war in the middle of the Pacific Ocean. That is NEVER going to happen. **The Chinese are playing their game with their rules and have already out maneuvered the USA's Aircraft Carriers.**

Washington DC has no world Grand Strategy other than trying to make the Middle East safe for Apartheid and opposing any change because it is a change.

(527) In 1835 the Brilliant Grand Strategy Analyst Friedrich List Predicted that the then new Railroads would Eventual Neutralize Sea Power.

The Chinese are financing railroad expansions and connecting railroads in dozens of Nations. The Chinese are spending hundreds of Billions of dollars and building a new Grand Silk Road that is uniting Asia and Europe - and eventually all of Africa via railroad lines that would go under the Suez Canal.

The Chinese are following Sun Tzu's core Grand Strategy rule of NOT attacking your opponent's strong areas but always attacking your opponent's weakest areas. Rather than the Chinese building Aircraft Carriers to play the US Navy's game with the US Navy's rules. **The Chinese are building railroad lines and playing their World Grand Strategy game with their rules.**

Actually - the Chinese are building Aircraft Carriers - several of them. They are called Air Bases in the South China Sea Spratly Islands. The difference between the Chinese Aircraft Carriers and US Navy Aircraft Carriers is the Chinese Aircraft Carriers cannot be sunk.

The US Navy knows where the Chinese Islands are located. However **unlike WW2 the Chinese today know exactly where the US Aircraft Carriers are located. Secrecy of location was the great decisive advantage of WW2 US Aircraft Carriers.** Today that decisive secrecy advantage does NOT exist - and thus the strategic value of Aircraft Carries in a shooting war with China does not exist.

(Historical Note) By the Spring of 1945 the US Navy had built 28 large Aircraft Carriers and 71 Escort Carriers (99 total Carriers). During WW2 Japan had 15 large Aircraft Carriers, 5 light Carriers and 5 Escort Carriers (25 total Carriers). All but 2 of the Japanese Carriers were sunk during WW2 Pacific.

The USA's WW2 manufacturing industry was able to overwhelm both Japan and Germany. US industry during WW2 built 300% more Aircraft Carriers than Japan (and in only 3 and 1/2 years). However - in the 21st Century most of the USA's WW2 manufacturing capacity is now either somewhere else in the world or dead. Over 60,000 USA manufacturing companies have closed their doors in the last 50 years. 100% of the skilled workers who made the USAs factories work during WW2 are now dead or far too old to work.

If there ever was a USA-China shooting war (God forbid) Washington DC would learn the very hard way that restaurants and sports stadiums and people on food-stamps do NOT win wars. **It takes factories to win wars and China has the factories**.

(528) Professor Nicholas Spykman's Strategy of "Containment" was Used to Defeat the USSR. In the 21st Century the Chinese could use "Containment" to Eventually Isolate the USA into mainly North and South America.

Professor **Spykman** (1893-1943) before WW2 ended formulated the Strategy of containing the USSR. That grand strategy took 45 years but succeeded in ending the USSR.

The same strategy could be used against the USA. The US Pentagon looks at a map of the world and sees two large oceans protecting the USA. **The Chinese look at a map of the world and they see two large oceans that isolate the USA.**

If USA-China relations deteriorate and stay bad - then China's long term Grand Strategy would be to isolate the USA into mainly North and South America.

The USA's Grand Strategy should be getting as close to China as possible without being taken advantage of by the Chinese.

(529) The German Geographer Friedrich Ratzel Analyzed Cultural Achievement and Winning Wars.

Friedrich Ratzel (1844-1904) a German college professor and Franco-German War hero wrote several books about the quality of a population's culture and how that quality (or lack of quality) **directly affects military successes or failure**.

The Chinese believe that **the quality of their soldiers is superior to the quality of the average US soldier.** Let us pray that idea is never tested in the real world.

One of the tested and proven results of the Vietnam War was that the quality of the Viet Cong soldier was superior to the quality of the average US soldier. For those who want to argue - then explain why the Viet Cong won the war.

There are many Pentagon excuses for the US Generals losing in Vietnam. The usual excuse is "The US Generals never really tried." FACTS - the US Generals dripped 5 Times the tons of bombs on Vietnam than the USA dropped on Nazi Germany. The US Generals used the biggest Chemical Warfare attacks in the world since WW1. The US Military killed 650,000 Vietnamese men, women and children. The US Pentagon killed 58,220 US Troops in Vietnam. That is a little more than "never really trying."

At the end of the day a 100 excuses mean nothing. You are either a winner or a loser. The US Generals must NOT be fighting shooting wars until they reorganize and retain themselves to be a winners.

(530) Sometimes NOT Having a Weapons System is Better than a Weapons System that can Get You Into Deadly Trouble but Cannot Get You Out of Trouble.

Before 1914 the German Nation was spending large amounts of National wealth (and work effort) towards building a large Military Navy. Kaiser **Wilhelm II** of Germany was an enthusiastic believer in the ideas of US Admiral **Alfred Mahan**. However in Grand Strategy terms the German Navy was big enough to deeply aggravate the British Navy - but NOT big enough to defeat the British Navy. **In both WW1 and WW2 the German Navy was the worst of all worlds - expensive in time, money and effort but less than useless against Germany's #1 competitor - the UK/England.**

When Germany attacked France in 1914 the Germans were sincerely and completely expecting the British to stay neutral as England had done in the 1870 Franco-Prussian War. However in 1870 there was no united "Germany" and there was no German Navy. **In 1914 the British correctly assumed that the Germans would defeat France if France received no help.** In 1914 the rankings of the world's Navies by size were: #1 England - #2 Germany - #3 France - #4 Austria-Hungary. The English could clearly see that when the Germans defeated France (and then confiscated the French Navy) the then combined French ships plus Germany's Ally Austria-Hungary's ships - **that would have made the German/Austrian/French Navy bigger than England's Navy.**

The English also did NOT want French's excellent ports on both the Atlantic and Mediterranean (and in the French Colonies) to fall into the hands of the Germans.

The British joining with the French (in 1914) to fight Germany came as a great shock and surprise to Kaiser Wilhelm and the Germans. England's Queen Victoria was Kaiser **Wilhelm's** Grandmother and the British Royal Family was mainly German. However - **the Germans forgot that Parliament rules England and Parliament was not German.**

Germany's WW1 Navy failures give the World 5 Grand Strategy lessons to remember and follow.

(1) **Do not build a weapons system big enough to get you into trouble but not big enough to get you out of trouble.**

The USA's Aircraft Carriers could get the USA into trouble but NOT out of trouble.

(2) **Major funding priority mistakes with your weapons systems can be fatal.**

Kaiser Wilhelm followed Admiral **Mahan's** advice and put much German wealth, work and manpower into building what was a less than useless German Navy. If most of the German money and manpower had been put into the German Army then the German **Schlieffen** Plan might have defeat both France (and the British forces) in the first two months of the war. The major disasters of WW1 (and their aftermath like **Hitler**, **Stalin** and WW2) would have been avoided.

Today the US Generals are wasting Billions of dollars building and maintaining obsolete weapons systems. The money, time and effort should to used towards building Artificial Intelligence Weapons systems, building the Space Force, building the Communications Force and reorganizing the Pentagon so it can stop losing everywhere and start to win everywhere.

(3) **Do NOT build a weapons system that will drive otherwise neutral Nations into the arms of your opponent.**

The Kaiser's WW1 submarine fleet was big enough to draw the USA into the war with Germany but NOT big enough to stop a million US Troops from crossing the Atlantic to France. The German submarines also did NOT have the ability to break the decisive British Sea blockade of Germany.

The Germans made exactly the same submarine mistake in WW2 - and in classic form they made the mistake bigger. During WW1 the Germans lost 202 submarines out of 375 total. During WW2 the Germans lost 784 submarines out of 1,156 total. The Rule is - **"If you make a failed attack bigger what you will usually get is a bigger failure."**

(4) **An all or nothing Strategy of weapons systems should always be considered and analyzed.**

If you cannot build a major weapons system big enough to defeat your opponent and win a major war - then you might be correct to NOT build a particular weapons system at all.

Since 1946 the Japanese have maintained only a token small Navy and Army (and they have no nuclear weapons). Unlike the USA - Japan has NOT lost a shooting war since 1945.

(5) **Having Big Armies and Big Navies and many wars Does NOT make Nations happy.**

If wars made Nations happy then the USA would be the happiest Nation in the world. The top 5 happiest Nations in the world are: Finland, Norway, Denmark, Iceland and Switzerland. None of them have big Armies or Navies and none of them have troops in somebody else's Nation killing the local populations. The last time Switzerland had a war was 1815 - over 200 years ago. The last 6 generations of Swiss have NO memory of their Nation being at war.

The US Navy is definitely big enough to easily start a shooting war with Iran but it is definitely NOT big enough to win a shooting war with Iran. Let us pray that the US Generals and Admirals do not once again have to learn the hard and stupid way.

(531) **The Eight historical Reasons for having a National Navy are: (1) To protect your Colonies. (2) To protect your Economic Trade Routes and keep them Open. (3) To protect your Merchant Marine Ships. (4) To keep vital Foreign Ports Open for your Shipping. (5) To have the Ability to Land Troops and Invade Areas. (6) To Blockade an opponent's Nation. (7) To protect your Coast Line from Invasion. (8) To Prevent a Blockade of your Nation by an Opponent's Navy.**

In the 21st Century the USA has little to protect on the Oceans or anywhere else around the world. The last back-to-back Trade Surplus the USA had with the world was in 1975.

(The top of the USA's Bell Curve). In the 21st Century the USA Trade Deficit runs about $500 Billion dollars per year. **Today the US Navy is protecting other Nation's trade routes and other Nation's ships and other Nation's cargoes and other Nation's profits that are coming out of the pockets of all Americans.**

(532) Do NOT put Great Amounts of Money into Building Big Warships only Because you Can Build Big Warships.

The USA putting Billions and Billions of dollars into building yesterday's weapon system's (like super Aircraft Carriers) is classic Washington DC of the past 60 years - living in a WW2 mindset of yesterday's weapons for a replay of yesterday's wars.

The delusion is that if you spend Billions of dollars on updates of yesterday's weapons systems that "proves" the almost worthless status symbols have value.

The only winner in this game is the Military Industrial Complex that is playing everyone for suckers and is laughing all the way to the Bank.

Washington DC overspending on yesterday's weapons for yesterday's wars is like the very rich women who spend $40,000 to $75,000 on Hermes Birkin handbags and <u>sincerely believing the handbags are worth $75,000 because that is what they paid for them.</u>

In future wars (during the Artificial Intelligence Age of Warfare) it will NOT be big size that will be decisive. What will be decisive in battle will be large numbers of relatively small and highly intelligent weapons devices.

(533) Nations that were "Land Powers" Can Become Great "Sea Powers" in a Short Time.

Rome was a 95% land power until the First Punic Wars in the 2nd Century B.C. However - almost overnight Rome built a Navy and became a great sea power. A sea power that defeated its main rival Carthage that was located on the African South side of the Mediterranean opposite Rome.

Japan had little more than fishing boats in 1875 but by 1905 Japan's Navy defeated the Russian Navy that (at the time) was the world's fourth largest Navy in the world.

China in 1960 had almost no Navy. However in the 21st Century China has become a significant sea power with more total warships than the USA (715 for China and 415 for the USA when this Book was published). China has the world's biggest merchant Sea Trade and is expanding its Military Navy as **a logical step to protecting China's almost 4,000 merchant ships.**

(534) **Having a large Navy does NOT make you the World's Biggest Exporter of Goods.**

There was a time when having a large Navy made you a great economic power. However - in the 21st Century having a large Navy does NOT make you a great economic power. **What makes you a great economic power is selling much more to the world than you buy from the world** as the USA had been doing from 1775 to 1975.

(535) **Today's ICBM's have Pinpoint Accuracy. The USA's large Aircraft Carriers would be Floating Coffins in a Shooting War with China or Russia.**

The new Russian power dive nuclear ICBM's can reenter the Earth's atmosphere (about 90 miles above the Ocean) at a speed of 24,000+ mph (that is 6.66 miles per second - 35,000 feet per second). The time between the ICBM reentering the Earth's atmosphere and exploding 35,000 feet above a US Aircraft Carrier would be about eleven (11) seconds.

For comparison a US Army .50 Caliber Browning Machine Gun bullet travels at 2,800 feet per second.

Here is what would happen to a US Navy Super Aircraft Carrier in the middle of the Pacific Ocean in a shooting war with China or Russia.

1. The Chinese or Russian ICBM coming from outer space would power dive down onto the Carrier at 6.6 miles per second and explode 35,000 feet above the Carrier.

2. The carrier deck would be hit with intense radiant heat with a high probability of exploding gasoline and ammunition in the fighter jets on the deck. Anyone on deck with exposed skin would be badly burned.

3. The high pressure ball created by the nuclear explosion would come down onto the Carrier and push the Carrier into the water almost up to its flight deck. Personnel and equipment would be mashed up onto the Carrier's internal ceilings. Many personnel would have broken bones and there would be a high probability of ammunition explosions.

4. The pressure ball would then bounce off of the Ocean and the Carrier would bounce up and almost come out of the Ocean. Everything lose would be smashed down onto the deck. More broken bones and a high probability of ammunition explosions. The pressure ball bouncing off of the Ocean or land is the classic nuclear mushroom cloud going up into the sky and pulling debris up with it.

5. The pressure ball mashing into the Ocean would have forced all of the air out the area. When the pressure ball bounces up the air will rush back in at 650 mph. Anything lose on the deck of the Carrier (or loosened by the blast) would be blown away.

6. The ship would have been hit with intense radiation. With 5 to 6 hours anyone alive on the ship would go into convulsions and tremors - and then everyone would die.

7. If by some miracle the Super Aircraft Carrier did not sink. After 3 days it would be a smashed ship full of rotting dead corpses.

ICBMs are usually MIRVed (multiple independent targetable reentry vehicles). Russian and Chinese ICBMs can have 6 independent nuclear warheads per ICBM and can target every bigger ship in a typical US Navy Aircraft Carrier battle formation. The ICBM coming down onto the Carrier taskforce would NOT be one target to shoot down but 6 separate targets all coming down onto the US Ships at speeds 8 times faster than the speed of a high powered rifle bullet.

Russian and Chinese nuclear missile submarines can fire nuclear missiles at a US Aircraft Carrier from 1,000 miles away.

Russian and Chinese submarines have nuclear torpedoes. Some are stealth nuclear torpedoes that can lay on the sea floor dormant in the general path of a US warship until the Carrier is close enough for them to activate. Then they attack from the Sea floor upwards and sink the ship.

Nuclear bombs can be under 100 pounds in weight and delivered by small smart-machine speed boats or smart machine mini-submarines that are not much bigger than large fish.

Disguised warships are as old as Naval Warfare. A 25 foot sailboat could have a nuclear torpedo tube built into its hull. A fisherman in a 14 foot boat could have a nuclear bomb. The Chinese have 175,000 fishing boats and all of them are connected to the Chinese Navy. In a USA-China shooting war even if the US Space Force could take out all of China's spy satellites the 175,000 fishing boats (spread all over the world and easily disguised as from another Nations) would know where the US Navy's capital ships where located.

Over 1,000 large Chinese merchant ships are registered in several non-Chinese Nations (including the USA). They fly the flags of their registered Nation - including the US Flag. However - they are also spy ships connected to the Chinese Navy. In a shooting war they would be giving the Chinese Navy the location of US ships - and possible attacking US ships.

The bigger the ship the more effort an opponent will put into sinking that ship. During WW2 the motto of the US Navy in the Pacific was "Hit the Carriers First." The US Navy's Aircraft Carriers are very big #1 prime targets with great psychological value to an opponent after they are destroyed.

(536) **The Chinese have Built an Aircraft Carrier but that Could be Bait (and a Trap) for the US Navy.**

The Chinese have built a Super Aircraft Carrier (70,000 tons) to prove that (1) they could do it. (2) As a status symbol in Asia. (3) They wanted to understand exactly what are the weaknesses of Aircraft Carriers. (4) Against a 2nd or 3rd rate Nation an Aircraft Carrier might be highly effective. (5) China's Super Aircraft Carrier might be a trick (and a trap) to exploit and encourage the US Navy's delusions of a Chinese Aircraft Carriers versus US Aircraft Carriers battle in the middle of the Pacific Ocean.

A Pacific Aircraft Carrier war (like WW2) will never happen again. However the Chinese Strategy analysts know that reliving WW2 is the US Pentagon's greatest obsession and a major weakness. **A good strategist always exploits his opponent's weaknesses.**

No Nation has unlimited money, time and effort. The money, time and effort that the US Generals and US Admirals and Washington DC waste on worthless status ego-weapons from yesterday's Ages of Warfare means that much LESS money, time and effort is being used to build the needed weapons and systems for tomorrow's wars.

Remember the Rule - One of your objectives is to trick your opponent's forces into defending strategically useless locations and/or trick your opponent into wasting money and time on strategically useless equipment. The Chinese are very successfully using that strategy against the US Generals.

Four Star **US Admiral Hyman Rickover** (1900-1986) the longest active duty service person in US Military history (1918-1982) and "the Father of nuclear missile firing submarines" said in 1983 - "America's Aircraft Carriers would last longer if they stayed in port." Many decades after 1983 - **today's US Aircraft Carriers are much more vulnerable to a Chinese and/or Russian ICBM** than when Admiral **Rickover** basically said they were floating coffins.

During the Vietnam War (1959-1975) the US Navy (and US Aircraft Carriers) were **battle proven to be irrelevant**. The proven irrelevancy of Aircraft Carriers in the Vietnam War was a half Century before this book was published and it was already "the writing on the wall."

(537) **Always Work at Protecting your Collected National Ego.**

Washington DC needs to learn the lesson from the Russian-Japanese war of 1905. The Russians could have lost several land battles and the average Russian would NOT have been concerned. However - when the Japanese fleet sank the Russian fleet in the Battle of Tsushima that revealed (in a crystal clear way for everyone to see) the intellectual bankruptcy of the Tsarist Military. That produced a bloody revolution in Russia. The Tsar survived the 1905 revolution but was permanently weakened and 12 years later did NOT survive the 1917 revolution

In warfare (as in life) you have to protect your ego from crushing defeats. If the US Navy lost 3 or 4 destroyers the voter's would be unhappy but their reactions would be "well - losses happen." However - if the US Navy lost 6 or 7 Super Carriers (and 20,000 sailors) in a week's time then the USA voters would be both profoundly depressed and angry and the voters would turn against the war and the Pentagon and whomever was President.

(538) In the 21st Century there is a Powerful Argument for having a Navy Built Around Many Smaller Ships that have the Capacity to Fire Nuclear Weapons.

For over 500 years the warships that had the biggest cannons had the advantage. To have ever bigger cannons you needed ever bigger ships. However - **with nuclear weapons the basic equation has fundamentally and permanently changed everything "big" become a target.**

Smaller boats have also been a proven winning formula for Centuries. Probably most US Navy Officers do not know that many of **the real "Pirates of the Caribbean" used dozens of small sailing boats** that were faster (and easily built) against the big fat and slow Spanish Galleons. It was very difficult for the Galleons firing round balls from smooth bore cannon to hit 25 or 30 small pirate sailing boats bobbing in the ocean. If 5 or 6 small boats were hit by cannon balls and sunk - so what? When all of the small pirate boats had surrounded the Galleon they would all board the Galleon at the same time. The Black Power rifles of the Galleon's crew could fire only 3 rounds a minute and were very inaccurate. It was like 30 wolfs attacking a rhinoceros from all sides.

Sometimes in warfare what is old is new again.

When the world has fully gone into the coming Artificial Intelligence Age of Warfare the many small boats will all be controlled by on-board smart computers (the boats will have nuclear and/or very powerful weapons) and there will be no humans.

What is the WW2 history of "Big Ships?" The three biggest warships of WW2 were the Japanese Navy's Yamato and Musashi Battleships - and Shinano Aircraft Carrier. All Three Super Ships fully loaded were 73,000 tons. That was 19,000 tons larger than the US Navy's biggest WW2 ships - the Iowa Class US Battleships at 54,000 tons fully loaded. The Japanese ships were more than twice the size of the American USS Essex Class Aircraft Carriers at 33,000 tons and the American USS Enterprise class Aircraft Carriers at 27,000 tons.

Battle is the ultimate proving ground for weapons systems and all three Japanese "Super Ships" were <u>proven in battle</u> to be less than worthless. None of the Japanese Super Ships ever fired their 18 inch guns at an enemy ship or launched any attack airplanes at an enemy ship. They were all sunk in battle with great loss of Japanese life. **The Japanese could have used the same time, effort, materials and money that went into the 3 Super Ships and built 85 Heavy Destroyers. The history of WW2 Pacific proved that the Japanese Destroyers would have been many times more effective than the Super Ships.**

In the Great Asian-Pacific war the largest Battleship versus Battleship fight was the battle of Siargao Strait (October 24th 1944). The battle was two Japanese Battleships versus 6 US Navy Battleships. However - the US Destroyers were the first to attack the Japanese Battleships and they definitely sank one Japanese Battleship (and probably fatally damaged the second Japanese Battleship) before the US Battleships opened fired. If none of the US Battleships were at the battle - the US Destroyers and Cruisers would have won the battle by themselves.

Bigger has many times in battle been proven to NOT be better.

Chapter 27

Nuclear Weapons Have Made the "Draft" Obsolete

The Strategist must assume that Tactical Nuclear Weapons will be used in the future. There will be many future wars and the use of Tactical Nuclear Weapons is only a question of when and where.

Nuclear weapons have made "Big" obsolete. Big Armies with dense collections of concentrated troops, Big Warships and Big Military Bases. **Every Big item that has military value becomes a prime nuclear target - and that means giant WW2 Armies in the Millions are obsolete.**

(539) The USA's best Deterrent Is an Opponent Understanding that a Nuclear Attack Against the USA would Bring Massive and Horrible Pain to the Attacker.

Dr. **Sigmund Freud** (1856-1936) was the founder of psychoanalysis. Dr. **Freud** said that insane people do understand pain. People with the most serious mental illnesses nevertheless know that if something will cause then physically pain then 99% of the time they will NOT do it.

Dr. **Freud's** axiom is proven by the **"Broken glass scattered over a sidewalk"** fact. There are many crazy people who were (are) significantly insane and who were (and are now) National Leaders. However the fact is that none of them have ever crawled naked over a sidewalk covered with broken glass. WHY?

The "WHY" is because **even the most insane person understands what will cause real pain. When National Leaders completely and truly understands that doing a specific action will hurt them badly - they will usually refrain from doing that action.**

Most criminals are psychopaths and insane. But they all understand the pain of going to jail.

This is a salesmanship factor. The USA's communications must always make it absolutely 100% clear that any attacker using nuclear weapons against the USA will suffer devastating destruction.

Any National Leader thinking of using nuclear weapons against the USA must be told clearly and sincerely that if he does so he will be the ruler of a destroyed Nation full of rotting dead corpses.

(540) Having many Nuclear Weapons is a good Defense but for Offensive Wars Nuclear Weapons are mainly Useless and Phantom Firepower.

Having many nuclear weapons does not mean you win non-nuclear conventional wars. The USA had thousands of nuclear weapons during the Vietnam War (1959-1975) but the USA lost that war.

Remember the Rule - (1) **If you cannot envelop a Nation and** (2) **that Nation will not surrender. And** (3) **using nuclear weapons is world public relations impossible then** (4) **you have no way to end the war.**

There is a high probability that a USA nuclear war with Russia and/or China could officially go on for years with neither side firing any additional nuclear weapons but also neither side surrendering.

In the USA that could mean many **years** of no satellites and thus many years of no cell phones and no internet.

(541) Any Nation that Believes Acquiring Nuclear Weapons will be the Ultimate Answer to all of their Problems is Going to be Very Disappointed.

Nuclear weapons are in many ways White Elephants that: (1) **cost massive amounts of money.** (2) They are **a constant world public relations problem** for their holders. (3) Nuclear weapons do NOT grow crops. (4) Nuclear weapons do NOT build automobiles or cell phones or any type of useful products. (5) Nuclear weapons do NOT provide medical care for the sick. (6) Nuclear weapons do NOT build schools and Universities. (7) Nuclear weapons do NOT raise the average worker's standard of living. (8) Nuclear weapons are a money draining detriment to the standard of living for average workers in Nations that have Nuclear weapons.

What have nuclear weapons done for North Korea? Is the average North Korean worker richer? North Korea's nuclear weapons put North Korea into the USA's and Russia's and China's Computer Nuclear Attack Targeting Systems.

What North Korea's nuclear weapons have done is start the Japanese thinking about building their own nuclear weapons and ICBMs. **Do the Chinese (or the USA) really want North Korea to create a nuclear Japan?**

In 1950 the USA had a 97% monopoly on nuclear weapons (only Russia had a few). However the USA's nuclear weapons did NOT defeat North Korea in the Korean War (1950-1953). America's nuclear weapons did NOT prevent the USA's defeat in the Vietnam War (1959-1975) or the de facto defeats in the USA-Iraq War (1991 to ?) and in the USA-Afghanistan War (2001 to ?). The USA's nuclear weapons did NOT prevent the USA from becoming the biggest debtor Nation (by gross amounts of money) in the history of the world.

Nuclear weapons have the classic-flaw of being able to get their holders into serious trouble but NOT out of trouble.

If you use your nuclear weapons against another Nation that also has nuclear weapons your nuclear weapons CANNOT prevent a massive retaliatory counter attack against your population.

(542) Nuclear Weapons have an Upside. Any National Leader who is Only A little Sane is NOT going to Start a Nuclear War for the Fun of Having a Nuclear War.

If there had been no nuclear weapons after 1949 the probability of a major shooting war between Russia and the USA would have been very high. **Nuclear weapons on both sides made the defense of both sides immensely strong and nuclear weapons also made an aggressive attack profoundly problematic.**

Nuclear weapons prevented World War 3 between the former USSR and the Western Nations.

(543) A Nation with 50 Nuclear Weapons has as much Psychological War Prevention as a Nation with 2,000 Nuclear Weapons.

If an opponent is insane enough to NOT fear 50 nuclear bombs then he is insane enough to not fear 2,000. In the 21st Century advanced Nations are 80% urbanized and usually have 50% of their populations in their top 15 urban areas. If you have 50 nuclear bombs you can nuclear bomb 50% of your opponent's population.

(544) In a War with Tactical Nuclear Weapons your Troops must be Spread Out over a Large Area. The Depth of your immediate FRONTLINES (against an Opponent using tactical Nuclear Weapons) will have to be 100 to 150 Miles Deep.

With tactical nuclear weapons "big anything" becomes the target. The bigger the mass of troops the bigger the target.

(545) Tactical Nuclear Weapons being Used Somewhere in the Future is Only a Matter of Time.

The world history of weapons is a history of increasing destructibility and the eventual use of that destructibility. There are no reasons to believe nuclear tactical weapons will be an exception forever. **The highest object of US Grand Strategy is to NOT be part of a Nuclear War when it happens.**

Alfred Nobel (1833-1896) the inventor of dynamite (and creator of the Nobel Prizes) believed that dynamite was so destructive it would put an end to warfare. As we all unfortunately know that was NOT what happened. Dynamite made it possible to kill more soldiers faster.

Ironically the Nobel Peace Prize was originally financed by the sale of explosives.

Nuclear weapons will be used somewhere in the future. The US Pentagon must be reorganized now so when a nuclear attack happens somewhere outside of the USA the US Military is not in the middle. **The US Pentagon must break its sick addition towards want to be in the middle of every war in the world.**

If cutting the Pentagon's budget by 50% is what it takes to break the Pentagon's War Monger addition of wanting to be in every war everywhere than maybe that is what is needed?

(546) Nuclear Weapons have Greatly Reduced the Direct Wars between Nuclear Great Powers (something to be very grateful for). However the Reduction in Big Wars has Increased the Number of Smaller Wars and Proxy Wars.

Small wars that can be contained have some upside. They release tensions that might otherwise build up to the point where a big war starts. Proxy wars also give great powers a way to have a shooting war without actually directly fighting each other.

(547) The Proxy Wars of Nuclear Powers have a Different Set of Rules and Standards from the WW2 Industrial Age Warfare. Proxy Wars are by definition <u>Limited Warfare</u> with usually very <u>Complicated Political Underpinnings</u>.

Since the end of WW2 the USA has been part of many proxy wars. **The proxy wars that the US Pentagon loves are highly dangerous** because the USA's Bell Curve (relative to the world) topped out in 1975 and the USA (as a great power on the downside of its Bell Curve) **cannot afford to be fighting non-vital wars.** The more wars Washington DC enters into the faster the USA will decline relative to the world even if the US Military wins those wars.

(548) Prussian General Carl von Clausewitz (1780-1831) is Famous for his Saying "War is a Continuation of Politics by Other Means." That Mindset is now Dangerously Obsolete.

The 21st Century rule is - **"Economic warfare and Cultural warfare are Political warfare by more productive means than Shooting warfare."**

Economic warfare is warfare by other means. As China has been proving since the 1980's with China's Economic war against the USA

General Clausewitz is popular in the US Pentagon because Clausewitz dummied-down Military Strategy to being a large barroom fight. Every military small minded General can understand **Clausewitz** because it is a simple minded Strategy for lazy thinkers. **Clausewitz's** Book can be summarized as "You win the battle and you win the war." However - the US Military won every battle in the Vietnam War and lost the war.

In the 21st Century - if the USA had a nuclear war (with Russia or China) in a few days 150 million Americans would be dead and half of America would be destroyed. **That is NOT "politics by other means"- that is National Suicide by very fast means.**

(549) The Highest Objective of the USA, Russia and China in the 21st Century Must Be the Objective of Getting Through this Century Without a Major Nuclear War.

There will be no "winners" in a big nuclear war - only losers.

SUMMARY

Nuclear bombs were invented in 1944 and nuclear bomb technology is already old. Many Nations have the ability to build nuclear bombs. **If you keep threatening other Nations with military attack (as Washington DC does) then the threatened Nations are logically motivated to build nuclear bombs.**

It has become a circular loop of insanity. The US Warmonger Lobby threatening Iran encourages Iran to build an atomic bomb. Then the Warmongers say that is the reason for why the USA must attack Iran.

Here is a solution for Iran. The US Military loves being outside of the USA. With good reasons because most US bases are horrible. A solution for everyone is a 10,000 US soldier base in Tehran Iran. The US Military is not going to nuke Tehran with 10,000 US soldiers in the city. This is a reverse "trip wire" tactic.

Chapter 28

There Will NOT be a USA-China World War Why?

There will NOT be a USA shooting war with China unless Washington DC starts a shooting war with China. WHY? Because out of the 37 different types of warfare China has already won the only war it wanted to win - the Economics War.

There is a major difference between the Western Christian mindset and the classic Chinese mindset. The Western Christian mindset is always looking for the ultimate complete victory. The Western mental-set demands "perfect" and "complete" victory and is NOT comfortable with 75% victory. However the classic Chinese mental-set can be satisfied with "Good."

China is already the #1 largest hard goods product manufacturer in the world. China continues to be on the upside of its reset Bell Curve and that will probably continue on its upside for all of this Century. China has nothing of decisive value to gain from a shooting war with the USA but China has much to lose in a shooting war.

(550) The USA-China Economics War Happened and is now Over - and China Won.

The Great Age of Exploration (1492-1850) was the Age when shooting wars paid for themselves. Shooting wars sometimes made the aggressor fabulously rich. However starting in the 20th Century (and the growth of highly destructive military weapons) shooting wars usually did NOT make the aggressor richer and often greatly (and permanently) damaged the aggressor.

In the "good old days of warfare" soldiers were paid very little. If they were wounded and permanently crippled - that was their problem. Military organizations were NOT welfare states inside of a welfare state.

When this book was published there was much talk coming from the American chattering class about a USA-China war. The USA's chattering class should have been talking about "the war with China" in the 1980's. Today **the USA-China Economic War is over and China is now the #1 manufacturing Nation in the world.**

Today if you go into a Super Walmart 96% of all of the manufactured hard goods are made in Asia. Walmart (the largest retailer in the world with 11,500 stores in 27 countries) is a de facto colony of China.

In 1940 if you went into a Sears Department Store (the USA's largest retailer in 1940) 98% of everything in a Sears store (both soft and hard goods) would have been made in the USA.

In 1940 - 99% of the cars and trucks on the USA's roads were "American Made." **Today the 100% "USA Made Car" is extinct. The minority of cars on the road that are labeled as "USA Cars" are full of foreign made parts (many are Chinese and Japanese parts) and the parts might have been assembled into a car in Mexico or Canada.**

In 2017 the USA assembled about 13 million vehicles in the USA, Canada and Mexico. China manufactured 30 million vehicles in China with an average of 96% Chinese parts.

China's Universities are graduating per year five times the number of engineers, scientists, mathematicians and technology specialists as the USA is graduating. China is graduating scientists-engineers and the USA is graduating Lawyers.

Lawyers (and a service/entertainment economy) do not win shooting wars. That is only one of the many reasons for why the USA must avoid as many shooting wars as possible.

The USA's industries during WW2 overwhelmed both **Hitler's** Germany and **Togo's** Japan with manufactured war products. However most of the USA's WW2 war winning manufacturing capacity has already been moved to China, Japan and other parts of the world.

(THE POINT) The point is that despite the Pentagon's enormous budget (that is mainly wasted money) the USA does NOT have the ability to fight **and win** significant sized shooting wars.

(551) Washington DC needs Plans and Programs to Make Sure that Someday Economic Sanctions are NOT Used by China against the USA.

If the USA becomes a difficult problem for the Chinese their Grand Strategy would be to envelop the Asian part of the USA's banking system and keep the US Dollar's influence mainly limited to North and South America.

The Chinese Yuan, Japanese Yen and Indian Rupee would become the main Reserve Currencies particularly in Asia. The USA's Banking system and US Dollar are already becoming progressively less important in Asia with each passing year. This is making US economic sanctions against places like North Korea, Iran etc a little less effective every year.

China could use their ever growing financial power to progressively restrict the use of the US Dollar in the World. As each year goes by it becomes easier to sell the idea of using something other than the US Dollars that increasingly becomes a bigger Madoff type of accounting fraud.

The world is no longer in year 1946 when the USA had 37% of the world's advanced hard goods manufacturing and was the financial creditor to the world.

At the beginning of the 20th Century the British Pound occupied a stronger position in the world then today's US Dollar. The British Pound was the world's reserve currency. However - everything has a Bell Curve and the predominance of the British Pound is long gone. The US Dollar is already on the downside of its Bell Curve and will continue to be less important with each passing year.

Washington DC must understand that the US Dollar as a weapons system is NOT a forever weapons system.

Rather than stupid wars in stupid places. What Washington DC should be thinking about is what will happen to the USA when the Federal Reserve can no longer live on fabricated debt - and the USA has to produce things of real value to pay for the items of real value that it imports and consumes?

(552) The Solution for the USA's "Trade Deficit" with China is to Concentrate on Rebuilding the USA's Industries so America can Once Again Produce Most of What it Consumes.

In 2019 the USA was 5% of the world's population and 15% of the world's advanced product manufacturing. (In 1946 the USA was 5% of the world's population and 37% of the world's advanced hard goods product manufacturing). If the current trends continue by 2035 with current trends the USA will be 5% of the world's population and 10% of the world's advanced manufacturing.

(Note) Increasing the USA's percentage of advanced manufacturing must be one of the USA's highest grand strategy priorities.

In 1870 the British Empire controlled 35% of the world's advanced manufacturing and 23% of the total world's population - but as is true with every Bell Curve the British Bell Curve eventually went into decline.

The decisive economic game changer for the USA will be when the world resists taking the US Federal Reserve's Madoff fraud-money. **Washington DC's Madoff phony money con game will keep on going for a long time - but it is NOT going to last forever.** When the US Dollar's Bell Curve is on its far downside there will be high US Dollar inflation that will be very painful for 70% of the US voters. Is Washington DC prepared? The answer is - NO.

(553) If your Opponent has a Strong Conventional Defense of Army, Navy and Air Force then your Strategy must Use the Non-Shooting Tools of Warfare.

The USA has a strong defense of the traditional old weapons systems and **the strategy competent Chinese are NEVER going to challenge the USA with Chinese Aircraft Carriers versus US Aircraft Carriers in a battle in the middle of the Pacific Ocean.** The Chinese understand the US Pentagon's greatest weakness - and that is the Pentagon's WW2 obsolete mindset. **The Chinese will continue to attack and continue to win big where the US Pentagon is NOT looking and NOT thinking.**

The first Economics War (EW1) against the USA was successfully won by the Japanese. The current Economics War (EW2) was successfully won by the Chinese.

The Chinese successfully used **Sun Tzu's** Rules of Grand Strategy against the USA. **(1) You never attack strong points you always attack weak points.** (2) You search for your opponent's weakness. (3) You attack your opponent's vulnerabilities. (4) You attack what your opponent is ignoring. (5) You attack decisive faults that your opponent cannot psychologically and/or emotionally change. (6) You exploit your opponent's delusions by manipulating the delusions of your opponent for your own ends.

Cultural Warfare has been a major Chinese weapon for over 2,500 years. The reason for WHY the Chinese culture is the only culture in the world that has survived from over 2,500 years is because the Chinese eventually defeated all of their conquerors by culturally enveloping them and turning them into de facto Chinese.

The correct winning Strategy for the USA with Iran would be for the USA to use the proven Chinese strategy and culturally envelop Iran in as many ways as possible. That is the correct strategy for returning Iran to being a USA Ally.

(554) America should Put the Idea of a Shooting War with China on the Far "Back Burner." Washington DC Spending Large amounts of Money on War Preparations Against China is a Waste of Money and an Unnecessary Aggravation of the Chinese.

China is using the non-shooting tools of warfare and winning the war against the USA. **China does NOT want or need a shooting war win against the USA because China is already winning everything China wants to win.**

The Chinese Leadership knows that it has time on its side. In 1955 the USA was 40% of the world's gross total economy. In 2018 the USA was 20% of the world's gross economy - **a 50% loss that continues to become an ever bigger loss with each passing year.**

In 1975 China was 5% of the world's gross economy. However - measured today by using Purchasing-Power-Parity China is 20% of the world's gross economy - **a 300% increase**

since 1975 and going up. Relative to the world China keeps going up and the USA keeps going down. Why? Because China continues to be on the upside of its reset Bell Curve and the USA is on the downside of its National Bell Curve.

China is already winning and will continue to win. China knows that it has time on its side. China knows that it has the flow of history on its side. **Why would China want to mess everything up with a shooting war?**

When a great power is in decline - as the USA is now. The correct strategy is for the American people to accept the flow of history and accept the world's new power-equation realities and make the best arrangements possible.

(555) It is VITAL that the USA Strategy Planners Understand that China's Definition of "Victory" is Different from the Western World's Christian Bible Influenced Definition of "Victory."

Chinese culture does NOT demand total victory. **The total absolute dramatic defeat of your opponent is a Western cultural idea** that originates in the Christian Bible.

The Bible talks about Armageddon and the cataclysmic battle of Good versus Evil. **The Biblical idea of the final great battle pervades down into the deepest levels of Western cultural mindsets** and also pervades down into Washington DC's and the Pentagon's military mindsets and planning.

Chinese culture is centered on the teachings of **Confucius** (551-479 BC) and Chinese military thought is based on the teachings of **Sun Tzu** (544-496 BC). Both **Confucius** and **Sun Tzu** advocated moderation in all things.

Chinese General **Sun Tzu** wrote that the supreme commander of the Army (and his senior officers) should never be hateful towards their opponents. **Sun Tzu** wrote you should damage in war only what absolutely needs to be damaged to produce a pacification. The classic Chinese objective was not to destroy forever the opponent but to use the opponent's assets and turn the opponent into an ally (if possible) but if not possible then turn the opponent into something useful.

It is important for the USA's strategy planners to understand that the Chinese mindset is very different from the mindset of **Carl von Clausewitz** and the USA's Generals. **Clausewitz** is classic Christian Bible with his insistence on ultimate victory in battle against "evil." **In Western Culture every opponent is "the ultimate in evil" and every war is a holy crusade that justifies any sacrifice.**

Chinese leaders do NOT want an ultimate shooting battle. Today's Chinese leaders want to make China stronger, safer and richer without shooting wars. However - the Chinese have been and will continue to use all of the non-shooting types of warfare. **The USA's best grand strategy would be to apply an Economic Warfare strategy towards the world.**

(556) The USA's Leaders should Learn from the Chinese. Washington DC Should be Happy with Peace and Happy with "Good" Even If Everything is Not 100% to Washington DC's Subjective (and often wrong) Desires.

"Good" is much better than destroying yourself for some form of subjective "perfect" that is in the real world impossible to achieve.

(557) When Analyzing USA-Chinese Military Strategy the Most Important Fact is that the USA CANNOT Invade and Conquer China and China Cannot Invade and Conquer the USA.

Neither China or the USA can invade the other and that makes the USA and China **natural allies**. China and Russia are NOT natural allies. France and Germany were not natural allies. India and Pakistan are not natural allies. Iran and Saudi Arabia are not natural allies. Vietnam and China are not natural allies - and so forth. However - **the US and China (plus the USA and Russia) are natural allies.**

(558) The Grand Strategist works hard to Completely Understand his Opponent's Cultural Traits Including the Time-Reference Factor.

"Time Reference" means understanding how your opponent thinks about time - in terms of weeks or years or decades or centuries? **Time reference is especially important when US Military Planners do analysis involving China and Asia** and the Middle East that have peoples and Nations going back 4,000+ years.

The US Pentagon's planners usually think in terms of months and years and are at a disadvantage when up against their Chinese Strategy counterparts who think in decades and centuries.

Washington DC has a ghetto time-line mindset - or a TV News time-line mindset. If it is not an immediate crisis then it is not a problem.

(559) China has Reset its Bell Curve and will Once Again be the World's Biggest Economy.

In the last 500 years the history of the world has come full circle around the world - from (1) China being the world's richest Nation from 500 BC to the 1500's AD. (2) China's Bell Curve going down and crashing from the 1830's to 1976. Then (3) China resetting its Bell

Curve after 1976 and (4) China once again returning to be the world's richest Nation somewhere around year 2040.

Western World aggression against China went into high gear starting with the British aggressive Opium Wars of 1839-1860. England forced China to take Opium grown in India and that drained China of its sliver coinage. **England sucked the sliver coinage out of China and the Chinese economy went into a long deflationary collapse.**

The British "Pound Sterling" (i.e. a pound of silver) by the late 19th Century was to a great extent Chinese silver that was taken out of China in exchange for opium. The British were the world's first big time drug dealers.

For 2,000 years Keynesian economics worked in "closed China." The Chinese invented Keynesian economics 2,000 years ago. China's money for 2,000 years circulated in big loops **inside of China**. Then the British blasted China's doors open and drained China's silver.

Keynesian economics does work in a closed system (as China proved for 2,000 years) but **NOT in an open system** - as Washington DC somewhere in the future will learn the hard way. (Note - the reason for WHY the USA does NOT have massive inflation with its Madoff Trillions creations is because the US Madoff Money flows OUT of the USA and into the world. When that stops - disaster for the USA).

Japan attacked China in 1894. Germany established colonies in China. Russia took over Manchuria. The Japanese attacked China again in 1931 and started an aggressive and brutal colonization of China. After Japan was tricked into attacking the USA and UK in December of 1941 the already extensive war in Eastern China spread to Western China and became bigger.

The great Asian-Pacific War (1931-1945) was followed by the 1948 Chinese Revolution. The Communist won the Revolution in large part because of the super high inflation in China that was long term caused by the British draining China of silver a 100 years earlier and paper money replacing silver. Most middle-class Chinese were ruined by inflation and said "can the Communists be any worse?" The eventual answered turned out to be "Yes" with the horrible "Chinese Red Guard Cultural Revolution" of 1966 to 1976.

By 1976 **China had crashed and burned.** China was a shambles with 300 million Chinese living in extreme poverty and nobody was "rich." However in 1978 Chinese Primer **Deng Xiaoping** (1904-1997) came into power and reset China's Bell Curve. Politically China retrained a Communist Government but economically China became a vicious Capitalist Economy aggressor.

When the English first started exploiting China seriously (in the 1830's) China's economy was 33% of the world's total economy. That percentage crashed down to only 5% in 1975.

China's economy today is about 20% of the world's total economy and continuing to go up. Today the USA is about 20% of the world's economy and continuing its slide downwards.

By 2040 the USA should be about 15%-17% of the world's total economy. **China's economy will have returned to being about 30% of the world's total economy.** However there will be grand strategy decisive differences between the Chinese and USA economies. The USA economy will be mainly services and entertainment and Wall Street Gambling Casino money manipulations. The Chinese economy will be mainly manufacturing of products the world wants to buy.

The Chinese strategists know that all of the above is going to happen without a USA-China shooting war. The Chinese have already won. **A shooting war would only "mess everything up" for the Chinese.** If there is a USA-China shooting war (God forbid) it will be started by Washington DC's warmonger military strategy morons.

SUMMARY

The Chinese economy (as is typical of economies on the upside of their Bell Curves) is a **much higher quality economy than the USA's downside of a Bell Curve economy.**

The Chinese economy is built on manufacturing products that are wanted by the world (what the US economy was for 1790 to 1970). Today's current US economy is built on entertainment, retailing, "services" and financial manipulations and lawsuits.

The USA in the 21st Century is the classic textbook description for a Great Power on the downside of its Bell Curve. The current US Culture is centered on pleasure, entertainment, money/financial manipulations, debt creation and on never ending stupid wars in stupid places for stupid reasons based on the completely wrong idea that it is wars that make Nations great. **The USA today is the classic textbook description of a great power in decline.**

When Western Imperial Rome was on the downside of its Bell Curve (from the 3rd Century AD until its final complete crash in 476 AD) Imperial Rome (like the USA today) was centered on pleasure, entertainment, debt creation and never ending stupid wars for stupid reasons. The same collection of actions and mindsets (if not reversed) will eventually do to the USA what it did to ancient Rome.

Most of the idiots in Washington DC are telling themselves "America is so powerful, and grand, and big it can never go down." The Roman Governments during Rome's decline had exactly the same delusions. However, the Roman Governments were so stupid and incompetent for so long most Romans arrived at the point where they said - "**We don't care what happens.** The Barbarians can't be worse than the scumbag government we have now." The Chinese middle-class said the same thing in 1948. "**We don't care what**

happens. The Communists can't be worse than the scumbag government we have now." The Russian people said the same thing in 1917. "**We don't care what happens.** The Communists can't be worse than the scumbag government we have now." Maybe someday the USA's voters will say the same?

Chapter 29

How Will the Current USA-Muslim World War End?

The USA has had continuous military operations in the Muslim world since the Kuwait/Iraq War of 1991. Thousands of USA Troops have already died in the Middle East. Over 3,000,000 US Troops had already been in the Iraq and Afghanistan wars (when this Book was published). The complete cost to the USA is going to be many Trillions of dollars. Washington DC has the USA in a slow motion USA-Muslim World War that the USA has zero chance of winning by force of arms.

The USA's wars in the Middle East are a slower-motion Vietnam War but with sand rather than jungle. The death of US solders is much less than Vietnam but the long term dollar costs are much higher.

Washington DC's Middle East Strategy has always been fundamentally flawed, unworkable and stupid. The most important lesson from the Vietnam War is that enormous amounts of money thrown at a War does NOT guarantee victory when your Grand Strategy is fundamentally brain dead stupid.

(560) A Thousand Years ago Christian Crusader Military Troops Invaded the Middle East and it was the Biggest Military Disaster in History Until WW1 &WW2.

In the 21st Century - Christian USA Troops Invading the Middle East are making the SAME mistake the Crusaders made a 1,000 years ago. The US Military in this 21st Century "Crusade against Terrorism" will eventually have the same disastrous defeat as the crusaders of a 1,000 years ago and for basically the same reasons.

(561) The US Pentagon (for the Last 60 Years) has Had an Insane Passion for Fighting Wars that Somebody Else has Also Fought and Also Lost.

The demented Pentagon has been fighting the same wars (in the same places) and LOSING like the original Western Nation invaders had lost. This is a deep sickness that must end.

(1) The US Pentagon proved that it could be defeated in Vietnam like the French in 1953 had been defeated in Vietnam.

(2) The US Pentagon proved it could de facto lose in Iraq much as the British had lost the Anglo-Iraqi Wars of 1930 and 1937.

(3) The US Pentagon is proving it can lose in Afghanistan exactly the same as the British lost the 1st Anglo-Afghan War of 1839-1843, and lost the 2nd Anglo-Afghan War of 1978-1880 and lost the 3rd Anglo-Afghan War of 1919.

(4) The US Pentagon is proving it can lose in Afghanistan exactly the same as the Russians lost the Soviet-Afghanistan War of 1979-1989.

(4) Currently the US Pentagon is in the process of proving it can be disastrously defeated in the Middle East like the Crusaders were disastrously defeated a 1,000 years ago.

The Muslim Holy Koran is clear - if Muslim groups are attacked by non-Muslims then all Muslims must help those attacked. **The more USA Christian soldiers "Surge" into the Middle East the more Muslims in growing numbers will join the fight to resist the invading US Troops.**

The US Pentagon has 26,000 full time employees - can anyone at the Pentagon do 4th grade math? The Muslim World has about 170,000,000 males in the 17 years to 30 years old group of potential fighters and Guerrillas.

The Holy Koran tells Muslims to help those who are attacked by non believers. This means that when attacked by US Christian Troops the Muslim Guerrillas will receive decisive help and support from the world's Muslims.

The only way the US Military can win any Guerrilla War in the Muslim Middle East is if the USA **attacks the reasons for WHY the civilian Muslim populations support Radical Muslim Groups.**

If Washington DC could eliminate the reasons for why Radical Muslim Groups are supported by other Muslims then the Sea of popular support that the Guerrillas swim in would dry up and the Radical Muslims will be gone. However it is psychologically impossible for Washington DC (and especially the Republicans) to support equal freedoms and equal rights for everyone in the world everywhere.

(562) The Grand Strategist is Never a Crusader and Especially NOT a Crusader in the Muslim Middle East.

All of the world's 1.8 Billion Muslims know that the Muslims won the Crusades 800 years ago. **Every Muslim in the world today knows that the Muslims are eventually going to defeat the current Washington DC Crusade in the Middle East - and they are correct.**

(563) If you are Contemplating a significant Shooting War -one of the MUST HAVE Factors is a Doable way to Envelop your Opponent. If you Cannot Envelop your Opponent then there Should Be Many Outstandingly Excellent Reasons for Starting a Shooting War.

The factor of SPACE is highly important for both attackers and defenders. One of the most important Rules is **the greater the amount of room an opponent has at his back the greater the natural protection your opponent enjoys.**

In South East Asia the Viet Cong always had a vast jungle at their backs to retreat into. As the Vietnam War proved - the US Military winning all of the battles and **pushing their opponent around on the map - that is NOT winning a war.**

If there was a USA-Iran shooting War the USA would definitely lose the war for the same reasons the US Military lost in Vietnam. **There is zero way for the US Military to envelop the vast land area of Iran** (636,400 sq miles or 1,648,300 sq kilometers). The land territory of Iran is bigger than the US States of: Texas. New York, Illinois, Ohio, Georgia and California combined - and the Iranian population is 83 million. In addition Iran is on the other side of the Earth from the USA. It is a classic "arrow shot too far."

(564) The Best USA Grand Strategy for Iran is the USA Maneuvering Itself INTO Iran (to the Greatest Extent Possible) and then Working From the Inside Out.

The USA winning in Iran means as many USA companies, services and entities as possible going inside of Iran and into positions to apply the many non-shooting forms of warfare and attitude change (in as many ways as possible) and working from the inside of Iran.

The 2015 Iran Peace Agreement (that Secretary of State **John Kerry** negotiated) was a first step on the road to putting the USA back inside of Iran so US entities could work on National Iranian attitude changes from the inside out.

Dumping the 2015 Iran Peace Agreement in 2017 was a highly amateurish (and wrong) Strategy decision. Iran's money (that had been mainly frozen in foreign banks since the late 1970's) had already been returned to Iran in 2015.

The USA returning to the Economic Sanctions against Iran that started in 1979 (and have NOT work since 1979) is a return to a proven failure.

The USA had a much bigger percentage of the world economy in 1979 than in 2019. If US sanctions against Iran did not work in 1979 then why are they going to work now?

Washington DC threatening Iran with a shooting war goes past complete military incompetence and is criminally insane. This is how mistakes happen and large wars start. And for what? To make the Middle East safe for Apartheid and colonial oppression?

Based on Centuries of examples of what has worked - American culture, ideas and business must be put inside of Iran to the greatest extent possible so the Roman (and British) proven methods of **attitude changes can be made from the inside out**.

(565) When Making Your Strategy Plans One Factor You Must clearly Understand Is If You Are Fighting (1) a Professional Army or (2) a Citizen Soldier Army.

The type of army you are fighting (and the type of army YOU have) makes a significant difference in what Strategy you must have to defeat your opponent.

The differences between a professional army and a citizen soldier army has been an on-going military strategy debate going back 25 centuries to **Sun Tzu.**

Most military philosophers through the centuries have agreed that a motivated citizen soldier army is the best. **The USA had a citizens soldier army in WW1 and WW2.**

The most effective citizen soldier armies are the ones that are fighting wars that are on the Moral High Ground. Ideally the citizen soldier armies should clearly and unquestionably be fighting wars for freedom and the good of humanity.

Professional armies and professional soldiers and mercenaries (what the USA has today) do NOT especially care about who they are fighting or why they are fighting. **For most professional soldiers war is an end in itself,** and for the Pentagon's many $100,000 per year mercenaries it is about the money.

For most of the US Pentagon's Officers any war anywhere is its own justification. Shooting wars are to professional soldiers what water is to swimmers.

During WW2 the USA was fighting for the good of humanity. The USA was fighting for freedom and democracy for everyone. In WW2 the USA was fighting against any Nation or group forcefully subjugating other people. **Every individual in WW2's American Military correctly thought of himself as being part of the "shining city on the side of the mountain" because it was absolutely 100% true.**

The Rule that should be followed (but is not) is that **if a war is NOT clearly on the objective (factual) moral high ground then the USA Troops should NOT be there and Washington DC should not be supporting the war.**

(566) In the 21st Century the US Pentagon Continues to make its Long Running Serious Mistake of Thinking the Middle East Conflicts are Leader-Driven.

The US Pentagon's WW2 mindset believes that every conflict is caused by some **Hitler** or **Tojo** or **Ho Chi Minh** or **Saddam Hussein** etc. When this book was compiled the USA had greatly expanded its Middle East Muslim War program of "leader assassination via drones."

There are three big flaws with the Washington DC's warfare via assassination robots:

(1) Sometimes when a USA drone assassinates a Muslim leader that leader is replaced by a younger, smarter more skilled and more aggressive leader.

(2) What goes around comes around. The other side (the Muslims) are going to be MORE MOTIVATED to keep fighting US Troops. **WW2 proved that the more a people are bombed the harder they fight.**

(3) **The current war between the Muslim world and the USA was never leader driven on the Muslim side - it was always idea driven.** The Muslim idea is that Western Power Nations and Christian Troops are NO LONGER going to control what happens in the Muslim world.

It is vital that the US Pentagon's Generals and planners learn that every Middle East opponent is NOT some version of Bonnie & Clyde. Every Muslim opponent of the USA is motivated by a set of ideas and some of those ideas (from some of the groups) are on the moral high ground.

The basic Rule is that **the more you refuse to understand your opponent's Strategy and motivations the more difficult it becomes to counter the Strategy and motivations you refuse to understand.**

Up to this point the US Pentagon's analysts have had a highly flawed understanding of what is truly motivating their opponents. Worst of all the US Pentagon's analysts do NOT want to know. They do not want to be confused by the facts.

The US Navy assassinated **Osama bin Laden** (May 2nd 2011). For the USA exactly and specifically what dramatically changed for the better in the Middle East? The answer is - "Nothing."

When this book was published the Washington DC Warmonger Lobby were screaming that the USA had to have a shooting war with Iran to "solve our Middle East problems." **The USA has ZERO vital interests in the Middle East. All of Washington DC's "problems" in the Middle East are problems of their own making.**

Ho Chi Ming died September 2nd 1969. What for the USA changed for the better in the Vietnam War (1959-1975)? The answer is - "Nothing."

The Vietnam war was an anti-colonialism war that 100% of the North Vietnamese supported and 65% of the South Vietnamese supported. The foreign aggressive invader (the US Military) was the reason for why the Vietnamese were fighting.

For the majority of Vietnamese the war was always an anti-colonialism war fought against foreign invaders - first the Japanese invaders 1942-1945 - then the return of the French invaders and finally the US Military invaders.

Saddam Hussein and **Muammar Gaddafi** were killed. So - is everything "wonderful" and perfect in the Middle East because **Hussein** and **Gaddafi** are dead?

The USA's problems in the Middle East are not "evil Muslim leaders." The USA's problems are US Troops in the Middle East who are in somebody else's Nation killing the local people.

(567) The Best Weapons System (and the ONLY system) for Permanently Defeating the Radical Muslim Groups are Ideas that are Better than the Current Ideas that Motivate the Muslim Radicals.

If you defeat your opponent's main ideas then you have won. As the USA's victory over the USSR in the Cold War (1946-1991) proved.

Remember the core rule - **the correct objective in good Grand Strategy is NOT your opponent's army - your correct objective is your opponent's belief system that is motivating the war.**

President **Ronald Reagan** in the Information Age of Warfare (1960 to ?) defeated and destroyed the USSR Empire with a group of better ideas than the USSR's ideas.

The most important message in this book (for Washington DC) is that there are ZERO military solutions for the USA in the Middle East. The USA will start to win the wars against the Radical Muslim Groups when (1) Washington DC comes up with better ideas (and superior arguments) than what the Radical Muslim Groups are selling and (2) when Washington DC starts successfully communicating its better ideas to the Muslim world.

When you defeat the ideas that are motivating the Guerrillas then you have won the war because the Guerrilla's cannot hide from their own minds.

(568) The 100,000+ US Troops That Have Been Constantly Fighting in the Middle East Since 1991 Are Never Going to Make the Middle East Safe for Apartheid.

There are 340,000,000 Muslims in the Middle East. Every year they become smarter and richer and more experienced soldiers (thanks to Washington DC's never ending wars). Washington DC's 100,000+ troops perpetually fighting in the Middle East since 1991 have

NOT (and are NOT) going to make the Middle East a safe place for colonialism. Washington DC has to **Stop killing US soldiers to protect brutal colonialism in the Middle East.** In the Internet Age everyone in the world knows what is happening and Washington DC's support for colonial oppression is a worldwide public relations disaster.

(569) It is Vital for the USA Planners to Understand Exactly When Time Is On the USA's Side and When Time is Working Against the USA.

Big parts of the Radical Muslim Groups (RMGs) want to return to the 18th Century - and time is on the side of the USA. **Over 80% of the world's Muslims do NOT want to live in the 18th Century - the majority of Muslims want to live in the 21st Century.**

Some groups like Hamas are fighting an anti-Colonialism war. **There are similarities between the current Hamas and the Viet Minh of the 1950s.** Time is on the side of Hamas for the same reasons that time was on the side of the Viet Minh.

Washington DC in the 1950's made the mistake of believing that the Viet Minh were "terrorists." Today Washington DC is making the same mistake of thinking that Hamas are 100% "terrorists." **Calling freedom fighters "terrorists" does NOT automatically win wars as the US Pentagon should have learned from their Vietnam War defeat.**

The US Pentagon keeps making the same old mistakes in the same old ways decade after decade and lost war after lost war. **If you (1) refuse to understand your opponent's complete motivations - and (2) you refuse to understand your opponent's Strategy for achieving his motivations - then (3) you will lose the war.** As the US Pentagon has been doing for the last 60+ years.

(570) The Best USA Weapon Against Radical Islam is Moderate Islam.

The Holy Koran gives Muslims permission to fight non-Muslims who are attacking Muslims. This is why the best weapon against Radical Islam are Moderate Islamic Nations, groups and their military's soldiers.

The USA can provide: weapons, training, intelligence and so forth but the USA's actions must be very low key and NOT in the news - and hopefully NOT reported anywhere. **The USA must have 99% of the ground combat done by Muslim soldiers otherwise there is a reason in the Koran for Muslims to be fighting US soldiers.**

(571) Growing Cultural Sophistication is always the Foundation (and usually the Predictor and Precursor) for Growing Military Power.

The world is currently in the Information Age of Warfare. Now more than ever **advanced culture and military power go hand-in-hand.**

The Middle East has reset its Bell Curve and has made great strides in cultural sophistication during the last 50 years. The Muslim Middle East is on the upside of its reset Bell Curve. That means **their real military ability is also growing.** The Middle East is advancing in many ways and at an increasing speed.

The Muslims (i.e.Pakistan) already have atomic bombs. The Muslims already operate significant airlines (Emirate Airlines has 3,700 flights per week to 81 different Nations). When this Book was published Dubai Airport had 89 million passengers a year and was the 3rd busiest airport in the world.

The US Pentagon's planners have to stop making plans based on yesterday's obsolete Middle East mindsets. The US Pentagon's planners must start making plans based on (1) the growing intelligence/skills in the Muslim Middle East. (2) Making plans based on the USA's ever declining power relative to the world. (3) Making plans based on the fact that the USA has zero truly vital interests in the Middle East. (4) Washington DC refuses to see that the Muslim Middle East is on the upside of a reset Bell Curve and will become more powerful with each passing year.

(572) The USA Had Zero Vital Interests in the Middle East From 1776 to 1983 (207 years). Today The Only USA Vital Interests in the Middle East Are the US Troops Fighting There.

One of Washington DC's standard tricks to the take the existence of wars (this time in the Middle East) as the justification for the wars continuing in the Middle East. This is what happened with the Vietnam War. **The Vietnam War quickly became its own justification for continuing the war.**

For the American people **the Middle East has become a circle of self-reinforcing insanity.** We are told that the Middle East is vital because of the US Troops fighting there. However if we take away the US Troops then there is NOTHING vital in the Middle East and Zero reasons for US Troops to be there and fighting.

(573) Manipulations and Practices that Hide and Make Secret the US Military's Lack of Progress Must Never be Tolerated.

The "Pentagon Papers" revealed and proved how the US Military's failures in the Vietnam War were kept secret from both the US Voters and the US Congress. Keeping the failures a secret prolonged an already lost war and murdered many additional young Americans. In the opinion of the author officials like **Robert McNamara** (Secretary of Defense 1961-1968) should have been executed for the murder of US Troops and the needless murder of hundreds of thousands of Vietnamese.

If **Robert McNamara** had been executed for war crimes probably the USA-Iraq War 2 and the USA-Afghanistan War would never have happened and/or would have been much smaller and faster.

(574) The Standard Con Game of the Warmonger Lobby in Washington DC is to Attack Muslims in Their Homelands - and Then Use the Muslim's Defending Their Homeland as the Justification for the US Pentagon Expanding its Aggressions.

Washington DC is currently using the trick of making the existing US wars against the Muslim World the justification for continuing the wars against Muslims.

Vietnam was the same con game. After the shooting war started and increased in size then **the war became the justification for continuing the war.**

A USA-Iran war would be the same. After Washington DC attacked Iran (based on con games, lies, tricks and deliberate provocations) then **the war would become the justification for continuing the war.**

(575) Starting in the 19th Century (and Going Through WW2) one of the US Military's Rock-Based Absolute Axioms was to never <u>NEVER go into a Ground Forces Land War on any Part of the Asian Mainland.</u>

From after the US Civil War in the 19th Century through WW2 the **US Military retained its Axiom of no US Troops in ground wars on the Asian mainland.** The US Military must return to that proven correct Axiom.

Back in the 1870's the US Military strategy planners were smart enough to look at a map of Asia and ask "How do you envelop an opponent when there is unlimited space at their backs?"

During WW2 there were at one point about 5,000 US logistic personnel in Burma mainly helping the "air lift" into China. Plus a few US Troops to protecting a few US Airfields in China. The biggest US Military combat unit fighting on the Asian mainland during WW2 were the 3,000 "Irregular Troops" in Merrill's Marauders in 1943 Burma.

The US Army during WW2 reached a size of over 8,000,000 but it was only Merrill's one Brigade of 3,000 US Troops that were directly engaged in combat on the Asian mainland.

The US Military was in an Asian mainland war with the Korean Civil War (1950 to ?). After three years of fighting (and 36,914 USA dead by 1953) the war came to a cease fire on almost the exact boundary lines where it had started. (Over 1,000,000 South Korean civilians had been killed by 1953). US Troops have been in Korea ever since the cease fire.

The US Military entered into another Asian mainland war in 1959. This time with the Vietnam Civil War (1959-1975). After the death of 58,220 American soldiers (and after US Forces killed 650,000 Vietnamese women, children, old people and men) the US Military lost the Vietnam War.

In 1991 US Military entered into another Asian mainland war (this time in Iraq). Almost 3 decades after the war started there are USA combat troops in Iraq trying to capture enemy held territory.

In 2001 the US Military entered into another Asian Mainland war - this time the Afghanistan Civil War. Twenty years after the start of that war the anti-USA Taliban controlled 50% of the countryside.

Are all of the 26,000 incompetents in the US Pentagon grand strategy brain dead? Based on the Post WW2 record of losing shooting wars the answer is "YES." **The old axiom of NO big numbers of US Troops fighting in a shooting war on the Asian mainland has been proven and reproven to be absolutely TRUE in many Post WW2 tragic US Military lost wars.** How many more lost wars on the Asian mainland does the US Military have to lose before returning to the correct 1870 grand strategy?

After 30 years of US Generals losing land wars in Asia there is no need for the US Military to lose another shooting war (this time in Iran) to once again prove that the old axiom is true. The US Military must return to its past proven successful Axiom. **NO USA Troops in land battles on the Asian mainland. Especially NO USA Troops in Iran land battles.**

(576) Washington DC in 2019 was Making the Same Mistake with Iran that Washington DC has been Making with North Korea since 1953.

From the Korean cease-fire agreement (July 27th 1953) to now Washington DC has been trying to isolate North Korea. From 1953 to 2019 that policy has NOT been a success.

Washington DC has the same proven-failed sanctions policy against Iran that has failed with North Korea since 1954, failed with Cuba since 1961 and has failed against Iran since 1979.

The USA has been on the downside of its Bell Curve since 1975. If economic sanctions did NOT work (starting 65 years ago with Korea) then why are they going to work now? After 65 years of failure it is time for Washington DC to change its rancid, stupid and NOT-working failed strategy.

(577) The Warmonger Lobby in Washington DC keeps Saying that a USA-Iran War would be Over in a Few Weeks and the USA would Easily Win.

Here are a few of the facts. (1) The Iranians have been preparing for an attack by the USA since 1980. (2) Troops that are well "dug in" are costly to dislocate and the Iranians are extensively "dug in." (3) **In a USA-Iran War the USA has already lost the factor of surprise.** (4) Iranian troops have many natural barriers to aid in their defense against USA troops. (5) Iranian troops have a vast space to retreat back into. (6) In a shooting war with Iran there is a 95% probability that both China and Russia would aid Iran. (7) Washington DC would have no way to stop the aid to Iran from China and Russia without risking a nuclear war that would kill half of the USA's population. (8) For te end of WW2 to now the US Military has had nothing but defeats and muddles with significant shooting land wars on the Asian mainland. There are ZERO reasons to expect anything different if there is a shooting war against Iran. (9) Nobody in the world would help the US Military in a shooting war with Iran other than token meaningless small numbers of what in reality would be only combat observers.

(578) General Dwight Eisenhower became US President in 1956 and had Exactly the Correct Post WW2 Grand Strategy. President Eisenhower said that the American People should NOT Support White Man Colonial Rule in Asia, or in Africa or the Muslim Middle East or Anywhere Else in the World.

Unfortunately for the USA most Presidents since **Eisenhower** (President 1952-1960) have done the exact opposite and have had wars to promote White man Colonial Rule - **all with disastrous consequences for both the American people and the world.**

In 1956 Egypt declared its complete independence from England and took control of Port Said and the Suez Canal. British and French forces attacked Egypt. US President **Eisenhower** said it was a military mistake for America to fight for White man Colonialism in Asia (or anywhere) and President **Eisenhower** did NOT support England and France fighting against Egypt's independence.

When Lieutenant **John F. Kennedy** replaced Five Star General **Eisenhower** as President (1960). **President Kennedy** reversed **Eisenhower's** military policy and supported White man Colonialism in Asia and expanded the Vietnam conflict into a major war - and that war eventually turned into a major USA defeat and world disaster. **The biggest casualty of the Vietnam War was the USA's world image of being "the good guys."**

Unfortunately for the American people Washington DC doubled down on its Vietnam mistake. Washington DC's enthusiastic (and moral low ground) support for White man Colonialism in the Middle East has continued to this day. What "good" has it given the average American? The answer is "None." **Washington DC's support for oppressive colonialism has been a world public relations disaster for the USA.**

(579) Time is on the Side of the Muslim World and Working Against the USA. The Muslim World's Overall Strategy for Dealing With the USA is a Waiting Game.

Every year the Muslim World becomes a little stronger, smarter, richer and more competent. **Remember the Rule - one of the first things you must do when formulating a grand strategy is to correctly ascertain if Time is on your side or if Time is working against you.**

Time is on the side of the Muslim World. Every Muslim in the Middle East knows that it was the Muslims who eventually defeated the Western Christian Crusaders. **Every Muslims knows that they will eventually defeat all US Military Forces and drive them out of the Middle East.** The Middle East Muslims are NOT defeated in their minds and are NOT going to be defeated in their minds. They ARE going to eventually win.

For the USA there are zero Military shooting war solutions in the Middle East. ONLY the non-shooting solutions will work.

Chapter 30

How to Make Long Lasting Peace in This Century

There will be many wars in the 21st Century. Warfare in the last half of the 21st Century is going to be as different from today's warfare as 1964 warfare was different from 1864 warfare.

In 1914 two of what had been the tactical mainstays of warfare for thousands of years (1) Horse Cavalry and (2) Mass Infantry Charges were both disastrously obsolete. However in 1914 none of the Generals (from any Nation) had the intellectual ability to see the obsolescence. All of the Generals had to learn the hard way - but only after they had murdered millions of their own soldiers.

In this Century many cherished weapons systems, tactics, beliefs, mindsets and "military traditions" will once again become disastrously obsolete. It is a certainty that many Generals and Admirals will once again acknowledge the new realities only after the death of many their troops and civilians.

The mass conscript armies of the Industrial Age of Warfare (that numbered in the millions of men) are obsolete. In the last part of this Century advanced Nations will have very few humans on the battlefields of advanced Nations and many of the battlefield decisions will be made by computers.

(580) The USA's Main Objective in the 21st Century Must Be the Avoidance of Significant Shooting Wars.

The USA will be in decline (compared to the world) for the remainder of this Century. The 3,000 year historically proven Rule for Great Powers is that **shooting wars that happen on the downside of a great power's Bell Curve are far more damaging than shooting wars that happened on the upside of a Great Power's Bell Curve.**

(581) The Current Information Age of Warfare (1960 to ?) is Already on the Downside of its Bell Curve.

The coming Artificial Intelligence Age of Warfare is on the horizon. America needs to be building weapons systems for future wars and NOT building updated weapons systems for the now far bygone Industrial Age of Warfare (1885 to 1960).

The more time and money the USA wastes on obsolete weapons the less money and time the USA will have to build tomorrow's weapons and build tomorrow's military organizations for tomorrow's wars.

(582) The USA's Priority Military and Diplomatic Efforts Must be Centered as Much as Possible on building Good Relations with China and Russia.

Only Russia or China have the potential to significantly destroy the USA and that means the top two Nations that American Strategy should be working to cultivate as Allies are Russia and China. That does not mean continuing to let China take advantage of the USA. It must be a partnership of equals were neither side takes undue advantage of the other side.

There are Grand Strategy incompetents in Washington DC who say "Russia is advancing in the Ukraine and we need to send in American troops to stop them." WHY?

What does the Ukraine (or any part of the old USSR) mean to the USA? The Ukraine was under Russian control from 1648 to 1991. **For over 300 years the Ukraine under Russian control did NOT stop the USA from becoming the world's greatest power.**

If the Ukraine had been an independent Nation in 1941 then **Adolf Hitler** might have won WW2.

The Washington DC military morons and the Warmonger Lobby say, "The Russians might take over Poland - or Latvia - or Estonian - or Lithuania and America needs to stop the Russians." The USA is supposed to have a shooting war with a major nuclear power to save Latvia?

Poland and the Baltic Nations **were under Russian control for centuries and it made zero difference to the USA becoming a great power.**

What Russia does in Europe is NATO's responsibility. **If the NATO Nations located in Europe do not especially care then why should the USA?**

Washington DC's Warmonger Lobby uses every change in the world's map as an excuse for a shooting war. The USA can no longer afford endless shooting wars - the insanity has to end.

(583) The Strategist must Above All Else be Mentally Flexible and Always Look at the Big Picture. We Must Also Remember that Everything in the World has the Word "Temporary" in Front of It.

One of Washington DC's biggest flaws (of the past 50 years) is seeing many irrelevant world changes as threats to the USA that must be challenged. The changes that the USA should be concerned about - like the change of the USA going from a creditor Nation to a debtor Nation - those are the changes Washington DC should be fighting.

(584) The ONLY way to Defeat Widely Held Beliefs and Ideas are with Better Ideas Communicated in Superior Ways.

In the current Information Age of Warfare the USA's Military Grand Strategy **toolbox MUST include**: (1) the Internet in all of its many forms. (2) Internet hacking and sabotage. (3) Social media massive and effective usage. (4) Commercials in all of their forms especially on social media in every language. (5) The use of movies. (6) The use of television shows. (7) The use of talk-radio. (8) The use of all of the above combined with large amounts of highly talented salesmanship.

In 2018 there were 20 billion devices connected to the Internet. In 2028 it is estimated there will be 40 billion devices connected to the Internet. **In today's world the US Army cannot defeat a popular idea with guns. The US Military battling ideas with guns is strategy that is out of the Dark Ages. You must have better ideas that are successfully sold with great skill and talent.**

The US Military MUST become media and communications experts using every means and form of communications in the world as their main weapons systems.

There must be a major change in the definition of what is the US Military. There must be a profound and gigantic transformation and it must happen as soon as possible.

Probably the current US Military is incapable of the changes needed. New organizations will have to be built from the ground up and then replace the current organizations. For the changes need EVERYTHING with the US Military must be on the table for a decision of "Keep" or "Discard" or
"Change To."

(585) In the current 21st Century Information Age of Warfare the Control of Land is now of Minor Importance. Control of the Sea is No Longer Decisive. The Decisive Vital War Winning Weapon is Who Controls the Minds of the World's People.

For 4,000 years of warfare (and up into the Industrial Age of Warfare 1885-1960) the possession and acquisition of LAND was the major objective in most wars. However - today the side that can control the most minds is the side that will win the wars.

(586) He Who Controls the Minds of the World's People Controls the World.

The old axiom "He who controls the Seas controls the world" is obsolete and no longer true. The current axiom **"He who controls the Air and Space controls the world"** is now true. The ultimate axiom is **"He who controls the Minds of the world's people controls the world."** The 2 new axioms of world power are tied together. If you control "Space" then you control the communications satellites of the world and that means you have the potential to control the minds of the world's peoples.

(587) The USA Used To Be Paramount for Having a Favorable Image with the World's Collective Mindset. However - that Once Great and Decisive Advantage Keeps Slipping Away at an Ever Faster Speed.

The USA used to have the respect of the world. Including the USA's enemies who privately admitted that the USA was doing many things brilliantly. However - all of that has been changing for the worse.

Internally Washington DC is into some profoundly stupid and insane political fights that are making the USA look like a joke to most of the world.

Externally The Warmonger Lobby has extensive control of Washington DC and supports anti-freedom fascism and White man Colonialism. The USA's Warmonger Lobby has moved the USA to the very wrong side of history.

The Rule is that stupidity is NOT cost-free. The Rule is that being on the wrong side of history eventually comes with a very big cost.

(588) One of the Grand Strategist's Prime Jobs (and Hardest Jobs) is Taking his Mind Out of Yesterday's Wars and Getting his Mind into Tomorrow's Wars.

Unfortunately many USA Generals and Admirals in the 21st Century continue to have their minds mired in WW2 concepts. The USA should do everything possible to avoid any shooting war until there is the much needed Grand Strategy Team reorganization of the US Military.

(589) You Might be Planning for a Fast War and a Fast Peace but Your Opponent might be Planning for a Long War and Never Surrendering.

It only takes one Nation to start a war - but it takes two Nations to end a war.

Japan's 1941 Grand Strategy plan towards the USA's and UK's oil embargo (put on Japan starting in July 1941) was (1) Japan attacking the USA and UK (December 7th 1941). (2) The Japanese taking by force the British and Dutch oil fields of Indonesia. (3) The Japanese hurting the Americans and British so badly in the Philippines, Malaysia, Burma and Hong Kong to the point where the USA and UK would agree to peace. (4) Japan would use for negotiations the Philippines, Malaysia, Burma, Hong Kong etc and in return for giving some or all of them back with Japan keeping the Dutch Indonesian oil fields and keeping its colonial empire in China.

The above Japanese Quick War and Quick Peace Plan was an excellent plan but it had a giant fatal flaw the Japanese never saw until it was far too late. It was US President **Roosevelt** and British Prime Minister **Winston Churchill** who deliberately tricked Japan into attacking the USA and UK. **Winston Churchill** used Japan to bring the USA into WW2. **Churchill** never had any intentions of making peace with Japan short of Japan's complete destruction as a Colonial rival to England in Asia.

The USA's Grand Strategy planners must be 100% aware of Russia and China wanting a USA-Iran shooting war because that would be a disaster for the USA. Today's Washington DC should NOT let China do to the USA what the USA did to Japan in 1941.

(590) The US Pentagon in the Middle East is Following the Clausewitz Strategy of Attrition on the Battlefield - But that Strategy is a Slow-Motion Disaster for the American People.

Using honest long term accurate accounting it costs the US taxpayers a fortune to kill ONE Radical Muslim Insurgent. Part of the real cost is the wars are being fought with borrowed money and **interest will be paid on the borrowed money forever.** Another part of the cost is about 35% of all US Troops that are in the Middle East have gone on (and will continue to go on) some form of disability that is a monthly payment to them for the remainder of their lives - in many cases that is going to be 50+ years of payments.

Do the math - the Muslim world has 100 Million males 17 to 25 years old who are possible Insurgents - and the USA is already the world's biggest debtor Nation by gross money amounts. The Federal Government's Debt is already increasing by over a Trillion dollars per year. It is 4th grade math - **the USA is going to run out of Madoff money accounting fraud long before the Muslims run out of fighters.**

Osama bin Laden said "Every member of al Qaeda knows the economic weaknesses of America like he knows the lines on his hands." Nobody in the world understood the USA's core decisive weaknesses better than **Osama bin Laden**.

Osama bin Laden openly told the world his Grand Strategy for defeating the USA. (1) Keep the USA fighting stupid wars in stupid places for stupid reasons. While (2) the USA financially takes on more and more debt to fight more wars. Until (3) the world value of the US Dollar collapses. Then (4) the debt-driven economy of the USA collapses. Parts #1 and #2 have been accomplished and Washington DC is on the road to having parts #3 and #4 become reality.

(591) The Grand Strategist Must NOT make the "Tar Baby Mistake" of Being Suckered Deeper and Deeper into a War He Cannot Win.

In the current war between the USA and the Muslim World (that started in 1991) the US Pentagon has proven and re-proven that it cannot decisively win wars. The US Generals in the Middle East are doing what they have been doing from the end of WW2 to now - winning the battles and losing the wars.

The USA could decisively win all of the wars in the Middle East. Exactly how to win the wars and win a long last peace is described in Chapter 31. **The reason for WHY the US Pentagon will never win in the Middle East is because the US Pentagon does NOT want to win in the only way the USA can win.** The US Pentagon (and especially the Warmonger Lobby) want "victory" by using ONLY Bombs and Bullets from the Military-Industrial Complex.

The US Pentagon NOT really wanting to win the wars (and win the peace) in the only way they can be won is the #1 reason for WHY the US Military Forces should be pulled out of the Middle East.

A USA-Iran war would be the same. Not only is a US shooting war with Iran not winnable. The Warmonger Lobby and Military Industrial Complex would NOT want to end a shooting war.

(592) The USA must NOT Remake the Vietnam Mistake of being Naive about the Amount of Material, Blood and Money it is Going to Take to Win a War Against a Determined Opponent who has Geographical Advantages.

A USA shooting war with Iran would be long, horrible and the US Military would eventually lose. All of the BS about how the USA has enormous amounts of weapons etc etc is BS. During the Vietnam War the US Pentagon had a 1,000 to 1 advantage in both gross firepower and money spent on the war and the US Pentagon lost the war.

The American people had an enormous loss with the Vietnam War. However - the Warmonger Lobby made hundreds of billions of dollars. An Iran war would be more profitable.

(593) The Grand Strategy Team Must NOT Underestimate the Russians. This was Adolf Hitler's Biggest Mistake.

During 1941 the Russians lost half of European Russia to the Nazis. Then the Russians doubled the amount of their war production. That would be like the USA losing the Eastern part of the USA up to an Indiana, Alabama North-South battle frontline and then the remaining part of the USA doubling the amount of the USA's war production. What the USSR did in WW2 is amazing.

Russia is part of the "Big 3 Nuclear Powers" (USA, China, Russia) and is NOT down and out.

(594) In the coming Artificial Intelligence Age of Warfare (the Smart Machine Age of Warfare) WW2 types of Tanks will be replaced with Intelligent Combat Machines that can Travel Cross-Country on Land, Fly in the Air and Swim Under Water.

During WW2 the allies produced 4,356,649 Tanks, armored fighting vehicles, and self-propelled Artillery. The Axis powers (Germany, Japan and Italy) produced 670,288 of the same types of vehicles. However by the Vietnam War the WW2 updated Tanks were already on the downside of their Bell Curve. The Viet Cong (who had almost no Tanks until the very last 1975 takeover of South Vietnam) eventually won the war.

In the coming next Age of Warfare (the Artificial Intelligence Age of Warfare) there will be no humans in most Tanks. There will be no human pilots in most Jet Fighters. The Tanks will be smaller, lighter in weight and faster but with a very large punch. The line between Helicopters and Tanks will merge. Armies will be composed of many types of highly intelligent specialized combat-machines that will be able to: fly, run, travel like cars and swim underwater.

The battlefield winning military equation for the last half of this Century will be many different small but highly intelligent combat-machines in vast numbers.

(595) The Internet is a Major Weapons System and the US Government MUST HAVE a Much Closer Partnership with Every Major Internet-Platform Company within the Worldwide Internet System.

Ironically it was the US Military in the early 1960's (with the ARPANET project) that was one of the inventors of the Internet. However the US Pentagon has been very slow to see the Internet as the highly effective (and in many ways decisive) 21st Century weapons system that the Internet has become.

A core problem is that the Internet is NOT part of the Military Industrial Complex and thus not inside of the US Pentagon's thought and planning and money system. **The US Pentagon does NOT use the Internet is a pro-active way and that must change.**

(596) The Internet is NOT a "Benign Platform." The Internet has the Power to Change People's Minds and that Makes the Internet One of the Most Decisive Weapons of the 21st Century.

If you can change people's minds then you can win wars with little or no shooting war combat. However if you cannot change their minds then you can only win wars with old fashioned overwhelming military force.

To maintain the USA as a great power the US Pentagon and the major companies of Silicon Valley (and the Internet) must become close partners. Each side is going to have to make deep mindset and emotional adjustments to build solid ground for the partnership. Much new major legislation will be needed. Large amounts of old legislation must be discarded. However at the end of the day the US Military must have the full co-operation of Silicon Valley. That means **profound changes in the US Military's mindset to build and then maintain a relationship with Silicon Valley.** This is WHY the US Military must change from a War Machine to a Peace Machine. All of the non-shooting types of warfare must be used and completely exhausted before even considering a shooting war tool.

(597) The USA's Economic Sanctions Against Nations that Do Little Business with the USA Have Been Proven to be Almost Useless.

Economic Sanctions in the form of tariffs will work (to a degree) against Nations that have positive trade surpluses with the USA. However - the USA's **Economic Sanctions against Nations that do little business with the USA - like North Korea, Cuba, Iran etc have been proven to be ineffective.**

The USA has had economic sanctions against North Korea since 1950 - and North Korea continues to advance its production of both nuclear weapons and long range rockets.

The USA's Economic Sanctions by themselves have a track record of 80% failure. Why? World history's successful strategies usually had a "Big Stick" but they also had a "Big Carrot." If you were a British Colony in the 1800's the English took "their cut" but the British also gave in return many economic, cultural and political advantages. (There were exceptions with China being the biggest). However - **Washington DC's usual stumbling strategies tend to be too much Big Stick and too little (if any) Carrot.**

(598) In the 21st Century the USA's Intelligence Agencies Continue to Suffer from the "FBI Curse." That is Collecting Massive Amounts of Data and then doing Little Correct Analysis of its Significance in the Big Strategy Picture.

US Government agencies MUST become a major world force on the Internet for selling ideas that will counter the ideas of today's Radical Muslim Groups and every other Radical

Group in the future. **The FBI and CIA are stuck in the mindset of skullduggery, tricks and double crosses. How about good old fashioned salesmanship out in the open and as high profile as possible?**

(599) The USA's so-called "Intelligence Agencies" Using the Internet to Only Passively Spy on Everyone is NOT Using the Internet to Actively Change People's Beliefs.

The M-16 Military Rifle is only a very bad club unless you use its full potential. **The objective of the USA's "Intelligence Agencies" must go from finding out what people are saying to CHANGING what people are saying - thinking and believing.**

(600) In Past Ages of Warfare the Clausewitz Idea of Attacking and Capturing your opponents Capital City usually had a Big War Winning Premium. However in the 21st Century the USA's Opponents (like the Radical Muslim Groups) have No Capital Cities.

In past ages if you lost your Capital City that was (1) a significant Strategy defeat. (2) A major psychological defeat and (3) usually an organizational decisive defeat. However in the 21st Century Information Age of Warfare the "Capital City" of the Radical Muslim Groups is the Internet Cloud.

The capital cities of all of the Radical Muslim Groups (RMGs) can only be destroyed with better USA ideas and actions that are communicated in a more skillful, persuasive and convincing way than the RMG's are communicating.

(601) Any Nation (and especially the USA) Must Make its Grand Strategy Plans Based on What that Nation Will Be in the Future Rather Than What It Was in The Past.

All of Washington DC (and especially the US Pentagon) must make their plans for the future based on the reality of the USA being on the Downside of its Bell Curve.

The USA's manufacturing base is in steep decline relative to the world. Without major changes by 2040 of this Century the USA will continue to be a great nuclear defensive power but the USA's offensive war creating powers will be substantially reduced relative to the world.

By 2040 all of Washington DC's current "sanctions" will have succeeded in destroying the US Dollar and US Banking System as a weapon and that maneuvering tool option will no longer exist.

Washington DC is failing to make appropriate long term plans that reflect the reality of the USA's (relative to the world) declining economy and ever increasing debts. Washington DC is assuming that the USA can continue to finance the most massive military budget (and the biggest by money amount welfare system) in the world forever with Madoff funny money accounting fraud and forever increasing borrowing (by over a Trillion dollars a year now and going up). **At some point in the future it will be 2 Trillion dollars of new US Federal debt per year.**

The mathematics is crystal clear and 100% certain - a Nation cannot live off of Madoff created debt forever without destroying the value of its currency. The day of the $30 Big Mac at McDonalds is only a question of "when"?

A big part of the US economy is based on the Wall Street gambling casino that is fueled with ever increasing Trillions and Trillions of debt and artificially low interest rates. The 2008 Wall Street crash showed how fast the gambling casino "Madoff wealth" can evaporate. However you cannot easily evaporate a Smart Phone Factory or Boeing Airplane Factory or a Computer Factory or an advanced Robotics Factory - etc.

(602) Economic Warfare is Shooting Warfare by Other Means.

Washington DC (and the US Pentagon) think of "war" as ONLY a repeat of WW2's type of actions with updated WW2 weapons systems. For Washington DC (and the Pentagon) if something does NOT have the 5 B's - Bullets, Bayonets, Bombs, Battleships and Blood - then it is NOT a "war." However there are 37 main types of warfare and Washington DC has little (or NO) defense against many of them because Washington DC refuses to see them as the new age's highly effective forms of warfare.

The USA absolutely must have a Grand Strategy Team of experts in ALL 37 types of warfare and it must be USING all 37 types of warfare.

The Japanese in the late 1940s (and then the Chinese in the late 1970s) both updated General **Carl von Clausewitz's** axiom to say **"Economic warfare is shooting warfare by other means."**

After WW2 most of Washington DC believed that history had ended and the world would stay the same forever. One of the bestselling books in Washington DC was **Francis Fukuyama's** 1992 book ***The End of History.***

The Washington DC power structure made the biggest economic mistake in the history of the world by believing that the USA could give away most of its heavy industry to the world and the USA would remain a world power as a "service economy" of hamburger stands and sports parks.

During WW2 the USA's heavy industry was able to decisively defeat both **Hitler's** Germany and **Toyo's** Japan with massive USA manufactured amounts of: guns - warships - airplanes and war materials of every type needed. However today's Washington DC (and US Pentagon) refuse to understand that hamburger stands, shopping malls, sports, entertainment, the over building of homes and the Wall Street Madoff money gambling casino are NOT what won WW2 for the Allies.

The next age of warfare (the Artificial Intelligence Age of Warfare) is going to be a hard goods manufacturing intense age of warfare. The problem for the USA is that Washington DC's post WW2 laws and regulations sent **thousands** of USA manufacturing factories to: China, Japan, Asia, Mexico, Canada and everywhere else in the world.

The rancid thinking US Pentagon looks at Japan and says "Japan can never be a treat to the USA because it has a declining population." By the end of this Century Japan could have a 20,000,000 soldier Army with each soldier 1000% more battlefield effective than today's US soldiers.

(603) The "Quality Control" of an Army is Seldom Constant. Yesterday's Brilliant Army might be Tomorrow's Failure. Yesterday's Defeated Army might be Tomorrow's Brilliant Success.

An example is the French Army in 1914 and in 1940. The French Army in 1914 was on the doorstep of defeat but never stopped fighting and eventually won. In 1940 the French Army on paper was stronger than the Germans. However in 1940 the French Army and Nation were demoralized and suffering from mass National depression. The French Army and Nation in 1940 had lost its will-to-win and surrendered.

Baron Antoine Henri Jomini (1779-1869) was a French General and Chief of Staff to one of **Napoleon's** best Generals, Field Marshal **Ney.** General **Jomini** later became the aide-de-camp to **Alexander the First of Russia**. General **Jomini** was the **Lindell Hart** of the 19th Century and wrote, "The well-known Spanish proverb, *He was brave on such a day,* may be applied to Nations and armies. The French at Rossbach were not the same people as at Jena nor the Prussians at Prentzlow as at Dennewitz."

The Spanish proverb was saying that soldiers can be brave on a particular day but NOT on another day - and you cannot say they will always be brave and always victorious. **History has proven many times that yesterday's big military winner is often tomorrow's big loser - and the other way around.**

The Russian Army that was defeated by the Germans in WW1 was NOT the same Russian Army that defeated the Germans in WW2.

The USA's Soldiers who fought the battle of Guadalcanal (August 1942-Febuary 1943) and the Battle of the Bulge (December 1944-January 1945) were NOT the same US soldiers that fought in Vietnam or are fighting in Afghanistan.

(604) It is Vital that the Grand Strategy Planners Understand that a Particular "Brilliant" Grand Tactic and/or Strategy is Often Time-Specific for a Particular Historic Period and War.

In the first years of the 1800's **Napoleon Bonaparte's** use of speed (moving fast both to the battlefield area and on the battlefield) plus the "Elan" of the French Infantry attack - those 2 factors won many battles. However - in 1914 when WW1 started the French Army's doctrine of "**Napoleonic** Elan infantry charges" almost destroyed the French Army in the first 2 months of the war.

By September 1914 the French Army was on the doorstep of collapse. However - the then greatly overextended German Army's attack ran out of momentum and started to stall. The British Expedition Force (although small) was able to attack the German's upper right wing flank. The French Garrison of Paris rallied under General **Joseph Gallieni** and attacked the lower flank of the Germans at the River Marne (and forced a German pullback) thus saving the French Army.

As pointed out in previous copy - if the Germans had taken the money they wasted on their less than useless Navy and had put that time, effort and money into their Army the Germans would probably have won WW1 on the Western Front by October of 1914. That would have profoundly changed world history - no **Adolf Hitler**, no end to the Austrian Empire and Russia controlled by Germany and no **Joseph Stalin**.

(605) The US Pentagon (with its Current Organizational Mindset) does NOT Understand the Non-Shooting Tools of Warfare and does NOT Know how to Use Them. That Puts the US Pentagon (and the USA) at a Decisive Disadvantage.

The US Pentagon's mindset is that you win wars by using what the Military Industrial Complex sells. **The profound problem is that the US Military Industrial Complex does NOT sell the types of tools that are needed to win today's wars in today's world.**

In the Middle East the US Pentagon has never effectively used the many non-shooting types of warfare in a systematic, organized and comprehensive strategy. Why? The 2 biggest reasons are (1) nobody is trying to sell the US Pentagon those tools and (2) the US Pentagon does NOT want to understand the non-shooting tools.

Companies like Google are NOT going to the Pentagon and saying that they (for a price) can take some significant steps towards changing the minds of the Internet users in the Middle East.

The biggest victory (using non-shooting tools) of the Information Age of Warfare (1960 to ?) was the USA winning the Cold War against the USSR. The second biggest non-shooting victory of the Information Age of Warfare was West Germany enveloping East Germany.

The East German Communist Government and Nation was enveloped because West Germany used every non-shooting tool possible to open up East Germany to the world and bring the East German people closer to West Germany. Eventually the East German people WANTED to be ruled by West Germany.

(606) The Grand Strategist is Often Fearful and Deeply Concerned but Uses his Fear as a Motivation to Work Harder.

Real world shooting warfare is not the movies where the hero is usually calm and assured and without any fear because the movie hero knows the script and knows it is all make-believe.

In the real world **there are times when the Grand Strategist is very afraid and with good reasons** in an age that has nuclear weapons. However the Strategist has the courage to make the best decisions at the time and the courage to keep moving forward with his plans despite being frightened.

(607) Large WW1 and WW2 Conscript Armies in the Millions Were Already Obsolete Long Before the Last Century Ended.

With the arrival of tactical nuclear weapons large-anything becomes a target. A large concentrated mass of troops is a large target that can be reduced by 50% in one minute. For the soldiers NOT killed in an immediate tactical nuclear blast if they are exposed to enough radiation they will in 3 to 6 hours go into convulsions and then comas. They will shortly after die.

The proven rule from WW1 to now is that if a fighting unit loses 25% of its personal in battle its fighting offensive ability is reduced by 60%.

In the age of tactical nuclear weapons armies must be small with highly trained professional troops who know how to spread out fast and come together fast. Any large concentration of troops (or large ships) become prime targets for nuclear weapons.

The USA dropped atomic bombs on Japan (August 6th and 9th 1945). On August 10th 1945 Sir **Liddell Hart** the best known Grand Strategy Analyst of the 20th Century wrote in his newspaper column, **"We cannot tell whether the atomic bomb will end the war or end the world - or both, but it should spell the end of conscription for it makes nonsense of that military system."**

Conscription - the "draft" was already counterproductive by the 1960's Vietnam War. **If there was no US Draft the Vietnam War disaster would have been forced to end many years before it actually did end.**

The US Congress should cancel all of the USA's draft laws. **That would help the US Pentagon towards getting its mindset OUT of WW2's now obsolete concepts** and into the soon to arrive Artificial Intelligence Age of Warfare where "the Troops" will be machines and "war winning success" will 90% depend on the non-shooting types of warfare.

(608) Starting with the Industrial Age of Warfare (1885-1960) Defense in Depth became the Needed Defense. However Defense in Depth has Changed Greatly from WW1 and WW2.

"Defense in Depth" in the 21st Century (for advance Nations that might use tactical nuclear weapons) means a very deep distance depth of the "front lines" and much more dispersion of troops. Defense lines will continue to have "strong points" but they must be spread much further apart.

The Chinese have already extended their "front lines" several hundred miles out into the South China Sea by making the Spratly Islands non-sinkable Aircraft Carriers.

(609) The Grand Strategist will have his many Forces: Infantry, Guided Missiles, Satellites, Strategic Bombers, Jet Fighters, Navy - All Organized Towards for the Purpose of Each Part being Integrated and Positioned for the Maximum Possible Ability to Aid and Protect Every Other Part of the Organization.

If one part of your military organization is attacked then as many other parts as possible must be trained and positioned (and have the mental-set) to help defend the part being attacked. The interconnected defense will create more strength-of-spirit for the total organization. **If organized and trained correctly the combined strength of your military organization can be much greater than the sum total of every part.** However with bad and obsolete organization (what the US Pentagon has now) the sum total of your military can be much LESS than the total of its parts.

(610) Rivalry between a Nation's Intelligence Agencies should be Discouraged. As much Co-Operation as Possible (while guarding secrecy) should be Built into Interconnected Systems.

The USA is a unique Nation in having many different law enforcement agencies. Unfortunately for the USA there has historically been too much rivalry, empire building and non-cooperation between many Federal Agencies.

In the current Information Age of Warfare accurate and timely intelligence information (plus closer cooperation between all of the vital groups) is the decisive first line of defense.

At this point in the USA's history the Pentagon, CIA, FBI, NSA etc etc must have a written and specific description of the ultimate objectives for the USA in the 21st Century. Then everyone has to get onto the same sheet of paper. At the top of the list I would put **protecting both the letter and intent of the US Constitution because <u>it is following and obeying the letter and intent of the US Constitution that makes a person a true USA American.</u>**

(611) "Profiling" is at the Heart of Effective Law Enforcement and Much International Intelligence Gathering.

The left-wing has gone hysterical over profiling and calls it "racist" and "fascist." However - profiling is NOT profiling individuals but profiling crime patterns and profiling behavior patterns.

When there is a fire you do NOT randomly spray water around on the street - you spray the water onto the fire - **that is "profiling."**

When you have Cancer of the Kidneys you do NOT apply the X-Ray radiation to your whole body - you apply it to the Kidneys. That is "profiling."

The USA's Police are about 1/3 of 1% of the USA's population (about one COP for every 300 people). It is completely impossible for the Police to be "equally" everywhere. For the Police to be effective then MUST concentrate their force onto the problems. **The crazy left-wing US politicians "reduce crime" by NOT arresting criminals. Not arresting criminals reduces crime on the paper-reports but NOT on the street.** The Police everywhere know how to reduce crime - let them do their jobs.

Concentration of force onto the objective is THE CORE OF GOOD STRATEGY. The Police (and intelligence agencies like the CIA etc) MUST concentrate their very limited resources onto the most needed problems. "Profiling" is applying Law Enforcement's and the Military's very limited time and resources onto the specific problems that are most in need of fixing.

(612) It is a Management and Leadership Mistake to Pit one Part of your Organization Against another Part with the Idea that Competition will make both Parts Better. Many Times the Inter-Group Fighting Makes All of the Parts Weaker and Less Effective.

The organizational objective is to have a military that can disperse quickly to cover many points and be less of a target but then come together quickly and concentrate its forces with the least amount of lost energy, lost time and the least friction between the different military branches.

Frederick the Great (1712-1786) said, "The aim is to turn the army into an instrument of a single mind and will."

General Sun Tzu wrote, "The aim of every leader is to maintain cooperation, coordination and commitment among all members of his organization. This comes through qualities of heartfelt respect for the leader and that respect comes from an honest, rational and a well structured system. The leader who has earned credibility can command his many forces to quickly concentrate their power onto a desired object and achieve victory."

(613) Military Officers must NOT Cling to Obsolete Weapons Systems only because the Officers enjoy Fantasizing about Refighting Current Wars with Weapons Related to Old Weapons.

One of the reasons for why **Adolf Hitler** was so nonchalant about declaring war on the USA (December 11th 1941) was because **Hitler** sincerely believed that with his greatly expanded German U-Boat fleet in the Atlantic the USA would have no way to send significant numbers of troops and supplies to Europe. **Hitler was dreaming of refighting WW1 but with better submarines** and many more German submarines than in WW1. However - **German U-Boats were already a proven failed weapons system in WW1. Adolf Hitler made the classic mistake of making a failed weapons system much bigger and the end result was a much bigger failure.**

The Germans built 300% more submarines in WW2 than in WW1 but the USA's shipyards built 400% more cargo ships. The USA and British Navies also greatly advanced the effectiveness of their submarine killing weapons compared to WW1. The axiom is that **you might double the size of your previous failed attack but what usually happens is your opponent will double the size (or triple the size) of his previous successful defense.**

Hitler made the same mistake with Russia. **Hitler** attacked Russia in 1941 with more than double the German Troops of the 1914 Russian invasion but the Russians had more than doubled their defenses.

Is the US Navy making the same mistake with their 10 Super Aircraft Carriers? The answer is YES. **The USA's Aircraft Carriers had already been proven to be irrelevant during the Vietnam War of the 1960's.** The USA's Aircraft Carriers were also irrelevant during the Iraq War and are completely useless in the Afghanistan War.

The US Pentagon is building more multi-billion dollar obsolete weapons that have been proven to be useless. While the US Pentagon build yesterday's weapons for yesterday's wars the Pentagon is starving the non-shooting tools that have been proven to win Guerrilla type wars.

The US Pentagon is making the mistake of believing that if you manufacture a crossbow at great cost then it is a valuable weapon because it has a big cost. **Always remember that in**

Vietnam War the US Pentagon spend $1,000 for every $1 the Viet Minh spent and the US Pentagon lost the war.

(614) The US Pentagon is Making the Classic Mistake of Driving Forwards by Looking into the Rear View Mirror.

One of the USA's highest priorities must be reforming as soon as possible the obsolete organization of the US Pentagon.

(615) The Type of War that you Trained your Army and Navy to Fight has a very High Probability of NOT being the Type of War You End Up Fighting.

When WW1 started in August 1914 - Germany and Austria had 22 Cavalry divisions. The Allies (Russia, Britain and France) had 50 Cavalry divisions. All obsolete weapons cost money and manpower that would be more effective if applied to weapons systems for the next war rather than the last war.

At the start of WW1 German, French and British infantry divisions had only 24 machine guns per Division (a Division has about 11,000 men) as their standard issue. In particular the French military mindset of 1914 believed that massed infantry bayonet charges would be the decisive weapon system.

In 1914 half of the Artillery pieces (of ALL of the Armies) could not be elevated any higher than 16 to 20 degrees angle from the ground because all of the Generals thought that all future battles would be like the US Civil War (or the Napoleonic Wars) were your opponent was always in eyesight.

By 1918 the Generals had learned their lessons the very hard way. Some frontline defensive divisions had over 600 heavy and light machine guns. Artillery was built to fire at 45 degrees of elevation or higher.

The USA is on the downside of its Bell Curve and does NOT have the money or time for the Pentagon to learn the lessons of 21st Century warfare on future battlefields. Reorganizing the USA's Grand Strategy planning process is decisive for the USA's survival.

(616) The Rule that Became Standard Starting with the Industrial Age of Warfare (1885-1960) is that Winning the First Battles in a War (or Winning Most of the Battles in a War) is No Guarantee About Who is Going to Eventually Win the War.

Starting with the Industrial Age of Warfare all advanced Nations had the ability to build mass conscript armies with millions of soldiers. The new Industrial Age of Warfare massive armies were of such a gigantic size they could NOT possibly fit onto one battlefield. That meant winning a particular battle did NOT destroy the opponent's total army. You had to win the overall Grand Strategy struggle (in all of its many parts) to win the war.

The ultimate example is the USA-Vietnam War where the US Forces won all of the battles and lost the war.

(617) Speed has Always Been one of the Most Important Factors in Warfare. Once you have Made Up your Mind to Strike - then Move Fast and Hit Hard.

The Strategist has to correctly analyze his opponent's objectives and Strategy and then move with the most speed possible to counter them. **Albert Einstein's** equation holds true for warfare E=Mc2. **Energy is created by the speed of a maneuver that concentrates and multiples its striking force.**

In the laws of military ballistics you could shoot a ping-pong ball through the steel side of a Tank if you could get the ping-pong ball moving fast enough. **Speed equals power.**

Speed was a major factor that contributed greatly to **Napoleon** winning many battles in his early years. **The French Revolutionary Reformed Army had been reorganized and retrained to march and move faster than the troops of any of France's opponents. Napoleon** was thus able to arrive at advantageous positions first - and once the battle started **Napoleon** could maneuver around on the battlefield faster.

General Sun Tzu wrote about the value of speed - "Be as unknowable as the dark but always move like a lightning bolt."

Many times only a little extra speed can make a decisive difference. In both WW1 and WW2 air-combat - if one airplane could fly 5% to 7% faster than the opponent's airplane (and if the combat talents of both pilots were about equal) then the faster airplane had a decisive advantage.

At the O.K. Corral gunfight (1881 Tombstone Arizona) if you were a half second faster than the person you were drawing your gun against - then you were a 100% winner and the other person was 100% dead.

(618) Through all of the Ages of Warfare there has Been a Direct Connection Between a Nation's Productivity and that Nation's Ability to Win Wars.

Going back to ancient times (and up to the arrival of the Industrial Revolution in the 1800's) a Nation that could produce an excess of food and increase its population was the Nation most likely to be the dominate regional power.

With the beginning of the Industrial Revolution (1760-1820) the center of gravity for what Nation had world power shifted to the Nations that produce the most manufactured products.

By the start of the Industrial Age of Warfare (1885) the Nations with advanced industrial manufacturing were either already world powers - or (as with Japan) they were on the upside of their Bell Curves and going towards being world powers.

It was the genius economist **Adam Smith** (1723-1790) who first saw the changing foundation of world power. **The foundation of power went from the possession of land that produced excess food, minerals and populations to the Nations that could manufacture the most hard goods products.**

Adam Smith in 1774 specifically said that a Nation's ability to successfully wage war would be directly related to that Nation's manufacturing production capacity. That analysis became 100% correct.

Adam Smith wrote - "Fleets and armies are maintained with manufactured goods." **If you are not a great manufacturing Nation you are not forever going to be a world power.**

The brilliant German-American economist and military Strategist **Friedrick List** (1789-1846) wrote in 1835 that a Nation's industrial productivity had become directly related to its military capacity to win wars.

Karl Marx (1818-1883) wrote in 1867 - "Military power is built upon economic foundations."

The USA was the big winner in WW2 because the USA (at that time) was the world's #1 biggest manufacturer of products. In 1940 over 98% of everything that was in America's retail stores was made in the USA. 99% of the cars and trucks on the USA's roads where made in the USA. The USA's vast industrial base constantly produced large positive trade balances. Between 1930 and 1939 - during the 120 months of the "Great Depression Decade" 102 months had a positive USA Trade Balances and only 18 were slightly negative. **By 1940 the USA's positive trade balance was enormous.** In 1940 and 1941 before the USA was attacked the USA was the most profitable large Nation in the world.

In the 21st Century the USA's manufacturing base continues its decline relative to the world. Every year since 1950 the percentage of USA manufacturing productivity (as a world percentage of manufacturing productivity) has declined. Every year the percentage of "Made in the USA" products declines in America's stores.

During the USA's part of WW2 the USA's shipyards manufactured 37,118 military ships and boats of all sizes and types. The USA manufactured 108,410 Tanks and Self Propelled Guns and built over 300,000 airplanes. For comparison Japan built 85,000 airplanes during WW2.

Remember - for the USA WW2 was only from December of 1941 to August of 1945 -3 years and 9 months. WW2 for Germany started in 1939 (and for Japan in 1931) but the USA far out produced German and Japan.

In 1927 the city of Detroit was the richest large city in the world. In 1940 Detroit was the largest automobile manufacturing center in the world. Today's Detroit is in ruins and mainly a collection of abandoned factories with blocks and blocks of open space vacant land where manufacturing factories and worker's homes used to stand. **Detroit is the USA's metaphor for the USA's now greatly reduced capacity for hard goods manufacturing.**

What is the probability of Detroit once again becoming the world's richest big city? NOT a high probability. It is long past the time when Washington DC's strategy incompetents should have checked out of Motel delusion and checked into Hotel reality.

(619) The Soon to Arrive Artificial Intelligence Age of Warfare will Radically Change Tactics and Strategy.

Battlefield tactics are on the edge of many changes. Is the US Military ready for the new mindsets that will be needed?

Picture this - you are a Tank Commander in a new American Abrams M8 Battle Tank. You look up and there are 50 small smart helicopters flying towards you. Each smart helicopter weighs 150 pounds and is carrying a 15 pound Hollow Charge Tank-Buster Bomb on its underside. A big enough Hollow Charge to blast through 6 inches of steel. Each helicopter is highly intelligent and operates in communications and coordination with all of the other smart helicopters and has technology so it cannot be electronically gammed.

All of the smart flying bombs will have sensors like driverless smart cars so they will not run into each other. Every helicopter will have the same "mission" - to land on your Tank and detonate their bomb. If your Tank has "reactive armor" one of the drones will sacrifice itself and clear an area of reactive armor and then another flying bomb will land on the now cleared space and blast through your layered-spaced armor sending a molten stream of liquid steel into your Tank.

You (as the Tank commander) cannot outrun the smart helicopter/bombs because they are faster than your Tank. The small helicopters would split up into four groups and attack your Tank from all four sides at once. Your computer controlled smart Laser Gun on top of your tank would shoot down several of the smart mini-helicopters. However some of the flying weapons would have their own machine guns and would disable your Tank's Laser Gun. Many of the flying devices would be destroyed but a few would reach your Tank (and explode their bombs) and you will be dead.

The point is that in the future - big Tanks are going to be as obsolete as big Aircraft Carriers. **In a USA shooting war with Russia and/or China big everything will be the targets - big ships, big troop concentrations big airports big everything.**

(620) If You Find Yourself in a Stalemated War - Expand and Acquire Positions that Cannot be Attacked and Envelop Your Opponent with Positions that Cannot be Successfully Eliminated.

During WW1 the Western Front (France) became a stalemate after the first four months. However the British Fleet that was blockading Germany could not successfully be defeated by Germany's smaller fleet. The British Navy's blockade took 3 years to fully take-hold but by the Summer of 1918 tens of thousands of Germans were dying (or sick) from the effects of malnutrition/starvation and much of the German population was defeated in their minds.

In the Cold War between Russia and America (1946-1991) the USA was in an un-conquerable position. **Russian's Grand Strategy was to envelope the USA in a world full of Communist/Socialist Nations.** However - the USA kept expanding its allies in the world until it was crystal clear that the USSR's Strategy was defeated. With the USSR's Grand Strategy defeated the USSR's Will-to-Victory was also defeated and the USSR fell apart.

The day after Christmas Day 1991 the last flag of the USSR to fly over the Kremlin was lowered forever. The great Communist USSR Empire was gone.

If you defeat your opponent's core Strategy then you have defeated your opponent. Defeat your opponent in their mind and you have defeated your opponent.

(621) The Nations that Understand and Implement the World's new Strategy Realities will be the Future's Big Winners. The Nations who Have a Death Grip on the Past are Going to be the Future's Big Losers.

The coming Artificial Intelligence Age of Warfare (the smart machine age of warfare) will create a revolution in Tactics and Strategy. Those who are the first (and the most comprehensive) with the new changes to their organizations and **especially the new needed mindsets** will be the biggest winners in future shooting and non-shooting wars.

(622) The Next Age of Warfare (the Artificial Intelligence Age of Warfare) is Going To Be a More Manufacturing Intensive Age of Warfare than WW2. Is the USA Ready?

Some of the decisive industries will be: artificial intelligence, robotics, bio-technology, nano technology, miniaturization, cyber domination, salesmanship using all of the modern forms of communications, space aeronautics, space weapons, new energy sources, super materials, super software, ultra high speed computers and much more. The Nation (or

Nations) who are superior in the new manufacturing-and-communications skills and can apply that technology to all 37 types of warfare will rule the world.

(623) Many Times in History the Changes in Warfare, Tactics and Strategy Happened with Great Speed and Dramatic Effect.

The changes in both Land Warfare and Naval Warfare between 1864 to 1914 were profound. In a period of 50 years there were unbelievable changes. In 1864 "advanced" Nations like England and France were building wooden "battleships" that had fully rigged sails and muzzle loading black power cannons basically like warships in the year 1564.

As late as 1877 Britain built two battleships that had fully rigged sails - the Alexandra and Temeraire. The world will see much bigger degrees of change in warfare technology between now and the end of this Century.

Military Technology is more than hardware. Military Technology at its core are changes in mindset and changes in worldview. Guns do NOT win wars - people win wars.

(624) The Washington DC "Rule Book" must be Rewritten. Washington DC's (1) Incompetents and (2) Con Game Warmongers and (3) Law Breakers must be Punished.

"Bad Judgment" and the belief that the "inside people" have a right to "a loyalty higher than to the US Constitution" - that must no longer be an acceptable excuse that automatically gives Washington DC's manipulating Insiders a Free-Ticket to lie, falsify evidence and use the full power of the government as an aggressive weapon against those they subjectively disagree with.

Clearly there are Federal Agencies in Washington DC that want to (and have been) operating as a government inside of the US Government. **They have deliberately subverted the intent of the US Constitution for reasons that were personal and wrong.** The people involved should go to jail.

(625) In Washington DC there are Few Consequences for Gross Incompetence and the Result is Much More Incompetence.

The USA's strategy planning will take a giant step upwards in quality and success when the incompetent hacks, ass kissers, report falsifiers, fact manipulators, spin doctors, military-industrial complex enablers and warmongers start ending up in jail.

For the end of WW2 to now Washington DC has been protecting its many criminally incompetent "insiders" by promoting them to bigger jobs thus "proving" that nobody ever made a mistake. That con game on the American people must end. **A declining USA no**

longer has the resources to support an army of Washington DC scammers, crook and creeps.

(626) The One and ONLY Safeguard that Makes the USA's Homo sapiens Different from Nazi or Stalinist Homo sapiens is the US Constitution. Take Away the US Constitution and the Worst Crimes that Happened in the World Can Then Happen Inside of the USA.

The USA's left-wing groups attack the US Constitution as "yesterday's piece of paper written by fat old White men." The groups that want to dismantle the Constitution by doing things like packing the Supreme Court - etc. Those groups and people are attacking the core of what made America great and what made America safe and what keeps America safe.

Without the US Constitution the worst crimes that have happened in the world could happen inside of the USA. **The groups that want to destroy the US Constitution are the #1 enemies of both the USA and the World.**

(627) If Washington DC Insists on Living in the Past then the USA will Become the Past.

The Middle East is changing. The Middle East is on the upside of its reset Bell Curve. But there are groups in Washington DC going hysterical because the Middle East landscape is changing.

Great powers on the downside of their Bell Curves have all viewed "change" as evil and something to be resisted. The USA must fight that weakness and stop trying to freeze history.

Every great power in decline (i.e. the USA now) went through the stage of trying to stop history. It always makes things worse for the Nation in decline.

(628) The USA is both (1) a Great Power in Decline and (2) in the Middle East a Nation on the Wrong Side of History.

The above two facts make the probability of the US Pentagon clearly and decisively winning a significant shooting war in the Middle East about zero.

The correct Middle East Strategy for the USA is to do what is needed to reconcile the USA's Middle East schizophrenic contradictions.

If Washington DC is morally and intellectually too weak to reconcile its contradictions then Washington DC should withdraw its ground troops from the Middle East. The USA survived very nicely for over 200 years with no US Troops in the Middle East.

(629) The Organization of the US Military is Obsolete on Several Levels. The USA Military needs (at a minimum) Two Distinctly Separate Types of Troops with Separate Mindsets as to Their Mission.

The old WW2 type of "one-size-fits all military" is as obsolete as bayonets. The 21st Century US Military must have different Divisions that have some very different mindsets for the different jobs that are needed for success.

In areas like Iraq and Afghanistan once a moderate degree of pacification of the local population has been achieved by US Combat Troops. At that point the hunter-killer troops must be replaced by troops that are there to make life better for the local population.

(630) In the 21st Century the USA's "Military Industrial Complex" has in Many Ways Become Counter Productive.

From the end of WW2 to now a small group of Military Industrial Complex companies have dominated USA weapons development and production. For many years defense contractors hired retired military officers to "sell" weapons systems to the active duty military officers who were the buyers for the Pentagon.

There are restrictions on the above practice but in many cases the restrictions are for only 2 years. With today's many interconnected corporations in the "defense industry" the rules and laws are hard to enforce and easy to side-step.

The current Military Industrial Complex system of old retired Military Officers (with old ideas) going to work for the military contractors and making decisions about what the US Military will need 20 or 40 years into the future is a system that is HIGHLY FLAWED and highly dangerous.

Yes, the **old US Military Officers have a perfect understanding of what was needed to win YESTERDAY'S Shooting Wars.** However in most cases they know little or nothing about what is going to be needed to win tomorrow's shooting wars.

What the USA needs to win tomorrow's wars are the non-shooting warfare tools. 80% of the USA's military Officers understand almost nothing about the non-shooting warfare tools and worst of all - **they do not want to learn.**

What the USA has today are enormously expensive weapons systems updates designed by yesterday's US Military Officers to win yesterday's wars.

SUMMARY

From 1775 to now the average USA fighting soldier has been the pride and treasure of the USA. **Today's brave and hard working enlisted soldiers, sailors and airmen of the US Military deserve a better grade of senior Officers than they have had since the end of WW2.** The objective of this Book is to give our US soldiers and sailors smarter Leaders who will start to win long lasting peace everywhere.

The Earth has a million years of history in its future. Most of the factors that people passionately believe to be all important are in the long run of little if any importance. Most disagreements should NOT be settled with shooting wars because in the long run they will mean little or nothing in the world's history.

Chapter 31

The Proven Successful Way to Win and Keep Long Lasting Peace

The ultimate objective of every correct Strategy and war should be the same - an advantageous peace for as many people as possible.

The Roman Empire was NOT great because it only won battles. The true Roman genius and discipline was the Roman's understanding of how to win the wars and then how to win and sustain the peace for many Centuries.

The glory and brilliance of the Roman Strategy was proven by the Pax Romana. **The time when there was almost constant peace in the Roman Empire from 27 BC to 180 AD - 206 years.** However - peace in large parts of the Roman Empire started much earlier: 210 BC in Sicily, 146 BC in Greece, 133 BC in Spain and 100 BC in North Africa. **In many areas there were no wars for over 300 years.** Everyone's borders were safe because Rome controlled all of the borders.

The "secret" to Rome's success was it made life better for everyone. Rome became the outstanding organizational benefactor to all of the areas it controlled. After a very short time 90% of the local populations that had been "conquered" by the Romans wanted the Romans to stay in their lands because **the Romans were overall a great benefit for the local people.**

The Romans succeeded because they gave everyone what everyone everywhere in the world has always wanted and continues to want to this day: (1) **Safety** from being attacked by another Nation from the outside because Rome controlled ALL of the borders and all of the local rulers. (2) A **Stable Court System** with honest and **Reasonable Fair Laws**. (3) An honest and predictable **Stable Government**. (4) Safety from pirates and bandits everywhere - i.e. **Safe Streets**. (5) Continuous **Economic Growth**. (6) Easy access for every business person to a **vast Mediterranean Common Market** with the same laws the same government and the same currency everywhere. (7) **Full Roman citizenship** to anyone who was loyal to Rome and the Roman Empire.

Today's European Economic Union of the last 30 years is basically a recreation of what the Romans were doing 2,000 years ago.

Washington DC needs to copy the actions and proven successes of the Romans - who successfully (and 95% peacefully) controlled a great empire for many centuries.

The key to winning a successful peace is in the Bible - "Do to others as you would have them do to you for this is the essence of the law." (Matthew 7:12 and Luke 6:31). Rome was already basically practicing that commandment before **Jesus Christ** was born. **Jesus Christ** saw and understood what the Romans were doing and that it was overall good for everyone.

The Romans were the greatest peace keepers in history. **Jesus Christ never criticized the Romans.** The Sermons of **Jesus Christ** were aimed at the Zionists that were Hell bent on destroying Israel - and did destroy Israel in 66-70 AD by declaring war against Rome. The Zionists were insane 2,000 years ago and nothing has changed.

If the USA's Grand Strategy for Peace turns into what Jesus Christ said to do - then how wrong can the Strategy be?

(631) The good Grand Strategist Spends most of his Time and Efforts Planning for How To Maintain Peace and/or Achieve Peace.

The ultimate objective of war should always be an advantageous peace for as many people as possible.

In the Iraq War of (1991- ?) and the Afghanistan War (2001 - ?) like the Vietnam War (1959-1975) the US Military won all of the battles. However the US Military in Vietnam was defeated and gone - like it will eventually be gone in Iraq and Afghanistan unless the US Military does a 190 degree turn around and adopts the Roman proven successful method of winning the peace.

(632) A Competent Grand Strategy Team Puts More Prewar Planning into How To End a Proposed War than How To Begin a Proposed War.

Any moron can start a war and hundreds of wars have been started by morons. An Nation with an unlimited ability to borrow money (i.e. the USA) can keep a shooting war going. However **the objective should always be ending a shooting war and having a sustained long lasting peace.**

When this book was published all Americans who were born in 1989 (and after) have no memory of living in a USA that did NOT have Troops in somebody else's Nation killing the local populations. If that is not a "moral crime" than what is?

Fighting shooting wars is by definition a failure. Washington DC always fighting never ending shooting wars is by definition a gross failure of both Washington DC's management and a gross failure of the US Military's Strategy. **Only stupid and incompetent Generals have never ending wars.** Only morally bankrupt National Leaders have never ending wars

Constant warfare is NOT normal - it is sick and insane.

Wars are highly damaging to the social and cultural fabric of any civilization. The USA's never ending constant wars are one of the main reasons for the USA having (1) mass killings, (2) increasing amounts of escapist drug addictions, (3) growing extremist politics, (4) increasing numbers of suicides, (5) increasing numbers of Americans just giving up on life and crawling into their video games and alternate worlds of doing ONLY the bare necessity of what is needed to keep living.

From 1900 to the start of 1950 (the period when the USA became the world's greatest Nation) the USA was "at war" a total of 5 years and 5 months (WW1 - 1 year and 8 months, WW2 -3 years and 9 months). However - now that the USA is on the downside of its National Bell Curve (like the Roman Empire in its decline) the USA is constantly having never-ending stupid wars for stupid reasons.

For 95% of the 21st Century (up to this point) Washington DC has had some type of shooting war somewhere in the world. **Constant wars are history's #1 textbook characteristics for great powers in decline.**

In 1907 the British Empire was the biggest in the world. In 1947 the British Empire was coming apart and collapsing everywhere. During that 40 year period the British lost 1,151,300 military killed and 2,346,801 military wounded. **Wars killed the British Empire and they are killing the USA now.**

(633) The Best Grand Strategist Seldom Looks for Absolute Surrenders. He is Always Willing to Accept an Imperfect Peace that Looks Like it Might Hold and Be Expandable.

You work towards absolute surrenders only if (1) you are sure you can crush your opponent completely and (2) it is vital to create an example to avoid future wars.

Good peace agreements are by definition NOT totally one sided. A peace agreement that is too one sided will NOT last. Both sides have to be getting something of value and **both sides must have their own vested interests in keeping the peace going.**

(634) Most of the World's Ultra Vindictive Peace Treaties have Historically Been the Raw Material for Future Wars.

One sided peace treaties where one side acquires almost everything at the expense of the loser have historically been proven to be dangerous and often the reason for future wars.

The only one sided peace agreements that are safe are the ones where your opponent is totally crushed forever and can never challenge you in a future war.

One of the moral high points in American history (and what was outstandingly excellent Grand Strategy) was after WW2 when Japan and Germany were totally defeated. **The USA did NOT impose harsh conditions on either Japan or Germany.** The end result was both Japan and Germany became dependable allies of the USA and everyone had a long advantageous peace.

What is the most successful (and longest lasting) Peace Treaty in the history of the modern world? It is the 1812 Treaty of Ghent between the USA and Canada saying the USA-Canadian border would not be fortified. The USA has put together common sense treaties in the past - we have to return to those days.

(635) Do Not look for (or Demand) a Spectacular Surrender. They are the Exception and Not the Norm.

American military history (which is very short by world standards) has three examples of unconditional surrenders that have distorted the US Pentagon's mindset. (1) The Surrender of the South in the Civil War (1861-1865). (2) The unconditional surrender of Nazi Germany (May 1945) and (3) the unconditional surrender of Japan in August 1945. The formal Japanese Surrender was "Hollywood Movie Perfect" on the deck of the Battleship Missouri in Tokyo Harbor (September 2nd 1945).

Most of the hundreds of bigger wars in the world's history did NOT end in anything close to being so neat and dramatic. **The Japanese formal ending of WW2 was the historical exception rather than the historical rule.** The way most wars eventually end is the way the USA-Vietnam War ended. The way the North Korean-USA War ended and the way the USA-Iraq War will end.

(636) If Your Grand Strategy has Failed and You are on the Downside of a Particular War's Bell Curve then Make Peace as Soon as Possible and Save What Can Be Saved.

All Wars have Bell Curves like everything else in the Universe. If you find yourself losing (and it is NOT a temporary setback blip but the true downside of that war's Bell Curve) **then you must end the war as soon as possible with the best terms you can negotiate.**

Always remember the basic Rule, "You have to know when you are defeated." **If you are on the downside of your war's Bell Curve and losing then the sooner you try to make peace the more assets you will have for negotiations and the more assets you will be able to salvage.**

The British did the above with the demise of their Empire after WW2. Rather than fighting never-ending stupid wars in attempts to stop history like the US Pentagon does. The British made peace with their Colonial rebellion leaders. In many cases the British were able to bring their former colonies into the "British Commonwealth." Rather than losing everything the British saved something. **Washington DC must learn from the British.** A 50% advantageous peace for the USA is much better than a never ending war.

(637) The Fatal Flaw with the USA's Military Industrial Complex is that it ONLY Knows How to Produce Shooting War Tools.

For General **Carl von Clausewitz** shooting wars were the one and only tool. The US Pentagon is a follower of **Clausewitz** and the Defense Contractors are giving the US Pentagon what it wants (shooting war tools ONLY).

What the American people need is half of the Defense Budget going to the non-shooting tools that will be decisive for winning the future's conflicts and most importantly for winning the peace.

(638) The competent Grand Strategist is always Working Very Hard to Maintain the Peace because Peace has Never Been (and Never will be) a Static State and a Forever Given.

Peace is like a flowering plant - it takes work, time, money and TLC (tender loving care) to keep everything alive and flowering. **Peace must never be taken for granted** or as a forever given that will always be there.

Business has a close Rule to the above. The Rule for business says that **sometimes keeping the money you make is harder than originally making the money.** Sometimes keeping the customers you have is harder than originally getting the customers.

(639) For successful Peace Negotiations (and a Lasting Peace) the Grand Strategy Team must Precisely Know the Vital Desires of their Opponent.

Unless a war is a total destruction of one side or the other the time for peace negotiations will always arrive. **If you do NOT understand your opponent's objectives and Strategy then you are at a great disadvantage during the peace negotiations**.

Often what is vital to you is NOT vital to your opponent and the other way around. You must look for the "diversity of vital interests." **If something is valuable to your opponent but of little real value to you then you have found a working-area for a successful agreement.**

The classic 4,000 year old historical mistake has been **Nations (and Leaders with an ego) classifying worthless ego status symbols as "vital interests."** In negotiations everything that is not a vital factor for your side is negotiable and a maneuvering asset.

What if your vital objective is exactly the same as your opponent's vital objective? Then you have to split-the-difference and each side receives something but not everything they wanted. Each side has to understand that **50% of something valuable (peace) is better than 100% of a never ending stalemated shooting warfare.**

The 2 biggest enemies in peace agreements are greed and arrogance. Do NOT become ego arrogant and/or greedy and demand everything to prove that your are "bigger" than anyone else. **A peace negotiation that fails means you have won 100% of nothing.**

(640) The US Military Cannot Bomb and Destroy a Set of Beliefs that Are in the Minds of Hundreds of Millions of Muslims.

In the Middle East the US Pentagon's guns and bombs obsolete WW2 Strategy is doing 2 things. (1) the Pentagon is pushing the Radical Muslim Groups around on the map of the Middle East but NOT destroying them and most decisively **NOT changing their minds.** (2) The more Muslims that are murdered by US Troops the greater the number of Muslims that will join the Muslim Groups in their fight against the invading US Troops.

History has show and proven that the best way to defeat ideas is with better ideas.

If French Field Marshals **Lyautey** and **Bugeaud** were commanding the USA Military today in the Middle East. They would achieve peace for the USA by using what worked for them (and what worked for Centuries with the Roman Empire) and what worked for over a hundred years with the British Empire. They would introduce better ideas and better actions to the point where the local people wanted the Americans to win and to stay.

(641) The Versailles Treaty of 1919 is the Prime Example of What NEVER To Do with Treaties.

The 1919 Versailles Treaty of WW1 was vindictive and harsh on Germany. **The Versailles Treaty had the seeds of WW2 in every paragraph.**

The US Congress had more common sense a Century ago than it has today. The US Congress refused to ratify the vindictive and disastrous Versailles Treaty.

The Versailles Treaty was constantly (and very successfully) used by **Adolf Hitler** as a major political tool during **Hitler's** 1920's and early 1930's rise to power.

(642) A Less than Perfect Iran Peace Agreement is Better Than Another War the US Military Will Eventually Lose.

This is where talent in the Art of War and Peace comes into play. (Or perhaps the ***Art of the Deal***). The competent Grand Strategist will settle for an imperfect peace if it has advantages for his side and is the best that can be achieved.

Remember - greed, egotism and demanding everything you want are the **enemies of lasting peace agreements**. You should think like a Chinese - be satisfied with good when good is doable - but 100% perfect <u>for only your side</u> is a killer of the peace and long term NOT workable.

For centuries the Catholics and Protestants in Ireland were killing each other. They finally made peace and **neither side was 100% happy.** However the end result was much better than the never ending murders and economic/emotional burdens of war.

The Washington DC military morons and the Warmonger Lobby who sincerely believe that Iran is going to become a West Bank like slave people to Washington DC are truly insane. The correct strategy is to negotiate a peace that puts USA interests <u>inside of Iran so Iran can be changed from the inside out.</u>

(643) It is a Good Sign when Both Sides Fight Hard and Long Over the Specifics and Details of a Peace Treaty.

If both sides fight hard during the peace negotiations it usually means they probably intend to keep their agreements. If one side gives in too easily it might be because they never intended to honor the agreement.

In International Relations (and business) if you (1) ask for much more than something is worth and the other side gives you everything you asked for without much of an argument. Or (2) offers you a very large asset for a job or project that is not worth the asset. YOU should be highly apprehensive. They could be offering you more than what the job is worth because **they never intended to pay you (or never intended to honor the treaty).**

You could believe that something you are being offered is a long term deal - but the other side could be planning to pay you for only a short time.

Before it was made illegal in the USA some ruthless companies would hire key people from their competitors by offering double or triple their current salary. After learning all of their competitor's secrets and contacts the person hired was fired. Their old company would not hire them back because they were traitors who could not be trusted. The instigator company (1) learned all of their competitor's secrets and then (2) destroyed one of their competitor's managers.

(645) You should Start Offering Peace Terms to your Opponent as Soon as it is Clear that Your Opponent is de facto Defeated.

Open peace proposals from your side will sap the fighting moral of your opponent's troops. Most of your opponent's troops will know when their side is losing the war. If your opponent's troops think there is a peace agreement they can live with then that becomes a plus for your side. **A peace proposal puts your troops on the moral high ground and clearly fighting for peace.**

Professional military Officers (most of whom know little about grand strategy) almost always want to keep fighting. Their standard BS line is "If we go after anything less than unconditional surrender that will sap the fighting sprite of our troops." In reality it does the opposite.

If you can communicate a peace offer to your opponent's troops and population then your peace-offer will undermine the support of your opponent's Leadership. **Your opponent's troops and population will start to fight half heartedly** and/or abandon the cause.

During WW2 the Allies called for "absolute victory" against both Germany and Japan. That was propaganda bravado for the Allies side but **the extremism motivated the German and Japanese troops (and their populations) to keep fighting.**

Sun Tzu wrote about an already defeated opponent.

"If your opponent is already defeated and your opponent is withdrawing then never interfere."

"If your opponent is already defeated and you have surrounding your opponent leave an opening for his men to retreat.

"If your opponent is already defeated and desperate then never press the attack."

General Sun Tzu was saying that if your opponent is already defeated then killing more of your own troops only to be killing more of your opponent's troops is morally wrong and thus bad Strategy. However if you are not 90% sure of the overall defeat of your opponent - then you SHOULD press the attack. This is one of the many places where Warfare becomes An Art.

(646) NEVER Assume that Your Opponent is Defeated and/or Neutralized Without Being Absolutely Sure.

During the 1940 battle for France the German forces had almost 400,000 British forces trapped at Dunkirk. That was about 80% of the British combat Army in May of 1940. **Adolf Hitler** ordered the German Army to not press its advance against Dunkirk at all costs. **Adolf Hitler** between May 16 and June 4th 1940 let 338,226 allied troops escape to England. About 40,000 British and 40,000 French troops were eventually captured by the Germans at Dunkirk.

The Germans won a great battle at Dunkirk but the British were NOT defeated. Four years later on D-Day (June 6th 1944 - and the months afterwards) many of the same troops who had escaped at Dunkirk returned to France and defeated the Germans.

Knowing when to press the attack (or when to let your opponent go) - that is part of the Art of War and Peace.

(647) At the Start of a War you Must Put Your Objectives Down in Writing and Constantly Refer to Them.

You must have very specific (and written down) war objectives before the war begins so you do NOT stumble into a situation where there are no objectives and the war itself becomes the reason for continuing the war.

Washington DC's war in Vietnam quickly degenerated down into being the justification for having the war in Vietnam.

In the Afghanistan War the US Troops fighting has become the justification for US Troops fighting.

The Post WW2 US Pentagon has developed a great talent for turning every shooting war into the justification for continuing the shooting war. When this Book was first published the shooting war in Afghanistan was continuing. The two REAL REASONS for the Afghanistan war continuing were (1) it continued to make Billions of dollars $$$ for the US Military-Industrial Complex and (2) it gave the US Pentagon a live-fire playground. Every other reason for the war continuing was BS.

There should be a new law in the USA with 2 options. (1) If a shooting war goes over 1 year then in the next general election there should be a National vote as to if it should continue. Or (2) with today's tecnology a National vote is relatively easy. If Washington DC has the resources to continue a war then it has the resources to have a National vote <u>at any time</u> about continuing the war. If the Warmonger Lobby cannot "sell" a war to the USA's voters with honest facts and logical reasoning then why should the USA be in that war?

(648) There is No Glory in going Down in Flames. There is No Glory in Destroying Yourself Because You Refused to Admit You Were Defeated.

If you are in a losing war (with no road to success) then Do NOT keep dragging the war on and on hoping that "Something good will happen."

Being a good Grand Strategist is having the courage to admit you were wrong and the courage to end a de facto lost war on the best terms and conditions possible.

The US Military's (and the US Warmonger Lobby's) shooting wars to make the Middle East safe for Apartheid and brutal colonialism never had a chance of long term success. The US Military is on the wrong side of history in the wrong geographical area and in the wrong Century. It is time for the US Military to get out of the grave-pit before it is filled in.

(649) Five of the Hardest Parts of Learning How to be a Good Grand Strategist (and good National Leader) are (1) Learning how to Discard Proven Failed Strategies. (2) Discarding Incompetent People. (3) Discarding Wrong Subjective Beliefs. (4) Discarding Useless Allies. (5) Discarding Unachievable Delusional Objectives.

One of the most valuable talents for success in everything is having the courage to LET GO of what is NOT working.

The smartest management teacher of the 20th Century was Austrian-American **Dr. Peter Drucker** (1909-2005) who said, "You cannot bring in the new until you first make room for the new by discarding the now obsolete."

Washington DC MUST LEARN these 5 axioms - (1) every challenge does NOT have to be answered. (2) Every worthless status symbol does NOT have to be defended. (3) Every National luxury and privilege does NOT have to be maintained at any cost. (4) Every war does not have to go on forever. (5) Washington DC must learn how to let the unimportant often less-than-worthless status symbols go and concentrate on what are the decisive essentials for the USA's 21st Century long term survival.

(650) Nations and Leaders must Guard Against the Sickness of "Security Addiction Disorder" (SAD).

Security Addiction Disorder has 3 other names: (1) Security Domination Syndrome, (2) Sick Ego Domination Need, and (3) Security Psychopathic Syndrome. They are all the same Mental Sicknesses.

The National Leader (and/or Military System) that is afflicted with SAD has an uncontrollable need to Dominate as many other people and things as possible. They are paranoiac and feel insecure and the only relief from their insecurity is dominating other people, groups or Nations.

Many people in Washington DC have (to one degree or another) SAD mental set disorders. One of many examples are those people in Washington DE who would feel secure with Iran ONLY if Iran was a West Bank type of de facto prison camp. Anything less for Security Addiction Disorder afflicted people is unacceptable.

Iran is NOT going to become a large West Bank prison camp with a long and bloody war that the US Military would eventually lose.

There are SAD afflicted people who are mentally sick (and sick Nations) who love to hurt people only to be hurting them. **Every person with extensive real world experience has run across Security Addiction Disorder people who worked at hurting other people for no reason other than proving they were dominant and could hurt others.**

SUMMARY

The ultimate objective of every war is always the same - it is always an advantageous peace for your Nation and for as many people in the world as possible.

The US Pentagon's cultural mindset (and the mindset of the USA's Warmonger Lobby) continues to be dominated by the war ideas of **Carl von Clausewitz**. The 6 facts about **Clausewitz** are (1) **Clausewitz** is completely obsolete. (2) **War is NOT politics by other means - war is the failure of politics.** (3) War is the **failure of diplomacy.** (4) War is the **failure of common sense** on one side or both sides. (5) War is the **failure of ethics** on one side or both sides. (6) **Clausewitz's** writings say absolutely NOTHING about how to create or sustain a successful long lasting peace.

Chapter 32

The USA's Must Have 21st Century Grand Strategy Peace Team

The idea of a Central Military General Staff started with the Prussian General Staff in 1814. The USA's Joint Chiefs of Staff was created in 1949.

The current USA Joint Chiefs of Staff is dangerously obsolete. In the 21st Century the USA must have a major expansion and reorganization of its war and peace planning Team.

There is an old saying "If it is not broken then do not fix it." The US Pentagon (with the largest by far military budget in the world) has NOT clearly and decisively won a significant shooting war since being part of the winning side in 1945.

The US Pentagon since 1945 has been a shooting war mediocrity at both decisively winning wars and winning a long lasting peace. The correct Saying is - **If something vitally important IS Broken then it MUST be fixed**.

The US Pentagon and US Military since 1945 have been shooting war losers. **The Pentagon wins all of the battles and loses all of the wars.** There is definitely a major problem that must be fixed as soon as possible.

(651) 21st Century Warfare has Become Enormously Complex and Requires Extensive Mega-Talents. The Current US Pentagon does NOT have Those Talents. The USA Must Create the New Grand Strategy Team as Soon as Possible.

The new Grand Strategy Team is a 21st Century decisive weapons system. Without it the USA (and American people) will lose more wars and start losing them faster. The USA's downward decline (relative to the world) will increase in speed.

There are 34 main analysis areas (with many sub-parts) to the 21st Century's Grand Strategy equation. The Grand Strategy Team must find the most talented and skilled analysts in each of the 34 specialized areas of expertise.

(1) The Team must have **proven talented experts in: National and world economics, finance, business, international trade and management.** Many wars in history have been Economic Wars and victory or defeat had a close connection to successfully (or not successfully) managing the economic/financial factors.

(2) The Team must be able to make the overall correct **Comparative Economic analysis** (short term and long term) for all sides.

(3) The Team must have experts who have **a cutting edge understanding of computer/software engineering** because in the soon to arrive Artificial Intelligence Age of Warfare many tactical and strategic decisions will be made by computers.

(4) The important weapons of the future will be highly intelligent machines. **One way or the other companies like Apple and Microsoft and Google etc must become USA major defense contractors.** To have that partnership work successfully the US Pentagon is going to have to put its actions and objectives back onto the moral high ground.

(5) The Team must have members with **proven talent in understanding comparative cultures and religions, human nature and motivations.** Team members must be able to provide absolutely **correct reasons for why your opponents are fighting** and what could change their opponent's motivations one way or the other.

(6) An accurate estimate of the overall **comparative Quality of National Leadership** and the proven ability of each side's Generals to win wars rather than (as with US Generals) only know how to win battles while losing all of the wars.

(7) The Team must have members with **an understanding of comparative natural resource advantages and disadvantages.** They need to know what natural resources your side lacks that will affect a long war and know what your opponent lacks that will decisively affect your opponent's war efforts.

(8) The Team must have **experts who understand military weapons production capacities** for all of the sides. If a Nation has an advanced high technology manufacturing base (like China, Japan etc) they can <u>manufacture many intelligent war weapons in a short time.</u>

(9) The Team must know **where are all of the sides located on their comparative National Bell Curves?** Nations on the downside of their National Bell Curve (like the USA) 95% of the time do NOT win significant shooting wars against Nations like China and the Middle East who are on the upside of their reset Bell Curves.

(10) There must be correct estimates for the **Ambiguity and Crisis Tolerance** of the Leaders, Generals (and where applicable the voters) for your opponent's side. Where are your opponent's "breaking points"? In the Vietnam War the USA's "breaking point" came long before North Vietnam's "breaking point." **In a US-Iran shooting War the "breaking point" of the USA's voters would come long before the "breaking point" of the Iranians.**

(11) There must be Team members who can correctly estimate for real world usable military shooting war **gross firepower** for all of the sides. Keep in mind that **in the Information Age of Warfare gross firepower is far from decisive** - as the Vietnam War proved.

(12) The Team must have members with **a deep understanding of the psychology of decision making. The psychology of logic and most importantly the psychology of illogic and insanity.** There must be an extensive understanding of the psychology motivations for going into wars and the many motivations for wanting to make peace.

(13) There must be Grand Strategy Team members who understand **how to analyze the percentage combinations of the 37 main types of warfare and what percentage applies with each particular war being fought** (and how that percentage changes during the war).

Wars are motivated by: revenge, economic gain, religions, semi-religions like Communism and Nazism, political ideologies, National border issues, emotional status protection, population pressures, relative power adjustments between Nations, stupidity, insanity and many other main reasons (see Chapter 3 for all of the main reasons).

(14) Every Grand Strategy Team member should have equal value because every type of warfare analysis is different in every conflict. There are 37 main types of warfare and the most valuable Team members are the experts in the particular type of war combination being fought at that particular time.

(15) The Team must have analysts who can do the **Population demographics**, **geographic and geopolitical** advantages and disadvantages for all sides - and then do running real time updates.

(16) The Grand Strategy Team needs proven **talented people who understand comparative National Characters and cultures.** Team members who know that **different cultures have different National decision making preferences and thought patterns.**

The thinking process and value systems are significantly different between Western Christian Generals and Chinese Generals and Muslim Leaders - know the differences.

(17) The Grand Strategy Team must include talented **analysts who can correctly measure the relative quality of the <u>National Character</u> of Nations.** National characteristics such as (1) are your opponents soft comfort addicted people (like the USA's voters in the 21st Century). Or (2) are they people who will suffer great pain and never surrender (like North Vietnam and Iran).

(18) The correct National **Will-to-Win estimates** (short and long term). Estimates of when the USA's voters would turn against a war. In the past it has taken a long time for the US voters to turn against a war but there are indications that time period might be

significantly shorter in the future. **The bigger the war the faster the USA voters will turn against it.**

(19) The Team needs **experts in world political history, military world history, the history of diplomacy, the history of economics and the history of finance in warfare and peace.** The USA Team in particular needs experts who understand the history of the rise and fall of empires and great powers and exactly why the former great powers went up and exactly why they eventually declined and either (1) crashed and burned or (2) went into a very long period of stagnation that was a slow motion crash.

(20) A correct analysis of the **quality, depth and accuracy of the National Intelligence Organizations** for all sides. NOT how much money they cost and waste - but how correct and accurate they have been. Are they honest or do they have their own agenda's for being a government inside of a government? In other words - can their own government's trust them?

Can a US President trust his own FBI and CIA? Can a US President trust the Pentagon's General Staff?

(21) A sub-team that can do the **Right-Side-of-History correct estimates** for all sides. Who is on the Right Side of history - and who is on the Wrong Side of history - (when this Book was published the USA in the Middle East is on the very wrong side of history).

(22) The **Moral High Ground estimates** for knowing who will have the World Internet advantage and the World Public Opinion advantage. In general the World Public Opinion advantage goes to those who are clearly on the moral high ground of supporting freedom and justice for everyone everywhere. (That was NOT Washington DC when this book was published).

(23) An accurate analysis of the historical and current **trends of all of the major Paradigms** that are in play for all sides.

(24) The Strategy planning Team needs talented experts who understand National Bell Curves. **A Nation's Grand Strategy must change significantly depending on where your Nation is at on your Nation's Bell Curve.** A Nation on the downside of its Bell Curve (the USA in the 21st Century) needs a very different National Grand Strategy from when it was on its Bell Curve upside (from 1775 to 1975).

(25) Analysts who understand the **True Objectives and Desires** of all of the sides in the disagreements.

(26) Analysts who understand the **Overall Main Strategy** of all sides for achieving their Objectives and Desires.

(27) **The Team needs tough thinking members who know how to use Allies and know how to avoid being used by Allies.** The USA Team must absolutely have members who are loyal to the best interests of the USA first - and ONLY.

(28) The Team needs talented experts who can write down on paper a highly specific definition of (1) exactly what is a win and (2) exactly what is a defeat. If you are unaware of either definition (like the Generals at the US Pentagon) then how do you know where you are going? How do you know when you have strategically won? How do you know when you have been strategically defeated?

In WW2 both Germany and Japan were strategically defeated by August of 1943 - they had zero chance of winning. However neither Nation had made a definition of defeat and kept on murdering hundreds of thousands of their own citizens for the next 2 years.

(29) The Team needs **experts in multiple scenario building and an understanding of Black Swan probabilities.** Experts who are scenario building all of the time for when Black Swans happen.

The ideal Team should outline as many Black Swan possibilities as possible so no matter what happens it should never be a 100% total surprise. There must be at least <u>a rough outline for action responses</u> for every possibility with a chance of 1 in 25.

(30) There must be Team Experts who are both skilled at managing unexpected Black Swans and **skilled at turning Black Swans into opportunities.** The Team needs members with the talent to see the opportunities and act very fast and boldly.

(31) The Team will have highly specific and written down **"Stop Orders"** for every shooting war. At pre-determined points of (1) cost in lives and (2) cost in time and (3) cost in money if the core objectives are NOT achieved then **the Stop Orders will activate and the best peace possible will be negotiated to end the shooting part of the war.** (Naturally the "Stop Order" points would be the highest level of secret).

(32) The Grand Strategy Team must have **a permanent and on-going process for psychoanalyzing themselves to know their own weaknesses.** The Team needs on-going programs to compensate for their weaknesses. **The Team would continually be psychoanalyzing all of their opponents and all of their Allies.**

(33) Within the Grand Strategy Team no-holds-barred debates must be encouraged. **The ultimate criteria for accepting plans for action must be <u>the ability of the plans to withstand intense cross examination by experts</u> in all of the above major groups of knowledge.**

(34) The good Grand Strategy Team should have **the ability to take ALL of the above and tie everything together into one comprehensive and dynamic Grand Strategy master plan** that has both short term (one or two years) and long term action plans (to the end of

this Century). **The Grand Strategy plan must be under constant refinement and updating because the world's major factors are both in (1) constant change and (2) changing at an ever increasing speed.**

The existence of the Team (and all of its member's work) must be kept as secret as possible.

The US Pentagon's General Staff will greatly resist the new Grand Strategy Team. Why? Because the Pentagon is a collection of grand strategy idiots who do not understand what is happen and what is needed in the 21st Century.

The US Pentagon is going to see a new Strategy Team as a direct threat to their current power status quo. When old and bloated organizations are on the downside of their Bell Curves (like the current US Pentagon) any real "change" no matter how badly needed is viewed as a threat and greatly resisted.

(652) The Chinese Probably Already Have a Grand Strategy Team as Outlined Above. If Not - We can be 100% Sure the Chinese are Working on the Project.

Every intelligent National Leader will build the above type of Grand Strategy Team to the greatest extent possible. Judging from the spectacular success that the China have had against the USA (using the many non-shooting tools of warfare) **we can be 98% sure that a Chinese Grand Strategy Team (as outlined above) already secretly exists in China.**

(653) The Builders and Managers of the Grand Strategy Team must Remember that the most Gifted Grand Strategists Have To Be Born and Sometimes they have "Non-Standard" Personalities.

Any intelligent person can be taught how to be a competent oil-on-canvas art painter but only **God** can make a **Vincent Van Gogh.**

Talent can always be greatly refined and expanded with good training but no amount of training can create a **Leonardo da Vinci.**

Some of the greatest artists were NOT recognized as great artists during much of their lives. Some individuals who were excellent long term Grand Strategy analysts (like **Friedrich List** 1789-1846 or **Homer Lea** 1877-1912) were ignored during their lifetimes because their ideas were too advanced and their personalities were not what was expected.

Some of the best analysts in world history were like **Galileo** who was 100% opposite the politically-correct "official thinking" on some issues. Many times in history advanced correct analysts were so opposite the "orthodoxy truth" they become heretics in the eyes of their own "officialdom." **Generals everywhere have never liked being told what they do not want to hear.**

In late 1943 General **Tojo** who ruled Japan assigned the two best strategy analysts in Japanese Intelligence to analyze the war. Their analysis conclusion was that Japan had zero chance of winning the war and should surrender on the best terms possible and as soon as possible. General **Tojo** thanked them for their work. Shortly after they were assigned to Northern Manchuria and never seen again.

In the last 3,000 years of warfare there was usually somebody who correctly understood the complete Grand Strategy picture and know the eventual outcome of a war. But most of the time they were persons who were outside of the official leadership loops and often "nobodies."

(654) Ninety-seven Percent of all US Generals and Admirals are NOT good Grand Strategists. The Grand Strategy Team should have Some Generals and Admirals but Not Many.

If you went to West Point Military Academy (at the time this Books was published in 2019) you would have found that **West Point had zero classes specifically and only in Grand Strategy.**

Students graduate from West Point as a 2nd Lieutenant at age 22. They spend the next 15 years basically being told to NOT think and to follow orders. **By the time they are 37 years old (and a US General) their Grand Strategy creative thinking ability was either never there - or had been atrophied years earlier.**

Officers in big military organizations like the US Military spend most of their adult lives in jobs where they are not supposed to think. **To become a General you have to "fit the mold" but the "mold" is the problem.**

The US Pentagon's "General mold" has NOT decisively and clearly won an significant shooting war since 1945.

The Grand Strategy Team should be constantly consulting with the Generals and Admirals as to their opinions about what is logistically and tactically doable (and not doable) and what military resources are available. However the final Grand Strategy plans should come from the strategy team and be based on the super big picture.

Professional militaries have never had the money and ability (and most importantly the need) to systematically teach every Officer what is needed to become a good Grand Strategist.

An army of six-million troops only needs one excellent Grand Strategist and about 400 General grade Officers who know how to successfully carry-out orders and "follow the book" and win the battles. **What is missing (and has been missing) in the USA is the one good Grand Strategist.** That is WHY the USA must have the new Grand Strategy Team as soon as possible.

Maurice De Saxe (1696-1750) the Marshal-General of France and one of the greatest soldiers of the 18th Century wrote about Generals saying "Many commanding Generals only spend their time on the day of battle in making their troops march in a straight line and in seeing that they keep their proper distance. They appear to me like men who are only able to do what they have done all of their lives which is to conduct troops methodically. How does this happen? It is because very few men occupy themselves with the higher problems of war. They pass their lives drilling troops and believe that is the only branch of the military art. When they arrive at the command of armies they are totally ignorant and in default of knowing what should be done. One of the branches of the art of war, that is to say drill and the method of fighting, is procedure and mechanical and the other is intellectual. For the conduct of the latter it is essential that ordinary men should not be chosen. **Unless a man is born with talent for war he will never be anything other than a mediocre General.** It is the same with all talents; painting, music or poetry. **Talent must be inherent for excellence. All sublime arts are alike in this respect."**

Leo Tolstoy (1828-1910) was an officer in the Crimean War (1853-1856 it was England, France, Turkey vs. Russia). **Tolstoy** wrote in his 1869 greatest war novel in the world ***War and Peace*** - "Why do they talk of military genius? A General is only called a genius because of the glamour and authority with which the military are invested and because masses of sycophants are ready to flatter power, and to ascribe to power realities quite alien to it. Most Generals I have known were stupid or absent-minded idiots."

Five Star US General of the Army (<u>and</u> 5 Star General of the Air Force) **Henry Arnold (1886-1950)** one of only six 5 Star Generals in US military history and the only 5 Star General (or Admiral) of 2 Branches of the Armed Services said - "The technical genius which can find answers is not in the military, or in the government bureaucracy, but is to be found in universities and in people at large who have gifted intelligence."

It seems like the last USA Generals who had a competent understanding of Grand Strategy were General **Douglas MacArthur** (1880-1964) General **George Marshall** (1880-1959) and General **Dwight Eisenhower** (1890-1969).

In the opinion of the author of this Book many of the USA's WW2 Generals and Admirals where excellent because (1) during their early years as Officers (in the 1920's and 1930's) they had very limited military budgets and **they were forced to think about strategies and plans that would get the most from their small military assets.** (2) There was no US Military Industrial Complex selling hundreds of Billions of Dollars of overpriced semi-obsolete stuff designed for the last war. (3) There was very little US Army and Air Force equipment when the USA Military entered into WW2. That was an advantage because almost everything could then be a new modern design. Example - every other WW2 major combatant used WW1 style bolt-action rifles until the end of the war because they were leftover by the millions from WW1. The US Troops during WW2 had a modern semi-automatic rifle (the M1 Garand) because the USA did not have many leftovers from WW1.

Having an enormous military budget as the USA has had for decades means you have a mountain of obsolete equipment, obsolete organization for the obsolete equipment and obsolete mindsets. **The US Pentagon is superbly positioned and equipped to win yesterday's wars.**

(655) Being a Battlefield Hero Makes You a Hero but Does Not Usually Make You a Competent Strategist. Any Heroes on the Grand Strategy Team Must also be Proven Competent Strategy Planners.

Adolf Hitler (1889-1945) was a battlefield hero who won the German Iron Cross for bravery during WW1. However at the end of the game **Hitler** proved himself to be a disastrous Grand Strategist and a world class loser.

Hermann Goring (1893-1946) was also a battlefield hero. One of Germany's greatest aces who had won in air combat Germany's highest award - the Pour Le Merite (the Blue Max) and eventually the Knights Cross of the Iron Cross. However **Goring's** Strategy during the decisive air battle of Britain was fatally wrong and a disaster for Germany.

Today in the 21st Century **the US Pentagon's cultural paradigm is NOT the solution - the Pentagon's culture is the problem.** The US Pentagon has NOT decisively and clearly won a shooting war since the end of WW2 but the Pentagon staff fanatically defends their proven failed mindsets and shooting war losing failed systems.

Dr. Peter Drucker (1909-2005) who was the world's greatest teacher of management in the 20th Century, and **Albert Einstein** (1879-1955) **both said that those who created the problems are NOT the ones who can solve the problems.**

Einstein's exact quote is - ***"No problem can be solved from the same level of consciousness that created the problem."***

The main objective of this Book is to change the US Military's "level of consciousness" to something much higher and much better for dealing with the 21st Century's challenges.

(656) Being in the Military (for a Short Time) is a Big Help Towards Becoming a Competent Grand Strategist - But it is Not Absoluetly Vital.

Established military organizations are NOT friendly places for advanced creative thought. **Established military organizations are very much about training people to only think inside of the approved steel boxes - or to NOT think at all.**

US Marine basic training is all about not thinking and only following orders.

The USA Grand Strategy Team never has to enter the Pentagon. The Team's conclusions will go to the President of the USA. The President will present the ideas to all of his top military leaders, Congressional leaders and presidential advisors for comments and criticisms. If any criticisms seems valid the strategy plan will go back to the Strategy Team for the Team's replies and so forth. Eventually the President of the USA will make a decision and the orders will come from the President.

If some of the experts on the Grand Strategy Team were a little crazy (or a little weird or had dyslexia) that might be an advantage for creative thinking and planning. **If a person has proven talent for Strategy analysis and they are 100% loyal to only the USA then everything else is of secondary importance.**

The weakness that almost all older military organizations have is that they put culturally expected appearance, attitudes and behaviors before talent. **For those doing the grand strategy planning it has to be the opposite - talented analytical abilities (and loyalty to only the USA) must be everything.**

(657) The Grand Strategy Team Should Examine all Grand Strategy Suggestions even if the Suggestions are from "Non-Standard" Outside Sources.

A classic historical mistake is for the National Leader (and old military organizations) to only consider Strategy suggestions coming from a standard "acceptable" military and government source. This is the classic "NOT invented here mistake." If it did not come from an "acceptable source" then it is not looked at or considered.

Henry Kissinger was one of the best USA grand Strategist of the 20th Century. Like all great grand strategists **Kissinger** was born with talent. In the US Army during WW2 his rank was Sergeant E5 - the lowest rank of Sergeant.

One of the many problems the US Pentagon (and the Joint Chiefs of Staff) have is that they do NOT want to let go of the past. **Dr. Peter Drucker** said **"If you want something that is truly new you first have to give up what is old and no longer working."**

Effective meaningful changes are NOT created by committees of "establishment people" from the rancid past.

The US Joint Chiefs of Staff are not going to declare themselves obsolete and accept the desperately needed Grand Strategy Team. **The Grand Strategy Team's creation is going to**

have to come forcefully from a US President with enough courage to do what is desperately needed.

(658) The Grand Strategy Team must have Members with the Ability to Proceed with Plans Without having Guarantees about the Exact Outcomes.

The Grand Strategy Team should not have nitpickers. Sometimes during a war if the odds are only 65% in your favor and you are only 75% ready you nevertheless must move forward and pray that your opponent is more unprepared than you.

(659) The Grand Strategy Team should Not be Micro-Managers.

The Grand Strategy Team must spend 80% of its time and energy on macro-planning and super Grand Strategy. The world has never been more tied together and complex then it is now and that makes 21st Century strategy analysis the most complex and difficult in history.

The USA's Generals are excellent at the details of winning battles and excellent at keeping overall US Troop deaths down. The Grand Strategy Team should concentrate on the mega Grand Strategy issues.

(660) The Grand Strategy Team must do Everything Possible to Keep Shooting Wars Short.

Wars need to be short and fast for 2 big reasons - (1) the other side always learns and becomes more skilled. (2) In Democracies like the USA the voters have (and should have) less and less patience with wars that drag on for no apparent reason other than fighting a war to be fighting a war.

The Rule for Democracies is that unless you are directly and large-scale attacked by the other side - your Military must win fast or lose the support of the voters and thus lose the war.

Somebody might say "Well, the war in Iraq started in 1991 and the US Military continues to have shooting battles in Iraq so that proves the USA can have forever wars forever." Yes the US Military has been involved in shooting combat in Iraq since 1991 - however there has been (and are) many enormous USA social costs on many levels. **The massive amount of free-floating frustration and anger in the USA today is caused in large part by Washington DC's never ending stupid wars in stupid places for stupid reasons.**

One of the many costs of never ending stupid wars are very few college graduates in the top 5% of their class joining the US Military. The US Military's reputation in the world (and in the USA) continues to sink lower and lower and lower.

Professional soldiers (and especially the Pentagon's $100,000 per year mercenaries) love never ending wars. However **the longer that the stupid wars for stupid reasons go on the more frustrated, aggravated and anti-American the USA's left-wing Press and left-wing voting public will become.**

The future of the USA is in great danger because every year more and more USA voters become anti-American. **Increasing numbers of Americans are becoming ashamed to be Americans.** The Pentagon tells itself (and sincerely believes) that their never ending wars are saving the USA when in fact **their endless wars are destroying the USA from the inside out.**

In the 19th Century (1800 to 1900 when the USA went from being a backwoods highly indebted nothing Nation to the greatest industrial power in the world) the USA had only 3 standard Wars. The War of 1812 that was started by the British. The Civil War of 1860-1865 when the Confederate States attacked the USA to protect slavery. The Spanish American War of 1898 that was over in LESS than four (4) months. There were "police actions" in Central America but **in the 19th Century the USA had 94 years of official PEACE out of 100 years. It was FEW Wars that made the USA into a great Nation and world power. It was NOT never ending stupid wars.**

When this book was first published every American born after 1989 could NOT remember a time when US Troops were not in somebody else's Nation fighting some type of stupid war. **For all USA citizens the 21st Century has been 100% shooting wars somewhere for things of no value to the average working-class American.**

Americans need to open their eyes and see what is in front of them. **The all important "social fabric" of the USA is being destroyed.** THERE IS TOO MUCH TENSION. **Nothing, absolutely nothing in the Middle East begins to be worth the USA tearing itself apart from the inside out.** For the USA's survival the Tension-Levels must go DOWN however possible and wherever possible.

(661) When There are Major Change in the World (like the Explosion of Worldwide Communications) that Also Fundamentally Changes World Military Tactics and Strategy.

The USA Strategy planners (in this 21st Century) absolutely MUST understand that **there are now three times the number of Non-Shooting Types of Warfare for neutralizing an opponent than existed during WW2.**

Today there are many more types of non-shooting warfare tools then existed even before the 1970's. However they are useless if the US Pentagon (and Washington DC) **continue their stubborn refusal to understand them and use them.**

At this point in history the Radical Muslim Groups have far more successfully used the new Internet and communications tools of warfare than the rancid Pentagon Generals. This is classic textbook military evolution. The Radical Muslim Groups are NOT burdened with Trillions of dollars of US Military-Industrial-Complex junk. So **the Radical Muslim Groups use the brand new tools that work.** It is like Apple Computer starting in a garage with a new simplified technology application and coming in under the giant IBM. It is like Amazon using the modern communications tools to come in under Sears. It is Uber coming in under the Yellow Cab giant.

The poor in money Radical Muslim Groups were forced into being innovative and progressive. However - the hardware rich Pentagon's Generals continue to believe that money wins wars. Exactly the same grand strategy mistake that the USA's Generals made in Vietnam and continue to doggedly cling to today.

All wars are run by people. Defeat your opponent in their minds and you have won the war. **Create a desire for peace in the minds of your opponents and you have achieved peace. Salesmanship is a tool that can and does win wars. What Washington DC refuses to understated is that salesmanship (and the non-shooting warfare tools) are the only way to permanently make peace with the Radical Muslim Groups.**

(662) When Technology has Major Changes than Tactics and Strategy must ALSO have Major Changes.

Around 1885 (the start of the Industrial Age of Warfare 1885-1960) Technology changed from what for 600 years had been single shot military rifles (that could shoot 3 to 5 rounds a minute) to machine guns that could shoot 500 rounds a minute. **The Technology changed greatly but tragically the tactics and strategy mindsets of all of the Generals in the world did NOT change.** That produced many of the disasters and horrors of WW1 and those disasters laid the groundwork for WW2.

Communications technology has already changed the world but the US Pentagon continues to have a WW2 mindset. The USA must have the new Strategy Team to bring the USA's Military mindset up into the 21st Century. **For the USA the new Grand Strategy Team is NOT an optional luxury it is a National survival necessity.**

(663) Most of the USA's Military Budget Must be Shifted to 21st Century Warfare Tools.

The only way to win a "terrorist war" is to dry up the sea of civilians that the terrorists swim in. The only way to dry up the sea of civilians helping the terrorists is to successfully sell the idea that the USA is really and truly giving the local Muslim people a better life and more long term hope than the terrorists.

The military concept of "fronts" in the 21st Century's USA-Muslim World War is obsolete. The "front" in the 21st Century is everywhere and everything. **The "frontline" in the 21st**

Century is the internet cloud. The "frontline" in the 21st Century is international business everywhere in the world.

The current con game from the Warmonger Lobby is saying and acting as if mud holes like Afghanistan are "the Front." Or Iran is "the Front." In a world where a person almost anywhere in the world can call somebody else almost anywhere in the world the "Frontline" becomes the whole world.

(664) The Grand Strategist Always Listens Carefully to the Critics

The best Grand Strategist never automatically dismisses anyone without taking a few minutes to hear the outline of their reasoning and perhaps asking them a few questions.

Ninety-five percent of the war criticism and/or "advice" in the 21st Century is useless and/or dangerous. However there is usually somebody somewhere who has the mainly correct solution about something important. The Grand Strategy Team must remember that during the last 2,000 years of military history the individuals who had mainly the correct analysis were mainly people not directly connected to the main decision making center of power.

Nicolo Machiavelli (1469-1527) wrote that most of the advice given to a Prince is usually (1) wrong stupid advice, (2) self serving advice that has an agenda, and/or (3) espionage disinformation, and/or (4) vindictive advice designed to deliberately hurt the Prince.

Machiavelli is famous today but when he was alive he was an obscure far down the line functionary.

One of the biggest talents an individual can have in the Art of War and Peace is being able to recognize and good advice from the (1) stupid advice and (2) the advice that has an agenda of its own.

(666) A Grand Strategy Team of Many Experts Operating In Different Locations Will Have more Resistant Against Becoming "Yes-Boys."

Organizations as they grow in size and age almost always become organizations that are full of "Yes Boys." The Yes Boys flatter their bosses and they are the ones who are promoted. **Then the flaw becomes institutionalized (and a part of the organization's culture)** until it is an organization of Yes Boys who have no creativity of their own.

The definition of "Yes Boys" are those that tell their Boss everything the Boss wants to hear and nothing about what the Boss does not want to hear - but should be hearing.

The Grand Strategy Team (that will consist of many small self-contained groups of 3 to 10 people) will NOT be as good for the growth of Yes-Boys as the current US Pentagon where "loyalty" is a premium. But those in the Pentagon are "loyal" to what? Loyal to losing stupid wars in stupid places? Loyal to defending Apartheid and brutal colonialism? Loyal to the US Pentagon being full of "yes boys" who are excellent ass kissers and **Vince Lombardi** profanity imitators but military grand strategy morons?

During the Vietnam War (1959-1975) Secretary of Defense **Robert McNamara** (Secretary 1961-1968) told President **Lyndon Johnson** (US President from 1963-1969) only what **Johnson** wanted to hear - and that was "total victory was right around the corner."

Being a Yes-Boy ass-kisser worked for **Robert McNamara** who was promoted to President of the World Bank where he stayed until 1981. **McNamara** lived an exceptionally wealthy life and died at age 94. However - the tens of thousands of young American soldiers who died in a war (that **McNamara** knew was already lost by 1964) - those young Americans did NOT live long and wealthy lives.

One of the highest priorities for the new USA Grand Strategy Team is for another Vietnam to never happen again.

(667) The new Grand Strategy Team will Understand that Using the USA's Military Power to Try and Stop History is a Mistake.

It is time for the USA to accept its relative decline in the world and make the needed adjustments. Washington DC must stop trying to freeze history with endless shooting wars against change.

(668) The new USA Grand Strategy Team is an Absolutely Must Have Now that the USA is on the Down Side of its Bell Curve.

When a Great Power goes into the downside of its Bell Curve the decline has 95% of the time been impossible to stop without crashing and burning. However - England did successfully **slow its decline to the point where it is almost stable** - but almost stable at a new much lower level of world power than before the start of WW1.

Realistic management and courage can GREATLY SLOW the relative decline of the USA to a very low speed. **President Trump is strategically absolutely correct when he says that everything has to be "America first."** If it is NOT clearly an advantage for the USA then the USA should NOT be doing it.

The British Empire from 1880 to 1914 was the #1 Super Power in the world and traveling over the top of its Bell Curve. The Great European Wars of 1914-1945 added speed to the downside of the British Empire's Bell Curve despite England being on the winning side. However **through brave and smart management the English Ruling Establishment has stabilized England's decline and reduced the speed of the decline to a very slow pace.**

Washington DC (that is roughly in the same position as the British Empire after WW2) must also learn how to successfully manage its decline.

(669) The new USA Grand Strategy Team is Needed to Jump Start the Revolutionary Changes in Military Training and Organization that are Already Overdue.

The world's military history is crystal clear. Staying up with change is NOT an optional choice. When an organization (or Nation) starts living in the past then it will become part of the past.

(670) The new USA Grand Strategy Team of Experts is Essential for Making the Transitions into the Next Ages of Warfare.

Professional career Military Officers become emotionally attached to old weapons systems that have become obsolete. One of the many advantages of the new Grand Strategy Team is they will 85% NOT be professional military. **The new Team will NOT have an obsolete mindset with emotional vested interests in yesterday's weapons/tactics for yesterday's wars.**

Senior US Military Officers say "Officers have the experience to know what is needed." **Experience in what?** No US Officer alive today has any experience in clearly and decisively winning a US Military significant shooting war.

Yes, the US Military has much "experience." Experience in losing wars since the end of WW2.

(671) All Established Old Professional Military Organizations Cling to Obsolete Thinking and Obsolete Grand Strategy.

(1) In 1914 all of the Generals were doing massive infantry bayonet charges that were in part sword and spear warfare.

(2) In WW1 most German, Russian and British officers wore swords.

(3) After WW1 British Officers were issued an improved model of sword.

(4) When WW2 started one of England's first orders was for British Officers to have their swords sharpened.

(5) In 1939 (at the start of WW2 in Europe) the Polish Army had 50,000 Horse Cavalry and many of them carried spears.

(6) In the American-Japanese war of 1941-1945 many Japanese Infantry Officers went into battle with swords.

(7) Today's standard US Army M4 Rifle comes with a bayonet that makes it a short spear.

As far back as the US Civil War bayonets were proven to be useless for deciding a battle's outcome. **The Rule is - do not bring a knife to a gunfight.**

The world's professional Military Officers have a long international world history record of NOT always being "cutting edge" for effectively dealing with warfare's changing times.

(672) The Grand Strategy Team (with its Many Diverse Experts) Will Be More Resistant to the US Warmonger Lobby's Con Games that Have Been Used So Successfully During the Past 60 Years.

The Warmonger Lobby (both inside and outside of the USA) is working very hard to start a war between the USA and Iran. Part of their con game is - "All America has to do is bomb a few targets in Iran and the Iranians will be screaming to surrender."

The new USA Grand Strategy Team will hopefully do its planning based on the truth (and moral high ground ethical principles) and NOT buy into the Warmonger Lobby's usual con games.

(673) The new USA Grand Strategy Team Will Greatly Advance the USA's Strategy Analysis.

One tool for analysis is "reductionism." You have a problem and you (1) start by taking away everything that is NOT truly connected to the cause of the problem. Then (2) what remains are the core-factors of the problem. Then (3) you take away "reduce" everything you know cannot be a cure for the core factors. (4) What remains is the solution.

Another analytical tactic is the reverse side of the coin - the "extension" of the main factors. Most of the time the big questions of Strategy are in "gray areas." You take the main elements of a question and **run that element out to its far extremes of possibilities.** This usually takes the core of the problem OUT of the "gray area" thus **exposing the bones of the issue** and that makes the essence of what is happening easier to see and easier to understand and thus easier to solve.

(674) The new Strategy Team would STOP Doing the Same Old Proven Failures in the Same Old Failed Ways.

The typical Washington DC behavior for over a half Century is to take proven failures and do them over and over and over again. **There is a clear and compelling need for a major updating of the USA's Strategy decision making organization.**

(675) A Strategy Plan where Only One Particular Outcome will Produce Benefits is by Definition a Bad Plan.

The new Grand Strategy Team will Position the USA so that No Matter What Happens the USA will Win Something.

Every significant Battle-Plan must have a Plan B and a Plan C and a Plan D. Most Battle Plans have four major outcomes: (1) a win (2) a draw (3) a defeat or (4) a Black Swan. There must be a Plan B, C and D prepared (at least in outline) before the battle starts. You must position yourself so no matter what the possible outcome might happen you will win something.

The 1942 **Jimmy Doolittle** raid on Tokyo is an example. After the US Carrier Force was in Japanese waters no matter what happened it would have forced the Japanese to act. If the Carrier Force had been discovered early and never launched any airplanes the Japanese would have nevertheless been motivated to attack and capture Midway Island and the US Navy's trap would have worked.

(676) All Old Military Organizations Acquire Bad Habits. The Bad Habits then become Standard Operating Procedures for Continuing Failures. The new USA Grand Strategy Team will Break the Pentagon's Cycle of Failures.

Israel's Generals brilliantly and stunningly won the Six-Day War (June 1967) fought by General **Moshe Dayan,** General **Yitzhak Rabin** and General **Ariel Sharon.** Israeli's Army also brilliantly won the Yom Kipper War (October 6-26, 1973) fought by General **Moshe Dayan** General **David Elazar** and General **Israel Tal.**

One of the big reasons for Israel's Generals being so brilliantly successful against greatly superior numbers of Troops and Tanks was because **Israel's Generals did NOT have decades of obsolete military organizational habits and mindsets they were trying to keep alive** (like the US Pentagon).

Israel's Generals did not have a US Pentagon with 26,000 employees pushing papers around to be sure every World War 2 mindset procedure was being meticulously followed.

Israel's Generals studied **Liddell Hart's** Strategy rules. Many of Israel's Generals had talked and studied directly and extensively with Sir **Liddell Hart** (1895-1970) and most importantly they actually followed **Hart's** advice. The result was that Israel's Generals clearly and decisively won the **two most brilliantly fought shooting wars** of the last half of the 20th Century and they won **2 of the most brilliantly fought tank battles** in the world's history of warfare.

Why in the Six-Day War - and again in the Yom Kipper War were the superior in numbers of Troops and Tanks Muslim forces so completely crushed and defeated? The Muslim forces had (1) Double the number of Troops of Israel. (2) The Muslim forces had many more Tanks than Israel. (3) In the Yom Kipper War the Muslims had the great advantage of complete surprise when they attacked. And (4) strategically Israel was already enveloped on three sides when the wars started.

One of the main answers for why Israel's Generals were so brilliantly successful (despite their many material and geographical disadvantages) is because **the Muslim forces were trained by the Russians and were using Russia's WW2 battle tactics.**

By the late 1960's WW2 was already a bygone Age of Warfare. As the US Pentagon learned the hard way in Vietnam.

One of the most important lessons from the 1967 and 1973 wars was that a WW2 mindset was already yesterday's dangerously obsolete mindset that would lose wars.

The 21st Century Grand Strategy Team will have the courage and talent to leave the old comfortably world of the past and bravely swim into the unknown waters where tomorrow's mindsets for winning tomorrow's wars resides.

(677) Sometimes the Strategist has to Burn Down his Comfortable Present so There is Nowhere to Go Except Into the Unknown Future.

One of the greatest soldiers in history was **Hernando Cortes**. **Cortes** landed in Mexico (in 1519) with 600 soldiers, 15 horses and 15 small cannons. **Cortes** faced an Aztec Army of 200,000. **Cortes** scuttled his ships so his soldiers could NOT retreat. **Cortes** built an Army of allied Indians that were not Aztec - and using European tactics defeated the Aztecs. Winning what was eventually 180 Tons of Gold and 16,000 Tons of Silver and a giant continent full of natural resources. (That is an example of a war that paid for itself).

(678) Before Starting a War the USA Grand Strategy Team will be Sure they have the Support of (1) the Voters and (2) the Nation's TV Press and (3) the Major Cultural Opinion Leaders. Most Importantly they will be Reasonably Sure they can Keep the Voter's Support.

Almost all wars are very popular when they first start. However - if a war drags on (and the war can no longer be sold to the Nation's voters) then the military will find itself fighting a two-front war. Front # 1 being the military opponent and front # 2 the political battle with the Press and voters.

The Vietnam War turned into a 2 front war and the US Pentagon was completely unprepared to fight on the 2nd front.

(679) One of the Hardest Jobs of the USA's new Grand Strategy Team will be Resolving the Major Contradictions in the USA's World Strategy Paradigm.

What destroyed the USSR were Contradictions that the USSR's leadership refused to resolve. The old USSR was like Washington DC is today. **Washington DC refuses to deal with its contradictions and keeps kicking all of the problems down the road.** The USSR used to also kick its Contradictions down the road until the USSR's cans went over the edge of the cliff and the USSR also went over the edge of the cliff and ceased to exist.

Washington DC and the US Pentagon have to make a Command Decision. The USA either stands for (1) freedom and justice for everyone everywhere. Or (2) the USA stands for Apartheid and White man brutal colonialism.

Washington DC's current massive moral contradiction (that all of the world sees except Washington DC) are NOT going away by themselves. **The sooner the contradictions are eliminated the better for the USA.**

(680) Modern Wars do NOT Pay for Themselves and in the Modern Age the Costs of the USA's Wars are Forever Costs.

The new Grand Strategy Team must always be asking - is this war worth the forever costs it will accumulate?

The human price of the Vietnam War (1959-1975) to the USA was 58,169 dead and 304,000 wounded (over 100,000 seriously and permanently wounded). The direct 1959-1975 monetary cost of the Vietnam War was over $2.5 Trillion (in 2016 dollar values). In addition **over a million Vietnam Veterans were receiving monthly disability checks/benefits** when this Book was published.

The forever cost of the Vietnam War (and all current USA wars) comes from the $2.5 Trillion used to finance the Vietnam War. All of the "war funding" was borrowed money and became part of the permanent Federal Debt. **The Federal Government has been paying interest on the Vietnam War borrowed money since 1960.** Every month of every year the USA tax payers are continuing to pay millions of dollars of interest payments for the Vietnam War. **The Vietnam War interest payments will continue either forever or until the USA's financial system crashes and burns and starts all over again.**

The current Iraq War and Afghanistan War are also being paid for with borrowed money that is added to the National Debt. The interest payments will also continue for the remainder of this Century and into the next 22nd Century. **Your grand children's children will be paying for the Vietnam War and the Iraq War and the Afghanistan War and maybe the USA-Iran War etc etc forever and forever.**

It is simple 4th grad math. Eventually the continued war costs will crush the US economy.

What did the average American worker receive from the Vietnam War? The answer is less than nothing. What did the average America receive from the current Iraq War and the current Afghanistan War? The answer is less than nothing.

The human price of Iraq War 2, (2003 to ?) to the USA are 6,845 dead and 900,000 injured (and continuing to go up when this book was published). <u>The 900,000 wounded, injured, and mentally impaired is NOT a mistake or misprint.</u>

Almost 770,000 of the 900,000 injured are already on monthly government disability payment and will be for the remainder of their lives. America's taxpayers will be paying for the Iraq War (and the Afghanistan War) into the next Century.

There are over 3,850,000 USA Veterans receiving disability payments. With the never ending wars in Afghanistan, Iraq and wherever in the world **that number goes UP every day**. (Yes - 3.8 million on monthly disability payments is almost unbelievable - you can look it up on Google).

(681) The Grand Strategy Team will Formulate the Overall Strategy and Objectives. Then the US President will Give his Generals and Admirals their Specific Objectives and Timelines.

Once the US President (in consultation with the Grand Strategy Team and Joints Chief of Staff) has decided that a shooting-action is worth the costs involved for the good of the USA. Then the President will have the Grand Strategy Team and Joint Chiefs of Staff formulate the specific and detailed logistics with one of the top priorities being a fast war. No forever wars.

(682) If the US Grand Strategy Team Cannot See a Doable Way to End the Opponent's Hope for Victory (Using Shooting War Tools) then the USA WILL NOT Go Into a Shooting War.

In the USA-Vietnam War (1959-1975) there were hundreds of battles and the US Military won all of them. However in 16 years of winning battles **the USA's Generals never came close to killing the Viet Cong's unshakable belief in final victory. The USA's Generals totally and completely failed to defeat the Viet Cong in the one and only place where they could ever be defeated - in their own minds.**

White man USA Armies in Asia (and especially in the Muslim Middle East) are now on the very wrong side of history. **Every Muslim in the Middle East knows that the Muslims will eventually defeat the US Military's White man Christian Army - and they are correct. The**

probability of the USA's Generals defeating the Muslim's idea of final victory over the outside invaders is less than zero.

The USA does have many doable grand strategy solutions in the Middle East. However they are all non-shooting war tool solutions. The core problem is that **the US Pentagon (in its current form) wants nothing (or very little) to do with any non-shooting war tools.**

(683) The Obsolete (and Disastrous) Mindset of the US Pentagon's Generals is that ONLY Bombs, Bayonets, Bullets and Blood are Part of Their Job.

In the 21st Century there are 37 different types of warfare. The USA must know how to win all 37 and must know how to defend the USA against all 37.

The Pentagon is called the "Defense Department" and NOT the "Bombs, Bayonets, Bullets and Blood Department." The USA's tax payers desperately need "Defense" from the many attacks that are ranging from the Mexican border immigration war to China's predator trade wars. The US People need MUCH MORE than a collection of loser Generals and Admirals who cost $700+ Billion dollars a year and all they do are lose current wars everywhere while dreaming about re-fighting WW2.

The USA's many military academies (West Point, Annapolis, US Air Force Academy, etc) should already be teaching how to win all 37 kinds of warfare. However - when this book was published every US military academy was lacking a pure 100% course in Grand Strategy.

The Book you are reading now is the most comprehensive history book in the world about old fashioned military Grand Strategy - plus the outline for the Grand Strategy needed for the 21st Century.

This book was not designed to teach you how to win battles it was designed to teach you how to win wars.

The US Air Force Academy should have already been renamed - The "US Air and Space Force Academy." There should have already been a brand new Academy for the "US Communications Force" new branch of the US Military. However none of the above was part of WW2 so the USA's Generals and Admirals are saying "who needs it?"

(684) One of the Top Priorities of the Grand Strategy Team will be Restoring the Reputation of the US Military.

The objective is to have college graduates in the top 5% of their class joining the US Military as a career because that is where exciting new things will be happening.

The US Pentagon losing war after war is more than the cost of dead soldiers and the wealth of the USA destroyed forever. The loss of dead soldiers and wealth can be mitigated with time. However **if your military acquires a reputation for being losers and failures and dead head dullards then the quality of both your Officers and enlisted men will take a big step downwards.**

Talented people everywhere (and at all times) want to be part of a winning organization. Most of the Pentagon's Generals will view the proposed Grand Strategy Team as a threat. However it is the solution for many problems including the re-vitalization of the US Pentagon's sinking reputation.

The American Military has been losing shooting wars since 1950. A smart, energetic, imaginative and creative 22 year old who is graduating from college looks at today's USA Military and asks himself, "Do I want to spend the next 35 years of my life in an organization that doesn't even understand the world changing power of the Internet?"

How many top 5% of US College graduates are saying - "I want to spend my life at the Pentagon?" That is like saying - "I want to send my life-career at Sears."

In December of 2017 **the Chinese Military openly stated their intentions to become the world leader in the soon to arrive Artificial Intelligence Age of Warfare**. Only a moron would not take them seriously.

The Artificial Intelligence Age of Warfare (AIAW) has NOT arrived yet but it is coming. There are many reasons to believe the Chinese will be dominate.

After the Artificial Intelligence Age of Warfare the next Age of Warfare (probably around 2060) will be **the New Species Age of Warfare.** The Chinese have already started human Gene Altering experiments and are already working on the Age of Warfare that follows the next AIAW Age of Warfare. Meanwhile the US Pentagon is building obsolete death trap Aircraft Carriers to re-fight WW2 in the Pacific.

(685) In the Last Part of this Century Smart Machines will have Replaced Humans on the Battlefields Between Advanced Nations. Only Poor Nations and Groups Will Have Human Soldiers.

It is much easier to train a Machine Army than a human Homo sapiens Army. Once you have perfected the software and machinery for one smart Machine Soldier you have it for the next 20 million Smart Machine Soldiers. None of the highly intelligent Machine Soldiers will ever forget any of their training and skills. They will be able to see in all directions at once and in all weather conditions. They will have hearing and a sense of smell far better than any Homo sapiens. They will never have to sleep. They will be able to function in conditions of extreme heat or cold and in high levels of radiation. They will have built-in radar, motion and heat detection. They will be able to stay underwater for long periods of time. The Machine Soldiers will all be able to make battlefield judgments 50 times faster than Homo sapiens soldiers and none of the Smart Machine Soldiers will fear injury or death.

The Chinese already build 30,000,000 cars and trucks every year. The day will come when the Chinese will be able to build 30,000,000 Smart Machine Soldiers every year.

(686) One of the Grand Strategy Team's Jobs will be Formulating the New Rules of Battlefield Tactics when the Number of Human Casualties is No Longer a Tactics Planning Factor.

Imagine how radically and profoundly the rules of battlefield tactics are going to change when the Commanders no long have worry about the number of human casualties?

(687) The Grand Strategy Team will be 100% Aware of the Fact that the Great Age of Western World Colonialism (1492 to 1945) is OVER and Dead.

The Grand Strategy Team will know that **any and all attempts at using shooting warfare tools to reestablish (or protect) White Man Colonialism anywhere in the world is now a major military mistake.** A major mistake that the US Military and Washington DC must stop making.

(688) The Grand Strategy Team will END Washington DC's current Classic Strategy Mistake of Building Strategy and Actions Around a Flawed Subjective Belief of What Washington DC Imagines the World Should Be.

For decades Washington DC has been making plans based on what Washington DC believes the world should be rather than making plans based on what the world is in reality. This has been a major USA strategy mistake.

The Strategy Team will NOT automatically dismiss widely held beliefs only because they do not fit into Washington DC's subjective view of the world. What that means is if 800 Million people believe something (even if that something is NOT true) for USA Strategy Planning it is a reality and must be treated as a reality.

(689) The new USA Grand Strategy Team will Work as Hard as Possible at being as Close as Possible to Those Nations that Can Destroy the USA.

The Grand Strategy Team will work for a new International Organization called "The Military Big Three" - the USA, China and Russia. Only the Military Big Three have the ability to destroy one or the other. **If the Nuclear Big Three form an alliance and/or understandings then all three are safe from mass destruction.**

The world's historical record of the last 4,000 years is crystal clear. **A Nation's government (and its people) have to work as hard as possible at being as close as possible to those who have the ability to destroy them.** Only China and Russia can destroy the USA and that means 70% of Washington DC's diplomatic actions should be centered on building working relations with China and Russia.

(690) The USA's TV Press "News" Knows How To Successfully Spin the "News" and Create Voter Public Opinion Either For or Against Any Issue.

In a Democracy when the military loses the support of the main line Press it will then lose the support of the majority of the voters. In all future proposed shooting wars the Grand Strategy Team must ask "Is the proposed war ethical and clearly on the moral high ground so there is deep and strong support for the war on Main Street USA? Or is the war supported only by the Washington DC Warmonger Lobby and the US Military Industrial Complex?"

(691) The Grand Strategy Team Will Work Hard at Always Being Optimists because Optimists Win More Wars Than Pessimists.

Optimists have far more success than pessimists in almost everything because the <u>optimists try harder and work longer</u> but pessimists give up too easily.

Be optimistic but **remember that being an unrealistic optimist will not by itself win wars** and could get you into deep trouble.

The balance between being optimistic but also knowing when you have lost is part of The Art of War.

The Grand Strategy Team as a group needs to be a collection of optimists and must work hard at being innovative, courageous and creative problem solvers. **The Grand Strategy Team must be optimists so during the times of crises (and there are always eventually times of crises) the Team will have enough optimistic faith to carry both the Team and the Nation through to success.**

There were many wars in history where you could have built a truly excellent factual case for the eventual winning side NOT being able to win. The same is true with many businesses that when started there was an excellent argument for them NOT succeeding. However they went on to become great successes. A person who had **true talent coupled with courageous optimism and perseverance** against all odds made it work.

The world history of business startups proves that the talent of the Leader is decisive for success. The world history of warfare also proves that the talent of the Leader is decisive.

Unfortunately for the American people and the world the US Pentagon has turned into a factory for mediocrity and for winning battles but losing wars.

(692) The US Air Force (and Especially the US Space Force) are the Future of Decisive Warfare. However Human Pilots in the Most Advanced Combat Airships Do NOT have a Future.

When this book was first published a human's world record-time for solving a Rubik's Cube puzzle was 4.7 seconds. The computer's record-time for solving the same Rubik's Cube puzzle was 0.6 seconds - less than one second and over 600% faster than the world's best human.

Do you as a human jet fighter pilot in year 2030 want to be in a dog fight against a smart machine fighter pilot? Only if you want to be dead. Not only will the computer pilot be many times faster in its thinking and reactions - in addition the computer pilot of a jet fighter would be able to take many times the G-forces a human could withstand. The computer pilot will be able to analyze, make decisions, dive, turn and maneuver faster than any human pilot. **The only limitations on the future computer combat pilots will be the limitations of the aircraft** and the computers.

The new U.S. Space Force will have Aircraft-Spacecrafts that can take off from Earth's surface and fly up into space and then return to Earth and land. With smart machine computer pilots all of the many difficult problems of: oxygen, heat, water, food, mental disorientation, vertigo, sleep etc are eliminated.

Future "great military powers" will be Nations that have the manufacturing ability to build many smart fighting machines including highly intelligent machine jet fighter pilots.

There might be a future "arms race" between China and/or Japan and/or India to build smart machine militaries. Also an "arms race" between China and Japan to bio-gene-engineer **the new super-smart species that will be the decisive weapons system of the late 21st Century.**

(Note) All of the "Star Wars" and "Star Trek" shows are factually wrong. In 300 years all Homo sapiens will be extinct.

(693) The Chinese will have Decisive Advantages over the US Military When the New Species Age of Warfare Arrives.

The yet to arrive Artificial Intelligence Age of Warfare will be followed by the New Species Age of Warfare. China has major legal and religious advantages over the USA for **gene-**

engineering the new super-smart species that will be the ultimate weapons system in the early 22nd Century.

(694) The USA's Grand Strategy Team will Constantly be Analyzing Ways to Apply all of the USA's Strong Points To All of the Opponent's Vulnerable Areas.

General **Sun Tzu** wrote "A winning Strategy should be like a pound of forethought balanced against an ounce of forethought. A losing Strategy is like an ounce of forethought balanced against your enemy's pound of forethought."

What **Sun Tzu** meant is that **you must put much more time and effort into planning than your opponent.** At this time the USA has NO central Grand Strategy Team to do that job. **There is a serious USA grand strategy planning deficit.** This highly dangerous weakness must be corrected as soon as possible.

Genghis Khan (1162-1227) **and his Mongol armies were excellent at intelligence.** The Mongols would always send out agents disguised as business traders or pilgrims into areas they were planning to attack. Their agents disguised as traders would map all of weak and strong points. Before the Mongol Army arrived and attacked the Mongols had a detailed and highly accurate picture of their opponent's military straights and weaknesses. The Mongols knew exactly where to apply their strongest attacks. The Mongol axiom was - **Few armies lost wars because they had too much accurate information and understanding about their opponent.**

(695) The USA's new Grand Strategy Team will have Specific and Detailed Objectives IN WRITING and will Constantly be Adjusting their Actions Towards Securing those Objectives.

One of the first steps in military analysis is to write down your objectives and make the objectives as specific and detailed as possible. **As the conflict unfolds you will constantly be adjusting your written objectives up or down.** The mental discipline of putting plans down on paper is a must for organized thinking.

You must have attainable well defined objectives otherwise your efforts from the start will be a failure. If your objectives are nothing but the usual Washington DC glittering propaganda slogans like "winning the war against terror" or "Peace in Palestine" then you will be on a road to muddled defeat.

Today we should be asking "what is Washington DC's objective with Iran?" Is the objective to have Iran become a de facto giant slave state West Bank? **Without highly specific and achievable objectives Washington DC will continue to run around in endless circles of undoable delusions.**

(696) The American Taxpayers are NOT getting Value for Their Money from the US Pentagon and There Must Be Improvements.

The US Pentagon has 6.6 million square feet of office space. That is equal to the office space of more than 6 Empire State Buildings. The Pentagon has 17.5 miles of corridors. The Pentagon has a full time staff of 26,000 personnel with many able to retire after 30 years and receive substantial monetary pensions that include 100% medical, dental and eye care for life. However the US Pentagon has NOT clearly won a significant shooting war since being on the winning side in 1945.

You have to ask the question - "Is the Pentagon a military organization that wins wars or is it the ultimate welfare state inside of the United States?"

The author of this book (**Andrew M. Molchan)** in 1964 predicted that America would lose the Vietnam War (1959-1975). Mr. **Molchan** correctly analyzed the outcome of the Vietnam War eleven years before the war's ultimate end of North Vietnam taking over all of Vietnam. In 1964 Mr. **Molchan** was 23 years old and one person doing analysis at an old desk in a small bedroom. If Mr. **Molchan** could correctly analyze the outcome of the Vietnam War with 100% accuracy then what were the 26,000 Military Officers, military staff, so-called analysts and employees at the US Pentagon doing?

We know what the US Pentagon has NOT been doing since 1945 - and that is clearly and decisively winning significant shooting wars. The Rule is if it is broken then it should be fixed.

(697) The good Grand Strategist Knows that One of His Biggest Potential Enemies (that Must be Watched all of the Time) is the Enemy He Carries in His Own Mind.

Military organizations in the 21st Century should have (at regular intervals of 3 or 4 years) all of their Generals (and everyone on the Grand Strategy Team) psychoanalyzed by a group of proven excellent psychologists and psychiatrists.

All Generals and Strategy planners must be aware of their own weaknesses and personal demons and know how to compensate for them in their planning and actions.

The Team should remember the historical fact that a person who might be "abnormal" in some ways might nevertheless be a brilliant military analyst. Also a person who can do a perfect imitation of **Vince Lombardi** in the locker room might be a military strategy moron.

The Pentagon's current organization has been proven and reproven to be obsolete. At this point the "need" is for analysts who do NOT "fit into" the US Pentagon's mediocrity-mold that has become standardized thinking and procedures for failures.

Three of the most gifted American military analysts of the 20th Century were "non-standard" and/or "abnormal" and/or had little respect for "established authority."

(1) **Homer Lea** (1877-1912) was a sickly soldier wannabe who joined the Chinese Army because the US Army would not have him. **Mr. Lea** wrote two books correctly describing the essential elements of the major campaigns of the first half of the 20th Century. **Mr. Lea's** predictions included correctly pinpointing the exact beach the Japanese would use when they invaded the Philippines in 1941.

(2) Major **Earl Ellis** USMC (1880-1923) was an alcoholic and a neurotic who died from the complications of over drinking. However Major **Ellis** in 1921 wrote Plan 712 that became the battle plan used by the USMC (with brilliant success) in the 1941-1945 USA part of the Great Asia-Pacific War.

(3) Colonel **Billy Mitchell** (1879-1936) was a WW1 war hero and the commander who during WW1 organized the first large (1,500 airplanes) air-ground coordinated attack in history. Colonel **Billy Mitchell** was a brilliant analyst who in 1922 could see that Battleships were obsolete and it was Aircraft Carriers that would win the next war at Sea. **Billy Mitchell** was Court Martialed and thrown out of the US Military for having "radical military views" and not accepting the "normal standards."

On the other hand - during the Vietnam War Commanding General **William Westmoreland** was Hollywood's Central Casting perfect image of what a US General should look like and how a General should act. Even his name was perfect for the war - more land in the West. However the Viet Minh in their black pajamas and palm leaf sandals won the war.

Now that the USA is well into the downside of its Bell Curve the flow of history is going to be MUCH LESS forgiving with all of the US Military's future dumb mistakes. There is a growing premium for Washington DC NOT making mistakes and the solution is the new Grand Strategy Team.

(698) The Grand Strategy Team should Have Some Input into what Senior Officers are Promoted into the Rank of General and Admiral.

The Pentagon's Generals and Admirals should have little impute into selecting new Grand Strategy Team members. If they do the Team will be a group of grand strategy mediocre dullards like the Joint Chiefs of Staff. **It should be the other way around. The Team should have input for the promotion of Generals and Admirals** and for those of lesser grade who show great promise and who should be in project leadership positions at a younger age.

There should be a new title for those young officers with proven talent. They should be given significant positions in new projects based on talent and NOT on seniority. A title of something like "knight" (to go with their regular rank of say Major) if that is Constitutionally legal (the US Constitution forbids royal titles).

The US Military must MUST be working closely with Silicon Valley but your classic dullard stuffed-uniform Pentagon type will never make that happen.

(699) The new USA Grand Strategy Team will Constantly be Reevaluating Both Allies and Enemies. Past Allies Who have Become Useless must be Downplayed. The Non-Shooting Warfare Tools Should be Applied Vigorously to Opponents (or Neutrals) that Could Be Turned into Allies.

History has proven that Allies and Enemies are always changing. Washington DC needs the new Grand Strategy Team to get out in front of the changes (and drive the changes) rather than the continuation of the current situation of Washington DC always stumbling to play catch-up after the changes have happened.

When this book was published many in Washington DC were saying that Iran was a forever enemy of the USA. A good Grand Strategist would "see" Iran as a Nation that was once an Ally of the USA and with skilled USA leadership (plus using the many non-shooting tools of warfare) could once again be a highly valuable Ally of the USA.

(700) The Grand Strategy Team must always be in a State of Upgrading. It should Bring in the Best People Possible and Discard the Least Productive Members.

One of the many blessings the USA had during its part in WW2 was the advantage of having General **George Marshall** as Chief of Staff of the Army. General **Marshall** (1880-1959) had an exceptional talent for recognizing and promoting the correct Army Officers into the correct jobs.

In August 1941 **Dwight Eisenhower** was a Colonel who had never commanded anything larger than a Battalion (about 1,000 men). However General **Marshall** liked what he saw. Less than 3 years later **Dwight Eisenhower** was a Four Star General and Supreme Commander of Allied Forces in Western Europe.

One of the reasons for the USA having a large number of excellent Generals in WW2 was because for 2 decades before WW2 the USA had a very restricted Defense Budget. **The USA's Officers of the 1920's and 1930's had to learn how to be innovative and creative and get the most out of what they had.** Like North Vietnam's General **Vo Nguyen Giap** had to do during the Vietnam War - and as the Taliban is doing now.

Before WW2 there was no USA massive Military-Industrial Complex peddling expensive (often useless junk) for yesterday's wars.

When General **Douglas MacArthur** was Chief of Staff of the US Army (1930-1937) and Congress wanted to see General **MacArthur**. He as a 4 Star General (and Chief of Staff) would take the Public Trolley Car (10 cents at the time) from the Defense Building to the Capital Building.

Too much money makes most people and organizations: soft, lazy, weak, afraid of pain, arrogant and incapable of saying they were wrong. The old saying is "The family goes from working-man to working-man in 4 generations." Many big corporations went from almost nothing to almost nothing in 4 generations.

(701) Some of the Most Important Factors in the Grand Strategy Equation Can ONLY be Completely Understood with Intuitive Talent.

The Grand Strategy Team must always be looking for the "beautiful minds" and if needed the group must ignore the unimportant things. If a person is (1) loyal to the USA first and only and (2) they consistently have had a very high percentage of the correct analysis then those are the two most important factors.

(702) The Grand Strategy Team Will Avoid the Common Mistake of Making Long Term Decisions Based on Short Term Circumstances.

Sir Basil Liddell Hart was one of the top five best Grand Strategy analysts of the 20th Century. After France had fallen (July 1940) **Liddell Hart** (1895-1970) knew that England and the UK alone could NOT defeat **Adolf Hitler**. **Liddell Hart** then publicly advocated for whatever form of peace agreement England could make with the Germans.

Liddell Hart in 1940 was an example of a brilliant Grand Strategy analyst who was only looking short term and temporarily lost the Art of being a good Grand Strategist. In 1940 **Liddell Hart's** analysis should have included the fact that the two potentially biggest land Army military players in the world (the USSR and USA) were not yet active players in WW2.

Liddell Hart did not see the long term possibilities for events greatly improving for England and the UK.

Liddell Hart's mistake of making the wrong long term analysis based on short term factors is a common mistake. The reverse common mistake (that the USA is currently making) is **ignoring long term threats because everything is okay in the immediate short term here and now.**

(703) Yesterday's Military Genius can be Tomorrow's Failure.

The Grand Strategy Team (that will have several members) will thus have some degree of "insurance" against a former brilliant analyst becoming a failed analyst. No person in history was correct about everything all of the time.

(704) The Grand Strategy Team must Speed Up the Transition to Smart Weapons Systems.

Professional military Officers are reluctant to leave old and nostalgic weapons systems. In June 1914 (before WW1 started) all of the world's biggest armies were spending massive amounts of money, time and effort on Horse Cavalry. At the same time they were spending less than 1% of their budgets on (1) Machine Guns - and less than 1% of their military budgets on (2) Airplanes - and 0% on (3) Tanks. The 3 Weapons Systems that became decisive for land victory on WW 1's Western Front.

Today - the USA's Military-Industrial Complex is making the pre-WW1 mistake of spending tens of Billions of dollars on obsolete weapons for yesterday's wars. **The very expensive updates of obsolete weapons ideas are blocking the progress towards building the needed new weapons systems (like the US Space Force) that will be decisive in a shooting war with China or Russia.**

(705) One of the Characteristics of 21st Century Wars Between Major Powers is that there Will Not be Large Amounts of Time to Correct Major Mistakes.

If the USA gets into a shooting war with China (God forbid) and the US Pentagon has the wrong mix of weapons **there is NOT going to be any time for major corrections.** If all of the USA's communications satellites are shoot down and there is no cell phone service and internet in the USA then it is too late to realize that the USA needed a Space Force.

(706) The Grand Strategy Team should have Great Humility for the Difficulty of Knowing the Eventual Outcome of Wars.

Nobody had the correct analysis of WW2 when it started in 1939 - not **Albert Einstein** not **Winston Churchill** not **Douglas MacArthur** not **Adolf Hitler** not **Joseph Stalin** not General **Tojo** of Japan not **Liddell Hart** not anyone. All of the world's Leaders and well known analysts in 1939 had the end result of WW2 very wrong.

In 1950 when the Korean War started General **Douglas MacArthur** was very wrong about how it would end.

In 1962 ALL of the USA's Generals and Admirals had the eventual outcome of the Vietnam War completely wrong.

In 2001 when the Afghanistan War started all of the Pentagon's Generals had the outcome of the war completely wrong.

In 2003 when the Iraq War 2 was started all of the US Pentagon's Generals said it would be over in a few months. All of the US Generals were wrong as to its eventual outcome.

Today's Washington DC has the same old failed system of "Pentagon Experts." Based on the past record of the last 70 years there is a 95% chance of the "Pentagon Experts" being wrong about the eventual outcome of the next big USA shooting war.

(707) Washington DC's Leaders Must Understand that the USA is Already Militarily at the Top in the World and with Any Future Significant War <u>There is Nowhere to For the USA to Go Except Down.</u>

In 1914 (when WW1 started) there was nowhere for the British Empire to go except down. Nations that are on the top militarily (like the USA today) must do everything possible to avoid large shooting wars because there is nowhere to go except down.

There is nothing in the Middle East that is of any decisive importance to the average worker in the USA. **Why is the US Pentagon constantly going back into the Middle East War Casino to roll the dice when the USA has nowhere to go except down?**

The answer is that having too much money can sometimes get you into trouble. If the US Pentagon had less money there is a strong argument for saying the average US voter would suffer fewer stupid wars in stupid places for stupid reasons that only benefit the Warmonger Lobby.

(708) If There Is a USA Proposed War and the Grand Strategy Team Cannot Build Excellent Reasons for the War. Reasons that can Stand Up to Vigorous Cross-Examination. Then the USA Should Not be In That War.

Inside of the Strategy Team every major decision must be "put on trial" and aggressively crossed examined by experts who understand all of the major factors connected to the proposed war. Secret transcripts of the "trial" should be kept for after the war when the ideas and reasoning and strategy axioms can be checked against reality.

(709) The Grand Strategy Team must Never Start a War that Does Not have a Doable Realistic Exit Option.

When Iraq War 2 started in 2003 the then US Secretary of State (a former US Army 4 Star General and former Chief of Staff of the Joint Chiefs of Staff and Secretary of State from 2001 to 2005) said - "there is nothing to worry about with attacking Iraq because unlike Vietnam we have is an exit plan."

When the 2003 second USA invasion of the Iraq turned into another Vietnam never ending war stalemate the Secretary of State DID have an "exit plan." He retired - and walked away leaving the Iraq War mess of horrible problems for somebody else to solve.

(710) The new USA Grand Strategy Team is NOT going to have Long Wars that Teach America's Opponents How to Be Better Soldiers and Smarter Leaders.

Aristotle (384-322 BC) was the teacher of **Alexander the Great** (356-323 BC). **Aristotle** wrote, "When an Oligarchy builds up forces of armed men from the masses it is sometimes building up a future challenge to itself. The reason for why the masses can defeat the Oligarchy is because the light and mobile forces of the masses (Guerrillas) can cope with the heavy cavalry and heavy infantry of the Oligarchy."

Strategists like **Aristotle** were writing about Guerrilla warfare over 2,300 years ago. **Aristotle's** advice to his student **Alexander the Great** was to NOT have long wars that teach your opponents how to become better fighters.

(711) The Post WW2 US Military is Infamous for NOT Even Trying to Analyze the Motivations of Their Opponents. The new Grand Strategy Team will Correct that Great Flaw.

When you (1) accurately understand all of your opponent's numerous motivations. (2) Understand their objectives. (3) Accurately understand your opponent's Strategy for achieving their objectives and desires. (4) Then you can formulate the correct Strategy to defeat your opponent's Strategy.

The US Pentagon can never defeat the USA's opponents in their minds without first understanding exactly where their opponent's minds want to go.

(712) The New Grand Strategy Team will END the Never Ending Wars.

One of the classic characteristics of great powers in decline (like the USA in the 21st Century) are never ending wars. The Grand Strategy Team will SLOW the USA's decline by ending the never ending stupid wars in stupid places for stupid reasons.

During the Roman Empire's Bell Curve highest point (30 BC to 180 AD) **Roman had 200 years of continues peace** with only one small war (the insane Zionists declared war on Rome in 66 AD). However during Rome's Bell Curve decline (280 BC to 476 BC) Rome was constantly having never ending wars - most of the wars were stupid wars in stupid places for stupid reasons that had nothing to do with the needs of the average Roman citizen.

The main reason for "why" the Western Roman Empire fell was because most of the average Romans wanted it to fall. Many average Romans became anti-Roman exactly like millions of American voters today have (and are) becoming anti-American.

The British Empire during its Bell Curve top (1870-1914) had only a few small colonial wars but during its Bell Curve decline had WW1 and WW2.

Sometimes wars can unit a Nation but wars also tear Empires and Nations apart. It is much safer to avoid wars.

(713) Up to this Point the Military Computer Programs of the US Pentagon have been Failures Because they have Been Programmed by Strategy Ignorant Yes-Boys Pandering to Grand Strategy Ignorant Generals.

The computer software rule has always been "delusional garbage in - delusional garbage out." During the Vietnam War the USA had tens of millions of dollars of IBM computers working constantly 24/7. Tons of "report information" from the battle areas went into the computers. Tons of paper computer readouts were produced. However **the data input into the computers was based on WW2 obsolete concepts and ideas.** The computers had software programs built around the Pentagon's delusions and obsolete **Clausewitz** concepts. The so-called "analysis" that came out of the computers was less than worthless because the software was wrong.

The Vietnam War was a classic example of the computer rule "garbage in - garbage out." **The US Pentagon turned the computers into "yes boys."** The computers kept coming to their deliberately programmed false conclusions that the US Military was winning the war and "there was a light at the end of the tunnel."

(714) The Grand Strategy Team will Oversee the Building of the Programs and Systems to Effectively Use the Internet as What It Is Already - a Major Weapons System.

The US Pentagon's Generals have a death grip on the idea that you can only win wars by bombing and killing people. However you cannot bomb the Internet out of existence. You cannot bomb ideas out of existence. **You can only defeat ideas with better ideas that are delivered to the world in superior ways.**

The Radical Muslim Groups have brilliantly used the Internet for their advantage. **Think of the irony** - the Radical Muslim Groups who want (in some ways) to return to the 17th Century know how to successfully use the Internet more effectively than the $700+ Billion Dollar per year US Pentagon.

The Pentagon learned nothing from its defeat in the Vietnam War. In the Middle East the too rich and too arrogant US Generals are once again trying to defeat ideas with WW2 obsolete battle tactics.

The Radical Muslim Group's Internet circulated ideas can only be enveloped and defeated with better ideas. The USA must be selling superior ideas and superior moral high ground actions that are sold with convincing salesmanship that uses all of the tools of modern communications to the greatest extent possible.

The new Grand Strategy Team will forget the obsolete Bombs, Bayonets, Bullets, Battleships and Blood of the Pentagon's loser obsolete Generals. The new Grand Strategy Team will completely understand that the new 21st Century decisive strategy is - **"He who controls the minds of the world controls the world."**

(715) The Grand Strategy Team will Understand that Winning Shooting Battles in Guerrilla Warfare usually Means Nothing. When your Troops have Left the Area the Guerrillas Will Return.

In Guerrilla Warfare winning battles means little or nothing. Winning minds means everything.

The father of modern successful counter Guerrilla warfare was French Field Marshal **Hubert Louis Lyautey** (1854-1934).

General **Lyautey** said that winning shooting battles in a Guerrilla war was useless. General **Lyautey** reproved that **the way to win Guerrilla Wars (and have a long lasting peace) was to envelop the local population's mindset.**

To win you must envelop the Guerrilla area's culture with new and better ideas. Introduce better justice and honest government systems. Greatly improve medical services, entertainment, introduce a modern education system for everyone. Use every type of tool available to bring a better and happier life for all of the peoples of the former Guerrilla areas. **Then - like the ancient Romans - you will arrive at a point where the local population wants you to say and they no longer support the Guerrillas.** Then (and only then) have you truly won a Guerrilla war.

(716) A Team of Grand Strategy Experts (that will be Heavy on Academic Talent) Will be more Resistant to the Football Locker Room Bravado that some Military Strategy Meetings Degenerate Down Into.

To win wars you need more than glittering profanity and screaming bravado. You need doable and imaginative planning based on proven axioms of success.

(717) The Grand Strategy Team Must Have Highly Talented Experts who Know How to Create World Public Opinion that is Favorable to the USA.

The USA's Grand Strategy Team could learn from the British who have been experts at influencing public opinion for hundreds of years.

When WW1 started one of England's first act was to cut the undersea cable from Germany to the USA. Then the only European "cable" to the USA was the one the British controlled. The only "cable news" from Europe was the "news" spin that the British Government wanted the USA's newspapers to receive and report.

The British Government also **paid cash** to key American newspaper and magazine writers and editors to publish favorable articles about the British and encourage the USA to join the war on England's side.

Foreign governments bribing key media and US government people (and fabricating falsehoods and lies) to influence USA public opinion, elections and actions is nothing new and continues to this day.

(718) The US Grand Strategy Team will Correctly Identify the True Motivations of its Opponents even if those Motivations are NOT what Washington DC Wants to Hear.

The Grand Strategy Team will NOT follow the usual US Pentagon practice of only talking to people who will tell them what they want to hear. Or **the standard US Pentagon practice of playacting at asking for advice but not really considering any advice from anyone who tells the Pentagon anything the Petagon does not want to hear.**

During the Vietnam War the US Pentagon asked **Dr. Peter Drucker** (the best management teacher in world history up to this point) for his advice about the Vietnam War. **Dr. Drucker** (1909-2005) gave the US Military excellent advice but the Pentagon's planning staff completely ignored **Dr. Drucker's** advice because it was not what they wanted to hear. **Dr. Drucker** was eventually proven to be correct.

After the war was lost and over **Dr. Drucker** said that he always knew that the Pentagon's Generals were only **pretending to ask for outside advice** and the Generals had already decided on their actions that were eventually proven to be completely wrong.

During the Vietnam War **Robert McNamara** (1916-2009) was US Secretary of Defense from 1961 to 1968. **McNamara** was an expert at playing the game of ONLY talking to people who he knew would tell him what he wanted to hear. Then **McNamara** went to **President Kennedy** (and later **President Johnson)** and said in all honesty that he (**McNamara**) had talked to dozens of people and they all said that the Vietnam War was succeeding.

(719) Rather than Understanding what Tools Are Needed to Change Opinions and Win in the Middle East the US Pentagon does its Battle Planning Ass Backwards.

When faced with a military challenge (like the Radical Muslim Groups in the Middle East) the US Pentagon **starts its Strategy planning** by (1) looking at their Trillions of dollars worth of shooting-war military hardware equipment. Then (2) the Pentagon tries to find ways to apply its mountains of semi-obsolete (or completely obsolete) military hardware to the problem.

Rather than correctly defining every aspect of the challenge using the formulas in this Book. **The US Pentagon first looks at its mountains of hardware and then tries to fit the often obsolete and inappropriate hardware to a superficially and only partly understood problem.** The process is ass-backwards. It is the US Pentagon's highly flawed standard procedure to justify the original multi-Trillion dollar purchases of the hardware.

Today's Middle East challenges require tools and mindsets that clearly do NOT exist inside of the US Pentagon.

(720) A Military or Political Leader can be Completely Sincere and Absolutely Loyal to the USA and also be Completely Wrong in his military Strategy Judgments.

People with bad ideas often fall back on the defense of saying "I have been absolutely loyal to the USA for all of my life." **It is not an issue of being loyal - it is an issue of having the correct ideas, mindsets and plans.** The US Generals in Vietnam were 100% loyal to the USA and 100% flawed in their planning.

(721) An Important Objective for the new USA Strategy Team is to NOT be the Victim of Another Iraq War2 Type Con Game.

The Strategy Team should be suspicious of all "weapons of mass destruction" information until they are absolutely sure it is true. The Team should be especially suspicious of the warmonger's usual con game of saying the USA must start a "preventive war" to stop something that is NOT absolutely proven.

Suppose Iran did build 3 or 4 nuclear bombs. The Iranians know that if they used them on the USA half of the Iranian population would be dead. In 2,700 years of history the Persians have NOT had any National suicides.

On the other hand the US Pentagon's "greatest ally" (the Zionists) HAVE committed National suicide in the past (66-70 AD). The Zionists are working on National suicide again. The new USA Strategy Team will work hard at NOT being dragged into the coming Zionist National suicide.

(722) National Leaders should Always be On Guard Against Becoming Megalomaniacal True Believers in their Own Military's Invincibility.

US Senator **Lindsey Graham** likes to say - "Where the US Military goes the US Military is victorious." Really? Since the end of WW2 (for winning significant shooting wars) it has been the opposite.

Adolf Hitler used to say - "Where the German soldier decides to stand nobody passes."

It takes more than "pride" to fly a 747 and it takes more than "pride" to win wars.

(723) The USA's Grand Strategy Team must Constantly be Working at Maintaining Four Lists of Advantages and Disadvantages.

The USA Grand Strategy Team will have staff members that maintain and constantly update four lists. The lists are:

(1) all of the advantages on the USA's side.

(2) All of the disadvantages on the USA's side.

(3) All of the advantages on all of the opponent's sides.

(4) All of the disadvantages on all the opponent's sides.

The Grand Strategy Team will constantly be working at ways to leverage the USA's advantages against the opponent's disadvantages - while doing what is possible to reduce the USA's weaknesses.

(724) Believing your Own Propaganda and Spin will Produce Highly Flawed Grand Strategy Planning.

No Nation can build good war analysis and plans based on delusions and myths. This is one of the reasons why the US Military has not clearly won a significant shooting war since 1945.

The lesson for the USA from both WW2 and WW1 (and from the Vietnam War) is that if the US Military wants to be on the winning side then it needs somebody else to do the majority of the combat fighting. If the US Pentagon is doing 90% of the fighting then winning the war is doubtful.

A protestor might say "Well the US fought and won the war in the Pacific against the Japanese by itself." In the great Asian-Pacific war the USA was fighting with China, the UK and the USSR. Those three groups were 45% of the world's total population. The Chinese lost over 3,000,000 soldiers in combat against the Japanese. Japan officially admits to having lost 480,000 dead Japanese soldiers in China. In WW2 Pacific the USA lost about 90,000 (ninety-thousand) in combat.

(725) General Sun Tzu wrote that Strategy Planners must Avoid the Five Flaws that will Cause Defeat.

(1) Never become overly reckless. Be bold and aggressive but always put the odds in your favor. Always guard against becoming overly extended in all vital areas. All of your pre-battle planning must be organized so you can (1) sustain a defeat if you lose the battle or (2) exploit a victory to the maximum if you win. Or (3) what often happens in a battle is you win some things and lose others and must be prepared for all eventualities.

You must build scenarios (and be prepared) for all of the most probable combinations of happenings.

(2) Never be excessively cautious. To win you have to take chances and sometimes put yourself in harm's way. You almost always need imagination and boldness to defeat a strong opponent. However - always keep the odds in your favor and always have a doable plan for retreat.

(3) Never let yourself become angry. Anger is both amateurish and dangerous for a professional Officer and/or National Leader. Anger interferes with rational judgment. If you become angry a skilled opponent will use your anger against you.

Osama bin Laden successfully used President **George W. Bush's** anger (after 9/11) and stampeded-tricked the angry President **W. Bush** into attacking Iraq. **Osama's** highly successful trick (1) more than doubled the 9/11 American death toll and (2) increased the 9/11 dollar costs to over a Trillion dollars and counting.

(4) Never be excessively nit-picky and fastidious. Nothing is perfect in warfare. There is never the perfect time to attack. <u>There is never the perfect time to make peace</u>. If you delay needed actions because everything is not absolutely perfect then you risk losing the war and/or risk having never ending wars.

(5) Never become emotionally attached to things (including Allies) that are (or have become) worthless status symbols and/or no longer have vital importance.

History is constantly moving forward. Both obsolete Allies and obsolete weapons system must be downgraded and/or discarded. Nothing is forever. **Sometimes what takes the most courage is leaving the beloved (but now obsolete) past - and moving on.**

(726) Sun Tzu wrote that Many People have Multiple Personalities and some of those Personalities can be Highly Destructive. The Grand Strategy Team must Put Great Effort

into Understanding and Mastering the Contradictions in their Organization and the Contradictions in their Nation.

Twenty-four Centuries before **Sigmund Freud** (1856-1939) - **Sun Tzu** emphasized the great need for National Leaders and Generals to understand and master (1) their own personality and (2) their own reasoning patterns and (3) their short comings so they (4) did not adversely distort their military planning and actions.

Sun Tzu wrote "Your personal growth depends on your level of self-control. Self-discipline is the road to challenging your inner contradictions. By cultivating the control of your own mind you will accomplish more with ease and be a successful Strategist."

Almost two-thousand and fife hundred years ago **Sun Tzu** was writing that Leaders (and key decision makers) should have themselves what was de facto psychoanalyzed so they could make better (and safer) decisions.

Within the Grand Strategy Team (that the USA absolutely must build) all of the members must be psychoanalyzed so they understand their own weaknesses.

If potential USA Grand Strategy Team members and/or US Generals are afraid to be psychoanalyzed then they should NOT be making war plans that if done wrong could cause the death of 50% of the USA's population.

(727) Two Vital Questions the Grand Strategy Team must Answer (in Writing) Before a War Begins are: (1) How will we Know When We Have Won? (2) How will we Know When We Have Lost?

The "how do we know when we have lost" question has NOT been as easy as it might first appear. When this book was published the USA had been at war in the Muslim Middle East starting with Operation Desert Storm in 1991. Was the US Military in the Middle East de facto defeated years ago?

The US Military was de facto defeated in Vietnam 11 years before the 1975 final takeover of the South by the North.

The 2001 USA invasion of Afghanistan was a defeat after the first year. However - in 2019 when this book was published US Troops were fighting in Afghanistan.

The new Grand Strategy Team will clarify the guidelines for when the USA has won and when the USA has lost and act accordingly.

(728) The Grand Strategy Team must Understand what General Sun Tzu wrote about Avoiding the Five Forms of "Disorders" that will Bring Defeat.

Sun Tzu's 5 disorders (that will bring defeat) are:

1. An organization with lax or non-existent discipline will bring defeat.

An army without internal co-operation and organizational discipline is a mob with guns.

2. A Nation that has wrong objectives will be defeated.

In the Vietnam War (1959-1975) one of Washington DC's objectives was to reestablish White man Western European de facto colonial rule in Southeast Asia (with the usual Native puppet government pretending to be in charge). However after 1945 White man colonial rule was the "Walking Dead" everywhere in the world.

Today Washington DC (in the Middle East) is making the same mistakes as in Vietnam. Washington DC supports repressive colonialism against Muslims and as in Vietnam is on the very wrong side of history.

3. Actions that are unprincipled and on the moral low ground will contribute to disorder and defeat.

During WW2 the USA was clearly on the moral ultra high ground. The USA was fighting for equal freedoms and equal justice for everyone everywhere.

In the Middle East Muslim-USA war Washington DC has **never** issued an Official Statement that clearly states that the USA's objectives are: equal freedoms, equal rights, equal justice and equal opportunity for everyone everywhere.

Washington DC's objectives in the Middle East are the same as the objectives during the Vietnam War - to bomb and shoot the world back into yesterday.

4. A Strategy that is overly indulgent will bring defeat.

The US Military's over indulgence of its collected-ego in the Vietnam War continued that war for 10 years longer than it should have been.

A serious over indulgence is the Pentagon's and Washington DC's belief that it is entitled to military victory everywhere.

Washington DC's over indulgence is the belief that God is always on Washington DC's side. God is always on the same side - the side that has the courage to successfully resolve the challenges that **God** puts in front of every individual person and every Nation.

5. You will be defeated if a war is overly restricted for political reasons and you do not bring those restrictions into you overall Strategy planning.

Many times there are clear political and/or military reasons for shooting war limitations. The solution is to bring into play all of the many non-shooting warfare tools and apply them to the maximum extent possible.

(729) The 21st Century USA Grand Strategy Team will have Experts in all 37 types of Warfare Constantly Upgrading the Master Software Programs.

Writing the first Grand Strategy military programs will be very difficult and time consuming. Once the first program has been started it will then be a never ending process of updating, refining and expanding the programs.

The integration of the many specialized areas of knowledge will take many highly trained and talented experts in their respective fields: economists, historians, sociologists, psychologists, computer engineers, geopolitical analysts, mega-trend projection experts, managers and individuals who have the imagination to think about winning the future's wars rather than only planning how to win the last war.

There must also be input from: Generals, Admirals, con artists, foot soldiers, politicians, statisticians, experts in gambling odds, religious experts, code experts, "skunk-works" experts, people who have a lot of common-sense street smarts and historians. Needed are the very best proven experts in how Muslims think and make decisions.

(730) The Grand Strategy Team will Discourage Rivalry and Empire Building Wherever and However Possible. It will Encourage Co-Operation and Team Unity of Objectives and Purpose.

Rivalry between the: Army, Navy, Air Force, Space Force, Communications Force and all of the Intelligence Agencies - and so forth should NEVER be encouraged. Rivalry should always be discouraged and as much co-operation as possible should be built into interconnected military systems.

(731) The Full Time and Energy of the USA Must be Focused and Prioritized onto Building ALL of the Future Defenses and Tools for the Future's Challenges.

The current never ending dumb shooting wars in dumb places for dumb reasons makes the USA weaker rather than stronger. The US Military must free its mind from the current WW1 like stagnation wars so it can concentrate on 21st Century warfare in all of its many forms.

The Old-Guard at the US Pentagon is going to come up with dozens of reasons for why the Grand Strategy Team should NOT be built. However - to paraphrase President **Trump** - "What do the American people have to lose?"

The US Pentagon (despite having a $700+ Billion dollar per year cost) has NOT clearly won any significant shooting war since 1945. So what do the American people have to lose with a new Grand Strategy Team?

It would be difficult for the new Grand Strategy Team to do a worse job with the USA's shooting wars than what has been done by the US Pentagon during the last 70 years.

The choice is (1) try a new organization that is very promising. Or (2) keep forever having the same old stupid wars in stupid places for stupid reasons that always eventually end up as de facto defeats.

(732) To Win Tomorrow's Wars the USA needs 2 new Military Branches. (1) The new US Space Force (USSF) and (2) the new US Communications Force (USCF).

The US Pentagon and Washington DC must take all of their dead-end wars off of life support - "clear the decks" - and concentrate on building a new military (and a new Grand Strategy Team) for the 21st Century.

Both the US Space Force (USSF) and the US Communications Force (USCF) are vital for the USA's survival as a significant Nation into the 22nd Century. **The USCF is absolutely vital for creating peace and maintaining peace because in the 21st Century "He who controls the mindsets of the world's peoples controls the world."**

(733) The new USA Grand Strategy Team will Constantly be Asking for Opinions and Advice from Proven Good Analysts Around the World.

The USA is a relatively new Nation - 1776 to now. Many of our close Allies (England, France, Italy, Japan etc) have military histories and strategy experience many times longer.

The Rule is - **Few Nations in history lost a war because they had too much correct advice.** The core problem is having people in Washington DC smart enough to recognize good advice when they hear it.

(734) The USA Must Become far More Creative in the Management and Use of it Talent.

The graduates of US Military Academy's who are especially gifted might go directly into the Grand Strategy Team as research assistants both "in the office" and/or out in the world.

An ancient Chinese tactic was taking the most intelligent young military officers and sending them out (as supposed civilians) into "areas of interest" both geographical and intellectual. Some of them would live double-lives for many years (or decades). The USA should be doing the same.

How many of the best and brightest young Chinese Military Officers are working around the world as supposed civilians? The answer is - "**Many**."

(735) The Grand Strategy Team will have a Section of Strategists who are Always Working on the Best Strategies for Defeating the USA.

Most people find it very difficult to change sides mentally. The solution is building a talented and competent **sub-Team of Strategists constantly working on the best strategies for defeating the USA.**

As the "Opponent Team" discovers news strategies that look viable for defeating (or tricking) the USA then the main line Strategy Team can closely examine the possible attacks so they are not blindsided. **The USA Strategy Team will put work and time into measures and precautions for guarding against all discovered weaknesses.**

(736) At the Start of Every US Shooting War the Grand Strategy Team must have "Stop Orders." At a Specific Point of Pre-Calculated Losses and/or Time Duration of the War (when the Stop Order limits have been reached) Washington DC Must Make the Best Peace Possible and Stop Fighting.

The existence and parameters of the Stop Orders must be Top Secret. Only the people at the highest leadership level would know the details.

Set the "Stop Order" points are where you think a war is not working. Your first action might be a "radical change in the plan"? However - in the Washington DC of the last 60+ years a "radical change in the plan" has always been taking the proven failed plan and making it bigger. Without a true change of strategy the war should be ended on the best terms possible.

A true "change in plans" would be taking a failed plan and doing mainly the opposite of that plan.

Stock Market traders have "stop orders" on all of their stocks. If they go down 10% or 15% the computer automatically sells the stock.

All of the USA's future wars must have preset "Stop Orders" before any shooting begins. For example - set the "Stop Order" for the USA-Iran war at one year of combat fighting and/or 10,000 US Troops dead. If at that "Stop Point" there are NO clear (and intellectually honest) paths to a fast victory for the USA - then the USA ends the shooting war with the best terms it can negotiate at the time.

During the last 60 years the US Presidents (and US Generals) have been like losers at a gambling casino. They keep telling themselves "One more Troop SURGE - one more year of war - one more invasion and forever victory will be ours."

SUMMARY. The war planners must agree on a "stop the war order" based on criteria of (1) time (2) casualties and (3) costs. At a certain point (and that point is kept top secret) an "exit plan" that was pre thought out before the shooting started is put into operation.

(737) **Everyone in the World has the Same #1 Job for the Remainder of this Century. That is to NOT Go Crazy over the Small Stuff and Start a Nuclear Shooting War that Will Mess Up the World. With a Million Years of History in Front of the World Absolutely EVERYTHING Today is the "Small Stuff."**

Thank You for your time. Hopefully this is the start of your understanding of Grand Strategy. It is a lifelong study because the details and specifics are always changing. It is the most important study area in the world because it is the study of how the World can survive.

Thank You for your Time.

Made in United States
North Haven, CT
04 November 2026

21766807R00236